Conservation of Andean Forests

Nicola Clerici
Editor

Conservation of Andean Forests

Springer

Editor
Nicola Clerici
Faculty of Natural Sciences
Universidad del Rosario
Bogota, Colombia

ISBN 978-3-031-80804-3 ISBN 978-3-031-80805-0 (eBook)
https://doi.org/10.1007/978-3-031-80805-0

This Springer imprint is published by the registered company Springer Nature Switzerland AG
The registered company address is: Gewerbestrasse 11, 6330 Cham, Switzerland

Foreword

Stretching from Venezuela to Chile, Andean forests are crucial ecosystems, representing outstanding reservoirs of biodiversity and habitats of unique endemic species. These forests provide vital services to society, by regulating global and regional climates, providing water, maintaining biodiversity, and supporting the well-being of millions of people and countless species. Yet, Andean forests are experiencing unprecedented threats in the face of the rapidly growing human population in the region. Increasing land use demands drive deforestation and habitat fragmentation, which is further exacerbated by agricultural expansion, mining, and illegal cropping. Furthermore, the converging effects of climate change and the spread of non-native plant species magnify the pressures on these fragile ecosystems. In light of these pressing challenges, it is imperative to recognize the urgent need for conservation efforts to protect Andean forests and their unique socio-ecological heritage.

This book provides an understanding of the ecological functioning of Andean forests, spacing from community to landscape ecology studies. It discusses the regional and global threats affecting them, and gives multidisciplinary reflections on their unique socio-ecological dynamics. *Conservation of Andean Forests* presents a synthesis of recent and relevant research on these key ecosystems, discussing in detail methodological approaches, and providing the complex background on Andean forest systems functioning.

The book is organized into 11 chapters, presenting research results and future implications, discussing case studies and research methods, and novel geospatial approaches. It is the goal of this book to stimulate new ideas for future work and help managers to take decisions, as well as to becoming a critical resource for students, new and experienced researchers, academics, and professionals working with forest ecosystems.

As Editor, I am indebted to all the recognized researchers who have contributed to this work, scientific reviewers, and Springer-Nature for agreeing to publish this book. Further, I thank Universidad del Rosario (Colombia) for supporting the Editor in realizing this initiative.

Faculty of Natural Sciences
Universidad del Rosario, Bogota, Colombia

Nicola Clerici

Contents

About the Editor

Nicola Clerici is Full Professor of Ecology at the Faculty of Natural Sciences, Universidad del Rosario, Colombia. He holds a PhD in Ecology and an MSc in Remote Sensing. He is also Associate Editor of the *Landscape and Urban Planning* and *Remote Sensing in Ecology and Conservation* journals. His research interests are focused on conservation ecology, spatial modeling of ecological indicators, and ecosystem services mapping.

The Andean Flanks: Montane Cloud Forests' Untold Biocultural Heritage and Ecological Legacies

Fausto O. Sarmiento, Manuel Oliva Cruz, Christiam Aguirre, and Héctor Esquivel

1 Introduction

Throughout all western South America, from Patagonia to the circum-Caribbean and Venezuela, the great Andean cordillera (8500 km) extends parallel to the Pacific Ocean, which allows the presence of different thermal floors in most of its extension, with numerous forest types and a great diversity of flora and fauna typical of tropical montane cloud forests (TMCFs). Often the foothills begin at 500 meters above sea level (masl); however, among tropical forests, the dry tropical forest, the humid tropical forest, the subxerophytic formation (desert areas), and the Chaparral formation (elfin areas) are distinguished through but at different elevations. They are located between sea level and 1000 masl in the mountainous circum-Caribbean. The elevation belt between 1000 and 2400 masl is known as Subtropical or "*Subandina*" (Cuatrecasas, 1968) and as "*Premontano*" o "*Montano bajo*" (Holdridge, 1967). In Colombia, it is also known as the 'coffee zone' for the importance of this crop (*Coffea arabiga*) wherein the "coffee axis" zone—between Quindío, Risaralda, Caldas, and north Tolima, in this altitudinal range of

F. O. Sarmiento (✉)
Neotropical Montology Collaboratory, Department of Geography, University of Georgia, Athens, GA, USA
e-mail: fsarmien@uga.edu

M. O. Cruz
Research Institute for Sustainable Development of the Cloud Forest, National University Toribio Rodríguez de Mendoza de Amazonas, Chachapoyas, Peru

C. Aguirre
Andean Cultural Subjects Research Group, Superior Polytechnic School of Chimborazo, Riobamba, Ecuador

H. Esquivel
Alexander von Humboldt Botanical Garden, University of Tolima, Ibagué, Colombia

N. Clerici (ed.), *Conservation of Andean Forests*,
https://doi.org/10.1007/978-3-031-80805-0_1

1000–2000 masl—produces the bounty coffee harvest, the basis of the Colombian agrarian economy (Esquivel & Nieto, 2003). In Ecuador, this zone is often referred to as "*montaña*" on the western flanks towards the coasts of the Pacific Ocean and to Upper Amazonia on the eastern flanks towards the Atlantic Ocean (Sarmiento, 1987). In Perú, they occupy several regions, including the "*selva alta*", "*ceja de selva*", "*rupa-rupa*" and "*yunga*" (Vidal, 2014). In Bolivia, the flanks are generally referred to as "*los yungas*" (Kesler & Beck, 1991), while in Argentina they use the feminine "*las yungas*" (Brown et al., 1991). Sometimes, people inhabiting the Andean flanks are known with the *Kichwa* term of "*yumbu*" or "*sacha runa*" because of their affiliation with wild mountainscapes. These areas often incorporate the montane, colline, and piedmont mountain provinces' that are landscapes with extensive forest cover.

Due to great pressures exerted by the expansion of livestock, agriculture, urban planning, mining, and timber extraction, forests of this sub-Andean altitudinal strip are the most threatened of disappearing (Esquivel, 2012). In this altitudinal range, with the most moderate and preferred climates for making a living, agriculture and urbanization developed since pre-Columbian times in the mountainous regions of Central America (Guatemala, Nicaragua, Costa Rica, and Panamá) and in Venezuela, Colombia, Ecuador, Peru, and Bolivia, since the mild temperature ranges between 18 °C and 26 °C are common. The rise of cities such as Medellín, Bucaramanga, Ibagué, Armenia, Pereira, Manizales, Palmira, Cali, and Popayán in Colombia; Santo Domingo de los Tsáchila, Guaranda, Zaruma, Tena, Puyo, and Coca in Ecuador; Tambopata, Chachapoyas, Jaén, and Cajamarca in Peru; Cochabamba, Sucre, and Tarija in Bolivia; Olmué, Temuco, and Valdivia in Chile; and Jujuy, Tucumán, and Salta in Argentina exemplify these favorable settlement conditions, with various of those cities considered as the "garden of the republic."

According to biotic inventories, the greatest diversity of species is recorded in these Andean flanks; not having to withstand extreme temperatures, these montane elements can adapt more easily in the range that corresponds to the tropical zone (Esquivel & Nieto, 2003). The presence of relict forests in the sub-Andean belt, despite the priority dedication of these areas to crops like coffee (974,000 ha in Colombia, 200,000 ha in Venezuela, and 96,312 ha in Ecuador), cocoa (189,000 ha in Colombia, 65,000 ha in Venezuela, and 601,000 ha in Ecuador), sugar cane (400,000 ha in Colombia, 50,000 ha in Venezuela, and 130,677 ha in Ecuador), banana, cassava, arracacha, beans, peas and others in Colombia, Venezuela and Ecuador, mixed with plantations of *Pinus patula*, *P. radiata*, *Eucalyptus camaldulensis* and *E. globolus*, all this biological richness has motivated the organization of Environment ministries, Regional Corporations, various NGOs (non-governmental organizations), ecological groups, and civic movements, keenly working in conserving the biota of these relict forests in all the seven countries traversed by the TMCFs on the Andes Mountains. With the proliferation of 'narcoeconomics' in the era of cocaine and poppy exports, many areas within the 'Andean crescent' have been dedicated to the harvest of coca leaves from *kuka* shrubs (*Erytroxylum coca*), outcompeting traditional subsistence agriculture as livelihood practice in the mountainscape (Andreas & Sharpe, 1992). Counter to the illicit market pressure, the increase of ethnotourism in the Andean flanks is being posited as a valid alternative pulse to benefit TMCFs conservation (see Textbox 1: *Kuka Umawa* Declaration to bridge science and society in the upper Amazon).

The ecological research of neotropical Andean flanks has significantly emphasized the then-dominant paradigm of systems ecology, solely on form, function, and change (Hamilton et al., 1994). Based on neutral models for ecosystem ecology, the Tropical Montane Cloud Forests (TMCFs) have been conceived as a 'black box' with unknown processes of throughput, receiving the inputs of intense solar radiation with high levels of UV^{B} radiation and abundant horizontal precipitation in the form of fog and mist, shrouding the mountainside or flank (Kappelle & Brown, 1991). As a result, the most significant outputs of the hydrological-based model were water and forest cover. Indeed, the TMCFs have always been taken as 'sponges' that capture humidity and accumulate hydric resources that are later utilized down slope when flowing with cascading rivers of white water (Bruijnzeel et al., 2010), which contributed to the establishment of the 'water tower' metaphor, particularly with the inclusion of the ice fields located on mountaintops, snow- and ice-caped 'towers', in the so-called "Third Pole," by the trifecta of polar research AAA (i.e., Arctic, Antarctic, Alpine). In the newer modeling approach of source-sink dynamics, the flanks have often been considered a transitional stage (Myster, 2023), with the presence of elevation-depletion gradients (either of temperature, precipitation, soil, or cloudiness, as we move down from the highlands towards the lowlands); this ecotonal characteristic of TMCFs, between high Andean and low coastal or Amazonian plains, has recently acquired scientific currency harboring biocultural diversity of microrefugia and offering a 'cultural bulge' (Sarmiento et al., 2022), mirroring Gentry's (1988) "tree-bulge "of diversity Fig. 1.

There is a new paradigm built into the premises of cultural landscapes on the TMCFs because of the evidence provided by functional ecology (*c.f.*, pull and push factors), paleoecology, archaeobotany (*c.f.*, pollen fossil and genomics), critical biogeography (*c.f.*, ecological legacies), and political ecology (*c.f.*, historicity and spatiality) of tropandean landscapes (Sarmiento & Sarmiento, 2021).The extent of the TMCFs on mountain flanks is still an active front of research, as different disciplinary angles provide either discrete boundaries of treeline, separating the montane from the alpine zones (Körner et al., 2011), or the joined transitional alpine treeline ecotone region, where forest to pasture transitions are often fuzzy (Malanson, 2023). The controversy about treeline becomes important in 'situated studies' as it depends on the location of the mountain flank (Körner & Hoch, 2023), whether in the northern or southern exposure aspect, the eastern or western orientation slope, or even the continental location of polar, temperate, or tropical locales, where a lot of influence can be exerted by human impacts creating a fire-prone or a teeth-prone boundary (Sarmiento & Frolich, 2002).

If the upper limit of the Andean flanks is hard to define to secure the extent of the TMCFs toward the highland grasslands, the difficulty is augmented when we approach the piedmont at the base of the mountain, surrounded by the lowland plains. This condition makes it necessary to consider two distinct boundaries on the flanks: the '*upper treeline*' and the '*lower treeline*.' There are more nuanced

Fig. 1 Morning view of the cerro *Puñay*, in Chimborazo province, Ecuador, as an example of the biocultural landscape of the *Kañari* and *Puruwa* nations, for whom the sacred mountain is equated as identity marker of their montane soul and origins, with one of the largest pyramids built on its summit. (Photo: Christiam Aguirre)

considerations for the classification of forests separated from the colline to the montane zones, especially in tropical and subtropical sites, containing either evergreen or moist forests, where a subclassification of the montane is in order (*e.g.*, the Andean, subandean, upper montane, lower montane, premontane, heath, peat, liana, bamboo forests, etc.) (Paulsch, 2002). The neotropical mountains also provide a considerable change in location according to the elevation/longitudinal gradient of the entire cordillera and the effect of atmospheric rivers and other air mass circulations that affect the location of the forest (i.e., the Massenerhebung effect, Föhn effect, Bernoulli effect, Venturi effect, Cascade effect, Barnum effect, etc.). Something that makes the entire neotropical flanks similar is the abundant plant and animal diversity harbored by the verdant, often associated with a green carpet that descends as a tapestry down the mountain slope with biotic vigor.

2 The Biological Richness of the Andean Flanks

The list of plant species is one of the most significant in the biotic inventories of the countries hosting tropandean landscapes. Whether in the Central American mountains or the northern and central Andes cordillera, there is a plethora of functional groups that allows a very complex system of natural elements to interact, creating a mosaic of rich green broadleaf and evergreen tropical fronds and foliage. The objective of the chapter is not to list the botanical inventories of representative sites, but it is clearly important to gauge the massive presence of the vegetal component in the Andean flanks. As an example of such a rich catalog, we are only listing the arboreal species of the three cordilleras with sub-Andean forests of Colombia (Table 1). We include the local name as well as the customary usage of the tree to add notions of utilitarian application of the forest resource in the area. Also, the altitudinal range distribution of the plant is indicated so that a better grasping of the spatial distribution of trees makes an impressive mosaic of different species at various altitudes. This is an important front of study due to the notion that increased temperatures due to global warming tend to move the upper limit of the forest towards more extreme heights (Sarmiento et al., 2023); we observed that the forest itself is not moving upwards the slope, but those species adapted to withstand warmer weather are individually migrating upslope, changing the community composition but not the community structure of the flanks. However, it remains to be seen whether community function has somehow been affected by such upslope migration with the 'escalator effect' (Dangles, 2023). In the tropical Andes, it is easy to identify flagship species, often considered to be the countries' emblematic 'national' plant, such as cockspur coral of Paraguay (*Erythrina crista-galli*), red quebracho of Argentina (*Schinopsis balansae*), cantuta of Peru (*Cantua buxifolia*), maga of Puerto Rico (*Thespesia grandiflora*), quina of Ecuador (*Cinchona succirubra*), wax palm of Colombia (*Ceroxylum quindiuense*), araguaney of Venezuela (*Tabebuia chrisantha*), or the ice-cream-bean *'pakay'* of Bolivia (*Inga edulis*). Even species that have been identified as prone to psychedelics for narcotraffic, such as coca (*Erytroxylum coca*) or *aywashka* (*Banisteriopsis caapi*), are indicative of a massive concentration of TMCFs flora.

Note that forests of these great mountain ranges have been studied since colonial times (eighteenth century), in such a way that several of the expeditions carried out in countries through which the Andes cordillera now crosses were key to register its rich biota, starting with the first French Geodetic Mission (1735–1739) and the rich collections that started the exploration of these mountains (Capello, 2018). In northern South America, the Royal Botanical Expedition of the New Kingdom of Granada stands out, led by the naturalist José Celestino Mutis, for 33 years (1783–1816) which allowed the dissemination of some 20,000 species of plants, 24,000 excised for Herbaria, and some 5000 plates or paintings of plants, as well as about 7000 animals and a collection of wood, shells, resins, minerals, and skins. Added to all this is the publication "*El Arcano de la Quina*" (Mutis, 1828). In Ecuador, the researcher José Mejía del Valle y Lequerica stood out (lived from 1775 to 1813), adding to the collections made in 1802 by Alexander von Humboldt, Aimé Bonpland, and Francisco José de Caldas. In Peru, the Botanical Expedition of Hipólito Ruiz and José Antonio Pavón of 1777–1788 was of great renown; more than 3000 duly processed excised specimens and more than 2500 botanical drawings were obtained, deposited in the Royal Botanical Garden of Madrid and in the Royal Cabinet of Natural History of Madrid, with a contribution of 141 new genera and more than 500 new species; in 1794 they published "*Florae Peruvianae et Chilensis Prodomus*",

Table 1 Main tree species of Colombian tropandean forests, their common name, uses, and altitudinal range of distribution

Scientific name	Common name	Family	Utilitarian function	Altitudinal range
Adenaria floribunda	Coralito	Lytraceae	Restoration	1000–2500
Alsophila engelii	Sarro, helecho arbóreo	Cyatheaceae	Restoration	1500–2800
Aiphanes aculeata	Palma de corozo	Arecaceae	Wild food	500–1800
Albizia carbonaria	Pisquin, dormilón	Leguminosae	Restoration	1400–2800
Alibertia patinoi	borojó	Rubiaceae	Edible	0–1200
Alchornea glandulosa	Algodoncillo	Euphorbiaceae	Fuelwoodand charcoal	1000–2800
Alchornea grandiflora	Carcomo	Euphorbiaceae	Fuelwoodand charcoal	1000–2200
Alchornea latifolia	Álamo, chopo	Euphorbiaceae	Watershed protection	1900–3000
Anacardium excelsum	Caracolí, mijagua	Anacardiaceae	Watershed conservation	0–1600
Anacardium occidentale	Marañón, caujil, chura	Anacardiaceae	Edible	0–1300
Attalea butyracea	Palma de vino, cuesco	Arecaceae	Restaur. Edible	0–1500
Bactris gasipaes	Chontaduro, cachipay	Arecaceae	Rest., Edible	0–1600
Beilschmiedia tovarensis	Aguacatillo	Lauraceae	Timber harvest	500–2200
Bixa orellana	Achiote, bija	Bixaceae	Foodstuffand medicine	0–1700
Bombacopsis quinata	Ceiba roja, pochote	Malvaceae	Restoration, frames	0–1200
Bravaisia integerrima	Palodeagua, naranjillo	Acanthaceae	Restoration	700–2000
Billia rosea	Cariseco	Sapindaceae	Restoration	1000–2700
Brugmansia candida	Borrachero, floripond	Solanaceae	Scopolamine	600–3000
Barnadesia spinosa	Espino, erizo	Asteraceae	Living fence	1900–3600
Bulnesia carrapo	Guayacán bola	Zygophyllaceae	Timber harvest	0–1300
Bunchosia armeniaca	Ciruelo de perro	Malpighiaceae	Wild fruit	500–2000
Calliandra haematocephala	Carbonero rojo	Leguminosae	Ornamental, watershed conservation	0–2000
Calliandra magdalenae	Carbonero, quiebracho	Leguminosae	Watershed conservation	1000–2500

(continued)

Table 1 (continued)

Scientific name	Common name	Family	Utilitarian function	Altitudinal range
Calliandra purdiaei	Carbonero	Leguminosae	Watershed conservation	100–1600
Calliandra schultzei	Carbonero	Leguminosae	Watershed conservation	0–1500
Calocarpum mammosum	Sapote costeño, mamey colorado	Sapotaceae	Timber harvest y edible	0–1500
Capparis odoratissima	Naranjuelo, tablón	Capparidaceae	Timber harvest	0–1200
Cassia grandis	Cañafistulo rosado	Leguminosae	Timber harvest	0–1300
Cedrela montana	Cedro de montaña	Meliaceae	Timber harvest	1700–3000
Cedrela odorata	Cedro rosado	Meliaceae	Timber harvest	100–2000
Ceiba pentandra	Ceiba, cotton tree, kapok	Malvaceae	Shades and canoes	0–1500
Cestrum nocturnum	Azahar de noche	Solanaceae	Restoration	0–1800
Cereus hexagonus	Cardo, cirio	Cactaceae	Foodstuff	0–1300
Cespedesia macrophylla	Lengua de vaca, pacó	Ochnaceae	Restoration	0–1500
Cnidoscolus aconitifolia	Papayuelo, chicasquil	Euphorbiaceae	Fodder, living fence	0–1300
Chloroleucon bogotense	Angarillo	Leguminosae	Fodder and fuelwood	0–1200
Cinchona pubescens	Quina	Rubiaceae	Medicinal	600–3300
Condaminea corymbosa	Quino	Rubiaceae	Restoration	200–1500
Cinnamomun triplinerve	Laurel colorado, chaviaco	Lauraceae	Timber harvest, fruit for birds	1000–2000
Cecropia angustifolia	Yarumo, guarumo	Urticaceae	Restoration	200–2400
Cecropia telenitida	Yarumo blanco	Urticaceae	Restoration	1800–3000
Coussapoa villosa	Uvo, matapalo	Urticaceae	Restoration, fuelwood	1000–2600
Ceroxylum alpinum	Palma de cera	Arecaceae	Fruit for birds	1400–2000
Ceroxylon parvifrons	Palma de ramo	Arecaceae	Fruit for birds	2000–3200
Ceroxylon quindiuense	Palma de cera, palma de ramo, palma blanca	Arecaceae	Wax for torches	1800–3300
Ceroxylon sasaimae	Palma de cera	Arecaceae	Wild fruits	1400–1800
Chamaedorea linearis	Palmicho	Arecaceae	Restoration	500–2500
Chamaedorea pinnatifrons	Molinillo	Arecaceae	Restoration	0–3000
Clusia alata	Chagualo	Clusiaceae	Foodstuff birds	1100–3500

(continued)

Table 1 (continued)

Scientific name	Common name	Family	Utilitarian function	Altitudinal range
Clusia brachycarpa	Chagualo, gaque	Clusiaceae	Restoration	1000–3000
Clusia cuneifolia	Chagualo	Clusiaceae	Foodstuff birds	1000–2700
Clusia decussata	Cucharo, chagualo	Clusiaceae	Restoration	1500–2700
Clusia ducu	Cucharo	Clusiaceae	Restoration	1500–2700
Clusia ducuoides	Monedero	Clusiaceae	Restoration	1000–2800
Clusia multiflora	Gaque, chagualo	Clusiaceae	Restoration	1500–3000
Clusia orthoneura	Flor de cera	Clusiaceae	Ornamental	0–2500
Clusia minor	Incienso, gaque	Clusiaceae	Restoration	0–1800
Coccoloba obovata	Buche gallina, uvo	Polygonaceae	Restoration	0–1400
Codiaeum variegatum	Croto, buena vista	Euphorbiaceae	Ornamental	0–1500
Cavendishia bracteata	Quereme, uvo de monte	Ericaceae	Landslide avoidance	1000–3200
Cavendishia cordifolia	Quereme, uvo de monte	Ericaceae	Landslide avoidance	1400–3000
Cavendishia engleriana	Uvo de monte	Ericaceae	Edible fruits	200–2000
Cavendishia quereme	Quereme, uvo de monte	Ericaceae	Landslide avoidance and edible fruits	1000–1500
Citharexylum subflBirdscens	Quimulá, gavilán	Verbenaceae	Timber harvest	1100–2800
Coriaria ruscifolia	Uvita de monte	Coriariaceae	Toxic fruit	1400–2600
Cordia alliodora	Nogal cafetero	Boraginaceae	Timber harvest	500–2000
Cordia lutea	Biyuyo, biyuco	Boraginaceae	Ornamental	0–1500
Cordia muneco	Muñeco	Boraginaceae	Ornamental	0–1600
Cordia sebestena	No me olvides	Boraginaceae	Ornamental	0–1300
Curatella americana	Chaparro, peralejo, carne de fiambre	Dilleniaceae	Restorator, Fuelwoodand charcoal	0–1100
Crescentia cujete	Totumo	Bignoniaceae	Medicinal fruit	0–1400
Croton gossypifolius	Sangregado, drago	Euphorbiaceae	Reforestation	0–1300
Cupania cinerea	Guacharaco	Sapindaceae	Reforestation, fuelwood	300–1500
Clethra fagifolia	Chiriguaco, cargagua	Clethraceae	Timber harvest	1500–3000
Calophyllum brasiliense	Aceite, cachicamo	Clusiaceae	Timber harvest, fruto foodstuff fauna	0–1200
Cyathea caracasana	Helecho gigante, palma boba, sarro	Cyatheaceae	Soil conservation, handscraft	1600–3000

(continued)

Table 1 (continued)

Scientific name	Common name	Family	Utilitarian function	Altitudinal range
Cyathea arborea	Palma boba	Cyatheaceae	Soil conservation	1500–3000
Citharexylum subflBirdscens	Quimulá, gavilán	Verbenaceae	Tool handlers	1000–2800
Dicksonia sellowiana	Helecho gigante, sarro, palma boba	Dicksoniaceae	Soil conservation, handcrafts	1000–3800
Daphnopsis cestrifolia	Pela mano	Thymelaeaceae	Restoration, fuelwood	0–1500
Dussia macroprophyllata	Ubre de vaca	Leguminosae	Restoration, Foodstufffauna	1700–2600
Delonix regia	Acacio rojo	Leguminosae	Ornamental, fodder	0–1500
Enterolobium cyclocarpum	Piñón de oreja, orejero Dormil6n	Leguminosae	Fodder furniture, medicine hemorroids	0–1300
Erythrina edulis	Chachafruto, balú	Leguminosae	Food, protein	1300–2800
Erythrina fusca	Cachimbo, ceibo	Leguminosae	Fodder and fences	800–2000
Erythrina poeppigiana	Cámbulo, minás	Leguminosae	Fodder, coffee shade	1000–2500
Erythrina rubrinervia	Chocho, siriguay	Leguminosae	Ornamental, fodder	1300–2600
Erythroxylum coca	Coca, hayuela, ipadá	Erythroxylaceae	Medicinal, energizer	0–1500
Erythroxylum ulei	Coca	Erythroxylaceae	Restorator	0–1600
Eugenia uniflora	Pitanga silvestre	Myrtaceae	Foodstuff and fences	0–1600
Euphorbia cotinifolia	Manzanillo de cercas	Euphorbiaceae	Living fence, caustic látex, honey	0–2000
Euphorbia caracasana	Manzanillo verde, Platero, lechero	Euphorbiaceae	Living fence, hney	0–2000
Ficus andicola	Caucho sabanero, uvo	Moraceae	Edible fruits, watershed conservation	1800–2800
Ficus dendrocida	Caucho matapalos	Moraceae	Restoration	0–1800
Ficus insipida	Higuerón, cachinguba	Moraceae	Látex antihelmíntic	0–1600
Ficus duquei	Caucho arepero	Moraceae	Ornamental, fence	0–1400
Freziera chrysophylla	Cerezo de monte	Pentaphylacaceae	Restoration, edible fruits	1500–3000
Garcinia madrunno	Madroño, machali	Clusiaceae	Restoration	500–2000

(continued)

Table 1 (continued)

Scientific name	Common name	Family	Utilitarian function	Altitudinal range
Genipa americana	Jagua, árbol tinta	Rubiaceae	Timber harvest, feet shaper	0–1300
Gliricidia sepium	Matarratón, piñón	Leguminosae	FodderLiving fence	0–1500
Guadua angustifolia	Guadua, guadua cebolla, paca	Poaceae	Erosion control, construction	0–2000
Guarea guidonia	Cedro macho, bilibil	Meliaceae	Timber harvest	0–1600
Gustavia speciosa	Chupo, jarapandiva	Lecythidaceae	Wild food	0–1600
Guazuma ulmifolia	Guazimo, aquiche	Malvaceae	Fire barrier	0–1300
Gynerium sagittatum	Caña brava, pintoc	Poaceae	Soil protection, constructions	0–2000
Guatteria lehmannii	Cargadero, garrapato	Annonaceae	Constructions, fuelwood, ties	1500–2500
Guettarda crispiflora	Huesito, frutepava, anís	Rubiaceae	Watershed protection, wild fruits	1000–3000
Gonzalagunia asperula	Frutepava	Rubiaceae	Restoration, wild fruits	300–2000
Gaultheria buxifolia	Mortiño, uvito	Ericaceae	Restoration, erosion avoidance	1500–2500
Gaultheria erecta	Mortiño, uvode monte	Ericaceae	Erosion avoidance	1400–3400
Gaiadendron punctatum	Platero	Loranthaceae	Restoration, living fences	1400–3800
Geonoma orbignyana	Palmilla	Arecaceae	Stem for canes, roofing support	500–3000
Geonoma undata	Palmicho	Arecaceae	Restoration	500–3500
Hamelia patens	Coralito, bencenuco	Rubiaceae	Restoration	800–1800
Hura crepitans	Mil pesos, tronador	Euphorbiaceae	Restoration	0–1400
Hymenaea courbaril	Algarrobo, pecueco	Leguminosae	Timber harvest, resine, ámbar courbaril	0–1300
Hedyosmun bomplandianum	Granizo y canelo	Chloranthaceae	Restoration, watershed protection	1800–3200
Handroanthus chrysanthus	Guayacán amarillo, araguaney, chicalá	Bignoniaceae	Timber harvest	0–2000

(continued)

Table 1 (continued)

Scientific name	Common name	Family	Utilitarian function	Altitudinal range
Hedyosmum goudotianum	Silvo-silvo, canelón	Chloranthaceae	Restoration, watershed conservation	1000–3000
Heliocarpus americanus	Balso blanco, melao majao, palo bobo	Malvaceae	Restoration, fuelwood	1000–2500
Hieronyma antioquensis	Candelo	Phyllanthaceae	Timber harvest, charcoal	1800–2900
Inga densiflora	Guamo machete,	Leguminosae	Restoration, foodstuff	0–1800
Inga edulis	Guamo rabo de mico	Leguminosae	Restoration. Food	0–1800
Inga lallensis	Guamo	Leguminosae	Constructions	1800–3000
Inga punctata	Guamo	Leguminosae	Wood and food	0–3000
Inga sierrae	Guamo peludo	Leguminosae	Fuelwood, charcoal	1500–2600
Ilex laurina	Cardenillo	Aquifoliaceae	Handcrafts, fuelwood	1500–3000
Ilex nervosa	Cardenillo	Aquifoliaceae	Wild foodstuff	1800–3000
Jacaranda caucana	Gualanday	Leguminosae	Ornamental	1000–2500
Juglans neotropica	Nogal, cedro negro	Juglandaceae	Timber harvest	1200–3200
Lafoensia punicifolia	Guayacán de Manizales, trompito	Lythraceae	Timber harvest, apícola, handcrafts	1200–3000
Leucaena leucocephala	Acacia Fodderra, panelo	Leguminosae	Restoration, fuelwood, fodder	0–1400
Ladenbergia magnifolia	Azuceno	Rubiaceae	Restoration, wild fruit	1500–3000
Ladenbergia macrocarpa	Azuceno	Rubiaceae	Restoration, wild fruit	1500–3000
Leandra subseriata	Niguito lanudo	Melastomataceae	Restoration	1500–2500
Lozania mutisiana	Laurelillo	Lacistemataceae	Restoration	0–2500
Licania cabrerae	(CR)	Chrysobalanaceae	Timber harvest	1900–2800
Licania octandra		Chrysobalanaceae	Living fence	600–1500
Machaerium capote	Capote, machetico	Fabaceae	Timber harvest	0–1300
Maclura tinctorea	Dinde, fustik	Moraceae	Fence posts	0–1500
Mammea americana	Mamey	Clusiaceae	Fruits	0–1500
Matisia cordata	Sapote	Malvaceae	Fruits	500–2000
Mauria heterophylla	Manzanillo	Anacardiaceae	Wild fruits	500–1800
Meliosma bogotana	Calabacillo	Sabiaceae	Wild fruits	1500–3000
Meliosma echeverriana	Calabacillo	Sabiaceae	Wild fruits	1900–3000

(continued)

Table 1 (continued)

Scientific name	Common name	Family	Utilitarian function	Altitudinal range
Matayba elegans	Zanca de mula	Sapindaceae	Restoration, fuelwood	1000–2500
Macleania stricta		Ericaceae	Restoration	350–2720
Macleania bullata	Chupe-quinde	Ericaceae	Talus protection	950–2900
Macleania rupestris	Uva camarona	Ericaceae	Fence, talus protection	1800–3300
Magnolia hernandezzi	Molinillo	Magnoliaceae	Timber harvest, fruto edible	1500–2600
Magnolia silvioi	Guanábano de monte	Magnoliaceae	Timber harvest	400–1500 (EN)
Magnolia espinalii	Molinillo	Magnoliaceae	Timber harvest	1900–2800
Meriania heptamera	Amarrabollo	Melastomataceae	Timber harvest y ornamental	1800–2800
Meriania nobilis	Amarrabollo	Melastomataceae	Restoration	1500–2500
Miconia asperrima	Niguito	Melastomataceae	Restoration	1500–2500
Miconia cladonia	Niguito	Melastomataceae	Foodstuff birds	1800–3000
Miconia jahnii	Niguito, tuno	Melastomataceae	Restoration	1500–2500
Miconia lehmannii	Niguito	Melastomataceae	Restoration	1800–3000
Miconia theizans	Niguito	Melastomataceae	Talus affirmation	1500–3000
Miconia albicans	Tuno	Melastomataceae	Restoration	1200–1900
Miconia caudata	Punta de lanza	Melastomataceae	Restoration	1000–2600
Miconia albicans	Canela de velho, tuno	Melastomataceae	Restoration	1000–2500
Miconia longifolia	Lengua de vaca, tuno	Melastomataceae	Watershed conservation	50–2900
Meriania nobilis	Amarrabollo	Melastomataceae	Restoration	1500–2500
Meriania peltata	Tuno rosado	Melastomataceae		1900–2800
Matisia lozanoi	Zapotillo	Malvaceae	Restoration	1900–2600
Montanoa quadrangularis	Arbo-loco, maguey	Asteraceae	Restoration, handcrafts	1.500–3000
Mollinedia tomentosa	Limonero silvestre	Monimiaceae	Restoration	400–2600
Muntingia calabura	Chichato	Muntingiaceae	Pionero y Foodstuffpara birds	0–1300
Myrsine guianensis	Cucharo, chagualo	Primulaceae	Para vigas y postes	0–2500
Myrsine coriacea	Cucharo blanco	Primulaceae	Timber harvest	1000–2200
Myrcia cucullata	Arrayán	Myrtaceae	Muebles y constru.	1000–2500
Myrcia popayanensis	Arrayán	Myrtaceae	Timber harvest y constructions	1500–2900

(continued)

Table 1 (continued)

Scientific name	Common name	Family	Utilitarian function	Altitudinal range
Myrcia splendens	Arrayán(hoja pequeña)	Myrtaceae	Restoration	1500–2500
Myrcia subsessilis	Arrayán	Myrtaceae	Restoration	100–2500
Myrcia complicata	Arrayán	Myrtaceae	Restoration	200–1800
Morella pubescens	Olivo de cera	Myricaceae	Restoration pioneer	1500–3.400
Myrcianthes rhopaloides	Guayabo de monte	Myrtaceae	Restoration, pioneer	1600–3500
Nectandra macrophylla	Aguacatillo	Lauraceae	Wild foodstuff birds	500–1500
Nectandra reticulata	Amarillo	Lauraceae	Timber harvest	0–1400
Nectandra acutifolia	Laurel amarillo	Lauraceae	Timber harvest	800–2500
Nectandra laurel	Laurel	Lauraceae	Timber harvest	500–3000
Notopleura macrophylla		Rubiaceae	Food	100–2500
Ocotea sericea	Quimula, amarillo	Lauraceae	Timber harvest	1400–3200
Ocotea micans	Laurel	Lauraceae	Timber harvest	1700–2500
Ocotea leucoxylon	Laurel	Lauraceae	Timber harvest	500–2500
Ochroma pyramidale	Balso tambor	Malvaceae	Erosion avoidance	0–1800
Opuntia schumannii	Tuna, penca, nopal	Cactaceae	Foodstuff	0–4000
Oreopanax cecropifolius	Mano de oso	Araliaceae	Protector	800–2900
Oreopanax capitatus	Cinco dedos	Araliaceae	Protector	200–2800
Oreopanax incisus	Mano de oso	Araliaceae	Protector	1500–2800
Oreopanax pes-ursi	Mano de oso	Araliaceae	Foodstuff birds	1000–2000
Oreopanax acerifolium	Mano de oso	Araliaceae	Protector	1200–1700
Oreopanax floribundum	Pategallina, papayo	Araliaceae	Musical instruments	1900–2900
Ochoterenae colombiana	Tambor	Anacardiaceae	Paper pulp	0–1500
Ormosia colombiana	Chocho, peonia	Leguminosae	Watershed protection	500–1600
Pachira insignis	Cacao de monte	Malvaceae	Food	0–1800
Palicourea acetosoides	Cafecillo, tinto	Rubiaceae	Wild fruit	1000–2500
Palicourea anceps	Tinto, bencenuco	Rubiaceae	Wild fruit	1000–2500
Palicourea angustifolia	Cafecillo, tinto	Rubiaceae	Fruto sivestre	1500–3100
Palicourea garciae	Aguadulce	Rubiaceae	Wild fruit	1500–3500

(continued)

Table 1 (continued)

Scientific name	Common name	Family	Utilitarian function	Altitudinal range
Palicourea perquadrangularis	Tinto, cafecillo	Rubiaceae	Wild fruit	500–2500
Palicourea pyramidale	Cafecillo	Rubiaceae	Wild fruit	200–2500
Palicourea thyrsiflora	Tinto, cafecillo	Rubiaceae	Wild fruit	1000–2500
Palicourea vaginata	Cafecillo, tinto	Rubiaceae	Wild fruit	1000–2400
Panopsis yolombo	Yolombó	Proteaceae	Fuelwood	1100–2600
Passiflora arborea	Gulupa de árbol	Passifloraceae	Foodstuff silviculture	500–1500
Pereskia bleo	Bledo, clarol	Cactaceae	Edible fruit	0–1500
Persea americana	Aguacate, cura, pelta	Lauraceae	Edible fruit	500–2000
Persea caerulea	Aguacatillo	Lauraceae	Soft wood	0–1800
Petrea pubescens	Chaparrito, flor azul	Verbenaceae	Honey	400–1500
Phyllanthus acuminatus	Grosella, cedrillo	Phyllanthaceae	Watershed protection	0–1600
Phyllanthus salviifolius	Cedrillo	Phyllanthaceae	Watershed protection	1700–3100
Piper aduncum	Cordoncillo	Piperaceae	Medicinal	0–2000
Piper archeri	Rabo de chucha	Piperaceae	Medicinal	1900–2800
Piper arboreum	Cordoncillo	Piperaceae	Medicinal	0–1700
Piper aequale	Cordoncillo	Piperaceae	Medicinal	0–2400
Piper artanthe	Cordoncillo	Piperaceae	Medicinal	1800–2900
Piper bredemeyerii	Cordoncillo	Piperaceae	Medicinal	0–1500
Piper crassinervium	Cordoncillo	Piperaceae	Medicinal	500–2400
Piper eriopodon	Cordoncillo	Piperaceae	Medicinal	1000–3000
Piper glabratum	Cordoncillo	Piperaceae	Medicinal	500–1500
Piper jericoense	Cordoncillo	Piperaceae	Medicinal	800–2100
Piper holtonii	Cordoncillo	Piperaceae	Medicinal	0–2400
Piper ibaguense	Cordoncillo	Piperaceae	Medicinal	500–2000
Piper imperiale	Cordoncillo	Piperaceae	Medicinal	0–2000
Piper tuberculatum	Cordoncillo	Piperaceae	Medicinal	0–2200
Piper sphaeroides	Cordoncillo	Piperaceae	Medicinal	1800–2200
Pithecellobium dulce	Payandé, chiminango	Leguminosae	Watershed protection	0–1400
Pithe. lanceolatum	Chiminango, gallineral	Leguminosae	Watershed protection	0–1500
Podocarpus magnifolius	Pino chaquiro, p. real	Podocarpaceae	Timber harvest	600–1900
Podocarpus oleifolius	Pino-chaquiro	Podocarpaceae	Timber harvest	1200–3000

(continued)

Table 1 (continued)

Scientific name	Common name	Family	Utilitarian function	Altitudinal range
Posoqueria latifolia	Jazmín naranjo	Rubiaceae	Timber harvest	0–1500
Pouroma bicolor	Caimaron, uvo	Urticaceae	Watershed conservation	200–1400
Prumnopitys montana	Pino de montaña	Podocarpaceae	Timber harvest	1800–3200
Pseudosamanea guachapele	Igua, cedrillo amarillo	Leguminosae	Timber harvest	0–1500
Psidium friedrichsthalianum	Guayabo agrio	Myrtaceae	Wild food	0–1500
Psidium guajava	Guayabo dulce	Myrtaceae	Food	0–2000
Pseudobombax septenatum	Majagua colorada, ceibo	Malvaceae	Watershed protection	0–1200
Psychotria carthagenensis	Cafeto, amiruca	Rubiaceae	Wild fruit	500–2000
Psychotria horizontalis	Cafetillo	Rubiaceae	Wild fruit	300–1500
Psychotria micrantha	Cafetillo	Rubiaceae	Erosion avoidance	200–1500
Psychotria poeppigiana	Beso de negra	Rubiaceae	Erosion avoidance	100–1500
Prunus opaca	Árbol bejuco	Rosaceae	Timber harvest	1800–3000
Quercus humboldtii	Roble	Fagaceae	Timber harvest	1500–3000
Rauvolfia leptophylla	Lechero	Apocynaceae	Timber harvest	0–2500
Retrophyllum rospigliosii	Pino chaquiro, hayuelo	Podocarpaceae	Timber harvest	1000–3000
Rhamnus sphaerosperma	Cabo de hacha	Rhamnaceae	Timber harvest	1500–2800
Rollinia muscosa	Anón cimarrón	Annonaceae	Edible fruit	0–1500
Rollinia edulis	Anón silvestre	Annonaceae	Wild fruit	0–1200
Ruagea glabra	Cedrillo	Meliaceae	Timber harvest	800–3000
Raimondia cherimolioides	Anón silvestre	Annonaceae	Edible fruit	900–2450
Salix humboldtiana	Sauce vela	Salicaceae	Watershed protection	0–3000
S. humboldtiana-f. llorón	Sauce llorón	Salicaceae	Living fase, barrier	1000–3000
Samanea saman	Samán, campano	Leguminosae	Timber harvest	0–1600
Sambucus nigra	Sauco	Sambucaceae	Medicinal	800–3000
Sapindus saponaria	Jaboncillo, chumbimbo	Sapindaceae	Washing clothes	0–1500
Sapium laurifolium	Lechillo	Euphorbiaceae	Reforestation	0–1700
Satyria warszewiczii	Uva de monte	Ericaceae	Erosion avoidance	1400–2600

(continued)

Table 1 (continued)

Scientific name	Common name	Family	Utilitarian function	Altitudinal range
Saurauia cuatrecasana	Dulomoco	Actinidiaceae	Erosion avoidance	1800–2900
Saurauia stapfiana	Dulumoco	Actinidiaceae	Erosion avoidance	1700–3200
Saurauia ursina	Dulumoco rojo	Actinidiaceae	Erosion avoidance	1500–2600
Schefflera quinduensis	Chaflera	Araliaceae	Ornamental, erosion control	1600–2800
Senna pistaciifolia	Alcaparro	Leguminosae	Fertilization	1500–2500
Senna reticulata	Martín Galvis, Dorance	Leguminosae	Medicinal	0–1800
Senna spectabilis	Vainillo, velero	Leguminosae	Erosion control	0–2000
Senegalia glomerata	Palo-bayo	Leguminosae	Fuelwood, timber harvest	0–1200
Styloceras laurifolium	Tijereto	Buxaceae	Wild food	1800–2800
Siparuna aspera	Limón de monte	Siparunaceae	Reforestation	500–3000
Siparuna lepidota	Limoncillo	Siparunaceae	Reforestation	1000–3000
Sloanea brevispina	Cadillo	Elaeocarpaceae	Reforestation	500–2000
Schinus molle	Falso pimiento	Anacardiaceae	Reforestation	1800–3000
Schinus terebenthifolius	Turbinto	Anacardiaceae	Wild fruit	1200–2500
Spondias mombin	Ciruelo hobo	Anacardiaceae	Wild fruit	0–1300
Solanum asperolanatum	Friegaplatos	Solanaceae	Wild fruit	500–3500
Solanum atropurpureum	Mancadera	Solanaceae	Wild fruit	500–2500
Solanum macranthum	Árbol de la papa	Solanaceae	Wild fruit	500–2000
Solanum ovalifolium	Friegaplatos	Solanaceae	Wild fruit	1500–3000
Solanum sycophanta	Arbol-tachuelo	Solanaceae	Reforestation	1000–2400
Solanum venosum	Lulo de monte	Solanaceae	Wild fruit	2000–3500
Stemmadenia grandiflora	Huevo de venado	Apocynaceae	Restoration	0–1300
Sterculia apetala	Camajón, camajorú	Malvaceae	Restoration	0–1300
Swartzia robiniifolia	Frijolillo	Leguminosae	Restoration	100–1300
Swartzia trianae	Frijolillo	Leguminosae	Restoration	200–1400
Swietenia macrophylla	Caobo	Meliaceae	Timber harvest	0–1500
Syagrus sancona	Chonta palma sancona	Arecaceae	Ornamental	0–1400
Tabebuia rosea	Ocobo, guayacán amarillo, flormorado	Bignoniaceae	Timber harvest	300–1800

(continued)

Table 1 (continued)

Scientific name	Common name	Family	Utilitarian function	Altitudinal range
Tabebuia rosea-alba	Ocobo blanco	Bignoniaceae	Timber harvest	500–1800
Talauma caricifragrans	Alpargate, quesocurao	Magnoliaceae	Timber harvest	1600–2400
Tara spinosa	Dividivi de tierra fría	Leguminosae	Fuelwood and charcoal	1600–2900
Tecoma stans	Chirlobirlo, fresnillo	Bignoniaceae	Ornamental	0–2500
Terminalia amazonia	Chipo, vara de león	Combretaceae	Timber harvest	0–1400
Tessaria integrifolia	Sauce playero	Asteraceae	Watershed conservation	0–2000
Tetrorchidium boyacanum	Palo blanco arenillo	Euphorbiaceae	Timber harvest	1200–2300
Tetrorchidium rubrivenium	Aguamoso, arenillo	Euphorbiaceae	Timber harvest	0–1600
Thevetia peruviana	Cobalonga, abeto	Apocynaceae	Restoration	0–1500
Tibouchina lepidota	Siete cueros	Melastomataceae	Conservation	1700–2800
Tibouchina urvilleana	Nazareno	Melastomataceae	Conservation	1000–3000
Toxicodendron striatum	Pedro hernández	Anacardiaceae	Conservation	900–2300
Thibaudia andrei	Quereme	Ericaceae	Erosion avoidance	300–2600
Thibaudia archeri	Quereme	Ericaceae	Erosion avoidance	450–1500
Thibaudia floribunda	Quereme	Ericaceae	Erosion avoidance	1600–3600
Thibaudia pachyantha	Quereme	Ericaceae	Erosion avoidance	200–1300
Trema micrantha	Raspador, surrumbo	Cannabaceae	Erosion avoidance	0–2200
Tithonia diversifolia	Mirasol	Asteraceae	Fodder	1000–2400
Tournefortia scabrida	Guazimo blanco	Borraginaceae	Erosion avoidance	1000–3000
Tournefortia fuliginosa	Verdenegro	Borraginaceae	Restoration	1380–3700
Turpinia occidentalis	Mantequillo	Staphyleaceae	Restoration	100–3000
Trichanthera gigantea	Nacedero, madre de agua	Acanthaceae	Watershed protection	100–2000
Trichilia habanensis	Cedrillo, siguaraya	Meliaceae	Timber harvest, restoration	0–1500
Trichilia pallida	Caratero, cedrillo	Meliaceae	Restoration	200–1800
Trichospermum colombianum	Balso blanco	Malvaceae, subf. Tilioideae	Restoration	0–1300
Urera caracasana	Pringamosa, ortiga	Urticaceae	Restoration	500–2000

(continued)

Table 1 (continued)

Scientific name	Common name	Family	Utilitarian function	Altitudinal range
Vaccinium meridionale	Mortiño, agraz	Ericaceae	Watershed protection	1800–3100
Vachellia farnesiana	Pelá, aromo, cují	Leguminosae	Restoration	0–1300
Viburnum lasiophyllum	Sauco montañero	Adoxaceae	Timber harvest	1200–2845
Viburnum triphyllum	Garrocho, chuchua	Adoxaceae	Timber harvest	1700–3200
Vismia baccifera	Lacre, puntadelanza	Hypericaceae	Timber harvest	1100–2300
Vismia guianensis	Carate, puntadelanza	Hypericaceae	Watershed protection	1200–3000
Vitex cymosa	Aceituno	Verbenaceae	Fuelwood and food	0–1500
Virola sebifera	Sangretoro	Myristicaceae	Medicinal latex	0–1200
Vochysia ferruginea	Gomo, horcán	Vochysiaceae	Timber harvest	0–2000
Vochysia lehmannii	Bayo, floramarillo	Vochysiaceae	Living fence	0–1200
Warszewiczia coccinea	Crestegallo, rojo	Rubiaceae	Woodworking	200–1400
Weinmannia pubescens	Encenillo	Cunoniaceae	Timber harvest	1800–3000
Zanthoxylum monophyllum	Moló, tachuelo, doncel	Rutaceae	Tools handler	0–1300
Zanthoxylum rigidum	Tachuelo, doncel	Rutaceae	Constructions	0–1300
Zamia huilensis	Palma de monte	Zamiaceae	Wild foodstuff	900–1600
Zamia mantana	Zamia	Zamiaceae	Ornamental	1800–2700
Zamia poeppigiana	Zamia	Zamiaceae	Ornamental	200–2010
Zamia tolimensis	Zamia, palma de yuca	Zamiaceae	Wild food	1300–2000
Zygia longifolia	Amé, suribio	Leguminosae	Watershed protection	0–1500

where they debated with Mutis about *quinology*. In Bolivia, the forested regions of the Andes are estimated at about 55 million hectares, with the regions of Santa Cruz, Beni, Pando, and Cochabamba as most floristically diverse. In Chile, during colonial times, the priority over plants' medicinal use stood out to such an extent that the aforementioned publication by Ruiz & Pavón, which includes Chile, was described as "*Yerbatero*" because what is known as a pharmacy today, it was managed by "apothecaries and herbalists"; the Jesuit "*Botica*" of Santiago was famous, until 1767 when the 'Society of Jesus' was expelled, where the pharmacist José Zeitler stood out for his extensive knowledge in ethnobotany.

Another important component of mountain jungles is the fauna associated with this unique forest cover. TMCFs have a great diversity of animals that are indicator species for the Andean flanks. The most iconic of those are birds and mammals, most of whom are considered 'national emblems' throughout the tropical Andes, including the clay-colored thrush (*Turdus grayi*) in Costa Rica, the quetzal in Guatemala (*Pharomachrus mocinno*), the keel-billed toucan in Belize (*Ramphastos sulfuratus*), the motmot in Nicaragua (*Eumomota superciliosa*), the eagle in Panamá (*Harpia harpyja*), the trogon in Cuba (*Priotelus temnurus*), the troupial in Venezuela (*Icterus icterus*), hummingbirds in Colombia (*Eriocnemis mirabilis)* and Ecuador (*Oreotrochilus chimborazo*), cock-of-the-rocks in Peru (*Rupicola peruvianus*), and condors throughout the tropical Andes (*Vultur gryphus*). As per their restricted distribution to cloud forest areas or previously deforested areas now occupied with shrubland and grassland (*i.e.,* paramo, jalca, and puna), the most representative mammals include the Andean bear (*Tremarctos ornatus*), mountain tapir (*Tapirus pinchaque*), cougar (*Puma concolor*), red brocket deer (*Masama rufina*), llama (*Lama glama*), Andean fox (*Lycalopex culpaeus*), white tail deer (*Odocoileus virginianus ustus*), guinea pig (*Cavia porcellus*), Andean rabbit (*Sylvilagus brasiliensis andinum),* and Andean topo (*Hesperomys rostellatus*). Of course, due to the horizontal precipitation and availability of solar radiation, the forest belt could become prime territory for tree frogs (e.g., *Dendrobates spp., Phyllobates terribilis, Cochranella pulverata, Hyloscirtus hillisi*), arboreal lizards (*Liolaemus andinum, Anolis heterodermus*), and iguanas (*Stenocercus various, S. bolivarensis*) as well as the habitat for the impressive giant earthworm (*Martiodrilus crassus*).

3 The Cultural Richness of Tropical Landscapes

There is now more evidence of abundant human groups inhabiting the Andean flanks throughout the area of the Andean crescent towards Orinoquia and Amazonia, on one hand, and the presence of cultures settled in the western flanks towards the Choco and Guayas watersheds, including towards the drier slopes of Peru and Chile further south. In fact, among the oldest pre-Columbian "urban" complexes, the Santa Ana-La Florida formative village (*Mayo Chinchipe-Marañon* culture) stands out, located in the southeastern foothills of Ecuador (5500 BP) (Valdez, 2018), and the city of Caral (*Caral* culture) located in the Supe valley in northwestern Peru (5000 BP), considered the Americas' oldest city and one of the oldest in humanity (Solis, 2006). The recent discovery of a lost city in the Upano valley, at the shade of Sangay volcano in Ecuador, is likely to represent the oldest human settlement in the whole of Amazonia (Yépez & León, 2015; Rostain et al., 2024).

In the territory of the *Yumbu* culture (1000 BP—400 BP), northwestern Ecuador, which is now part of the Choco-Andean Biosphere Reserve, extensive settlements of monumental architecture (*i.e., tolas*, truncated pyramids, and agricultural terraces) were established and widely distributed in montane forests (Lippi, 2001). While in the Chanchán basin, on the western flanks of Ecuador, the biocultural footprint of tropical landscapes is materialized in an agro-ritualized landscape of the *Kañari* culture (2700 BP—600 BP), made up of crop fields, agricultural terraces,

Fig. 2 Legacies of a long history of changes in land use/land cover in the Andes can be seen with critical geomorphological views of mountainscape features: (**a**) Cropping on former cloud forests sites of the western Andean flanks in the Chanchán basin, Chimborazo province. (**b**) Permanent fields with introduced trees (*chakra*) on the piedmont and the Terraces (*pata*) still visible under the Cloud forests and shrubland on the slopes of Mt. *Joyagzhí*, Chanchán basin, Chimborazo province, Ecuador. (Photo: Christiam Aguirre)

tolas, and *pukaras* (Aguirre Merino et al., 2021). It was possible to demonstrate socio-ecological legacies of more than 3000 years, through the use of certain crop systems and agroecological practices that allowed the transformation of montane cloud forests into agroecosystems with high resilience and ecological elasticity (Aguirre Merino et al., 2023) Fig. 2.

Many other sites in the northern Andean highlands show a few evidence of canals or irrigation channels, other than drainage structures to protect the mountainsides. Conversely, the sites of the central Andes include sophisticated hydraulic engineering and monumental constructions for water transport and storage. However, because of the location of the active volcanism of the northern Andes, the development of cultures and ancient civilizations was curtailed by eruptions and other disasters (Hall & Mothes, 2008), such as in "Joyas de Cerén" in El Salvador, with the molds of people preserved ash-casted in the so-called "American Pompei." Important human imprint is evident in the Central American highlands and throughout the tropandean landscapes of Guatemala, Nicaragua, Costa Rica, and Panama, where the mysterious "*bolas*" or stone spheres remain in the montane landscape. An impressive necropolis in San Agustín, Colombia, evidence a large, yet unknown ancestral tropandean culture. The location of *chiripampa* on the shore of lake *Kulta* is the site of the first European colonial town, now the city of Riobamba, built upon large *Puruwa* dwellings in central Ecuador.

This cultural imprint is particularly important to determine the fate of the oldest human settlement registered in upper Amazonia, the *Sangay* culture in *Wapula*, at the shadow of Mt. *Sangay*, the most active volcano in the hemisphere (Yépez Noboa, 2017). Even in recent times, the fact that entire populations have been erased from maps, such in Armero, Colombia, with the devastating eruption of Arenas volcano in Nevado del Ruiz should guide our understanding of ancestral signatures of humans in montane landscapes (Esquivel et al., 2016). In the Sierra Nevada de

Santa Marta massif, deep inside the forest, the lost city of *Teyuna* remains as testament of the large human presence in the cloud forest areas currently uninhabited by people. The most significant of these are located in the mountains of Huila, at the archaeological park of San Agustín, with megalithic funerary monuments and statuary, burial mounds, terraces, funerary structures, stone statuary and the *Fuente de Lavapatas* site, a religious monument carved in the stone bed of a stream, assuring the intricate human presence in montane forest areas of the tropical Andes. The same can be said about numerous other sites along the Andean crescent, including the terraces of the Jubones river, the mounts of the *shwar* and a*chwar* tribes in Ecuador, or the mega constructions of the *Chachapoya*, the "people of the clouds," as shown by *Kuelap* or *Karaji*a. Also, the ruins of *Inka* citadels in *Choquequiraro*, *Ollantaytambo* or *Machu Picchu*, the most widely recognized "modern wonder of the world" in the cloud forests of Peru.

4 Evidence of Nature-Culture Linkages from the Past Flanks

So, despite its significant presence, the current conservation narrative insists considering only the animals and plants of the Andean forests as factors to consider for the planning and management of protected areas. This emphasis on biodiversity alone has molded public perception of the Andean flanks as areas with a very great presence of trees and other plants that occupy the verdant, and the animals, particularly birds and mammals, that are able to prosper in this difficult topography with no 'cultural importance'. However, because of recent paleoecological evidence for the significant role of paleoindigenous people—who modified the high Andean landscapes in the Holocene (Schiferl et al., 2023; Bush, 2023) reportedly by the use of fire for clearing the TMCFs upper border and increasing the paramo grassland areas (Sarmiento, 2012; White, 2013)—the forests' people once considered as hunters-gatherers of the Pleistocene, have recovered their primacy in the configuration of montane landscapes as well. The pollen fossil extracted from the site of 'Baeza Viejo,' currently covered by cloud forests of the *Quijos* River valley has shown to contain pollen of corn, amaranth, and other cultivars associated with agriculture, and some of the structural bases of houses, channels, aquifers, and the terracing still evident under the thick layer of plants covering mounds in the montane site (Sarmiento et al., 2022).

Another technological application is found in teledetection by remote sensors with laser reflectivity, like Light Detection and Ranging (LIDAR) and Ground Penetrating Radar (GPR). Both systems have provided a geospatial tool that uncovered a complex network of structures underneath the once-thought-of pristine cloud forest; these TMCFs sites that, with the geovisualization power of new geographical models, have helped to secure our understanding of the biocultural diversity and ancestral knowledge kept by the people of these areas. In many cases, such as those in the *Eje Cafetero* in Colombia, or throughout the *Ceja de Selva* in Peru, colonization for cattle ranching, gold mining and timber extraction have already modified these Andean flanks towards a homogeneous pasture with African grasses, Monterrey pines, or Australian eucalyptus trees and many forbs associated with weedy fields therein (Esquivel, 2015). The trend of losing biocultural heritage of TMCFs in the few areas acting like remnants in a commodified matrix (Montagnini, 2022) moved the need to find innovative ways to grapple with the human-dominated expansion. Recently, an international workshop on Ecological Legacies of the Upper Amazon held in Coca, Ecuador, produced a joint document supporting the bridging of science and community for upper Amazonian conservation (Textbox 1:

Textbox 1: The document known as the 'Coca Declaration' with the phonetic alphabet for Kuk'a Omagua, the name of the original people that lived in these Andean flanks of the Upper Napo River, in Ecuadorian Amazonia, when Spaniards entered the uncontacted territory. Inclusion of the 'Coca Declaration' helps to reaffirm the narrative of ecological legacies of the Andean flanks and the decolonial scholarship that prompts Montology to explain how the Andean flanks are indeed a cultural landscape for sustainable and regenerative development. (Source: Sarmiento et al., 2024)

Kuk'a Omagua Declaration (Coca or Orellana)

On behalf of those attending the international workshop on *Ecological Legacies and Ethnotourism: Nexus between Community and Science,* both the members of original communities and tourism operators and the participating national and foreign scientists, after presentations and deliberations for two days (May 19 and 20, 2023) in the tropandean city of Coca, the "gate" to the Amazon, declare:

Whereas

- *That the Higher Polytechnic School of Chimborazo* (ESPOCH) Orellana Campus, the Pan-American Center for Geographic Studies and Research (CEPEIGE), and the Neotropical Montology Collaboratory of the University of Georgia (UGA) have organized a successful international workshop on **Ecological Legacies and Ethnotourism: Nexus between Community and Science**.
- *That the participants representing the points of view* of Amazonian, national, and foreign universities, and of the original peoples' communities located in the Napo River basin of Western Upper Amazonia, **identified the need to join efforts between scientists and communities**.
- *That the urgency of finding alternate decolonial mod*els, allowing for transdisciplinary scientific activity, that favors inclusion of ancestral knowledge and ending objectification of local researchers and community scholars as mere informants, which is imperative in this time of global environmental change.

- *That inventories of flora and fauna be complemented* with linguistic records of the communities to maintain the lists of species with vernacular names and that new species that are being discovered receive the appropriate binomial nomenclature naming the species with the vernacular territorial name or of the community that owns that biocultural element.
- *That scientific research is segmented by short-term financing*, hence when the financing ends, the generated science ends. This hiatus implies the needed activation of museums, herbariums, insectariums, serpentariums, university organizations, and civil society organizations, supporting long-term scientific activity.
- *That the influence of the prevailing religion* (i.e., Catholic, Evangelical, etc.) has penetrated ancestral practices and rituals, as a sign of accelerated acculturation tendencies of some groups that assume it to satisfy the visitor curiosity, instead of invigorating shamanism, with local animism and vernacular literacy. In addition, the religious division generated between missioners of foreign cults sometimes caused violent crashes in the local communities (e.g., between the *Enomenga* Waorani in the *Dícaro* community and the *Iromenga* Waorani in the *Toñanpade* community).
- *That the future of tropandean forests in the Andean-Amazonian flanks* depends on the sustainability and regeneration of cultural landscapes, considering them from a montological, integrative, and transdisciplinary perspective, for which it is necessary to prioritize geo-ethnotourism operations over mining exploitation activities. and ecotourism that focus only on the fauna and flora of the Amazon Hylea and not on its custodians, and in many cases its creators of the domesticated and manufactured "garden jungle" of yesteryears.

Declare

- *To make joint efforts to favor hybridity of Amazonian culture-nature* and thus, maintaining the ecological legacy of the original peoples who survived with their traditional knowledge and vernacular descriptors that explain, according to ancestral science and their habitual practices, the cohabitation of non-human entities with humans.
- *To advocate for improving the dissemination of traditional and indigenous knowledge* in a multicultural way, so that scientists who come to study the jungle put aside elitist feelings and superiority of scientific studies on vernacular literacy, the jungle worldview, and ways of accessing it, to knowledge through shared methodologies and results, both in its planning and in its execution and subsequent publication.
- *To require researchers being able to communicate in the national language* (Ecuadorian Spanish) and, preferably, in the local language of

native nationalities of these territories (e.g., *a'ingae* –Kofán; *paicoca* – Siona dialect; *sikopai* –Secoya dialect; *waotededo* and *waotidido*—Sabela, Guikita, Tiweno and Aushiri dialects; *Chicham* or *šiwar'a*—Shwar, Shiwiar, Achwar dialects); and *kichwa* -Amazonian (or Eastern) dialect. Whenever possible, have international approval for research on human subjects and sign the ethical code of the International Society for Ethnobiology that requires both prior and informed knowledge and consent of the communities, as well as the equitable distribution of the results of the research, be it intellectual, academic, professional, or commercial and industrial.

- *To insist on creating research stations in productive socioecological landscapes* in which the communities become custodians of the jungle and administrators of the facilities for long-term research whose financing guarantees continuation of basic and applied research, as well as the monitoring of environmental conditions in socio-environmental aspects of protected cultural landscapes, especially in iconic trees, food supplies, memory reserves, sacred sites, landscape reserves, indigenous territories, literary reserves, or biocultural reserves and micro-refuges of biodiversity and cultural and linguistic diversity.
- *To recognize the persistence of knowledge and the will of the people* who are in the territory. The country's legal and legislative frameworks must be integrated into the disciplinary curricula of universities so that they have a more authentic interculturality. There must be motivation of foreign researchers to expand and facilitate the process of integration of knowledge and the formation of local expert knowledge, be it from grade school to postgraduate degrees, or from the oral history of local scholars who can be non-western science teachers of researchers who come from abroad.
- *To demand that Andean-Amazon flank's schools* teach content required by the national curriculum, but in the languages and dialects of the area, with trained presential (face-to-face) instructors who teach *kichwa* of the Amazon –not *kichwa* from the mountains, or who teach radiophonic schools in *kichwa* not only in Shwar– favoring distance education with online, virtual connections.
- *To promote the integration of elements of modernity without prejudice* to maintaining the indigenous identity by crafting religious syncretism and strengthening the transmission of ancestral knowledge in a proud revival of interculturality and intergenerational transmission of knowledge from the elderly and older adults in their non-school special education and vernacular literacy.
- *To insist on the need for this type of workshops and seminars* that integrate ecological legacies and geoethnotourism to be replicated in many other Amazonian sites, so that the local authorities in charge of managing GADs and government institutions value and prioritize the integration of communities and scientists.

Recommend and Commit

- *To promote and demand that the Law* of Planning for the Special Circumscription of the Amazon Territory with respect to gender equality, preferred employment, environmental protection, and wise use of natural resources be observed. Also, that the Fund for Sustainable Development of Amazonia (FDSA) and the Common Fund (FC) coordinated by the Amazonian Technical Secretariat, serve to augment the collaborative activities of scientific research and ethnotourism in the Amazonian decentralized, autonomous governments, including administrators, scientists, and community members working on those territories.
- *To prevent economic powers from prevailing over the ecological powers* of the Amazonian cultural landscape by limiting extractives actions (e.g., illegal mining, drilling, and oil exploration activities in protected areas, biopiracy, and abuse of native practices for folklorized tourism purposes, disguising their authenticity to please ephemeral visitors.
- *To entering communities with the attitude of peers*, recognizing the collaborative effort between ancestral knowledge and scientific knowledge, and thereby depositing the information and not just extracting it without local benefit or published record of intellectual authorship of the research.
- *To confabulate innovation with ancestral practice,* to ponder to the world the beauty of the Amazon generated by the jungle gardeners, the domestication and familiarization of the biota in the Pan-Amazon ancient and modern cultural landscape and the Andean-Amazon flanks.
- *To show respect for the ethnic group and the ancestral knowledge* that they still cherish. It is necessary to reclaim and sustain the Amazonian identity, eliminating features that 'folklorise' identity markers such as hammocks, necklaces, earrings, ribbons, *chigras*, tattoos of various colors and symmetries that reflect origin and affiliation instead of a mere ornament attractive to tourists.
- *To become aware of atrocities of the "conquest"* that must be analyzed not only in their historical but also moral and ethical frameworks. But we must look ahead. We are of original descent and most mestizos and foreigners respect the culture of the jungle, including music, gastronomy, art, and legends. We must weigh the message of our endangered beauties, preventing them from getting lost in the maelstrom of globalization. In doing so, recovering the valuation of characters, geographical landmarks, civic dates of nationalities and erect monuments to these indicators instead of the "conquerors" or "discoverers".
- *To pay for communities' instruction in their quest to defend their ethnicity* and for financing sustainable and regenerative development of native peoples, since they are a treasured legacy and they do not have access to

the available facilities. Therefore, to ask the public powers and the corresponding private, civil groups to then carry out the diverse, equitable, and inclusive scientific advancement with a pan-Amazonian inspiration.

- *To eliminate the idea of the community as an "object"* and to integrate it as a "subject" of science with considerations of convergent and integrative montology. To seek that the concept of geoethnotourism has a public policy ordinance. The government and regional development units should benefit from the investment, not as white elephants but as proactive action-units for the conservation of biocultural heritage.

Given in the city of Coca, Ecuador, in the auditorium of the Orellana headquarters of ESPOCH campus, on Saturday, May 20th, 2023, on behalf of the workshop participants,

The Organizing Committee,
Fausto Sarmiento, Ph.D.
CMN_UGA – CMS_IGU
Renato Chávez, Ph.D. (c)
ESPOCH
Nelson Ortega, M.Sc
CEPEIGE

Kuk'a Omagua Declaration) as one of the recent efforts to put transdisciplinary mountain science to inform tourists and residents about the nature-culture hybridity of the Andean flanks (Sarmiento et al., 2020).

5 Conclusions

Assuming the newest narrative of the culture-nature linkages associated with these small remnants of mature forest areas, the story of cultural manipulation of the Andean flanks still requires more affirmative discoveries. Recent research efforts in bridging the sciences and the humanities have helped significantly to conform teams of diverse disciplines looking at the conservation of mountain forests (Sarmiento et al., 2019) with specific aims of bridging the natural sciences with the human sciences informed by different knowledge bases, to recreate a wholistic view of the Andean forests of the flanks, as priority areas for conservation because of their maintenance as potential nurse trees or biocultural heritage microrefugia for the next generations, mostly due to the nuances of globalization, including global climate chance.

An important first step in the right direction is to emphasize the need to rewrite the geography textbooks of the Andean countries, emphasizing the existence of a "fifth" region on its own right, making the Andean Flanks and their associated communities count in the future plans for conservation of mountain landscapes. Affirming that the verdant is not only a gradient of altitude and humidity that has to be crossed quickly from the Sierra to the Selva of from the Sierra to de Sea, the false premise of the ecotone of the cloud forest will disappear from the minds of young students (Wayusada, 2023). If more books on Andean Forests are published with the current trend of "decolonial" scholarship, we shall find young indigenous authors writing about their legendary mountain forests, with an intimate appreciation of the nature-culture hybridity that these sentient montane environments claim to protect.

Acknowledgments We are grateful to former students who contributed with a better appreciation of the montology turn and transdisciplinary understanding of the Andean Flanks. FS thanks the NSF grant 2148984 to Florida International University's UGA subaward RFLIT000151490P and to ESPOCH-UGA collaborative grant.

References

Aguirre Merino, C. P., Córdova Lliquín, J. D., & Piqué, H. R. (2021). Agriculturización: la alquimia de un paisaje precolombino Kañari en la cuenca alta del río Chanchán, Andes del Ecuador. *Estudios atacameños, 67*.

Aguirre Merino, C. P., Piqué Huerta, R., Parra Ordoñez, L. N., Guamán Cazho, V. A., & Valdez Bustamante, W. O. (2023). The archeological landscape of the Chanchán Basin and its Agroecological legacies for the conservation of montane forests in the Western Foothills of the Ecuadorian Andes. *Land, 12*(1), 192.

Andreas, P. R., & Sharpe, K. E. (1992). Cocaine politics in the Andes. *Current History, 91*(562), 74–79.

Brown, D. A., Grau, H. R., Malizia, L. R., & Grau, A. (1991). Argentina. In Kapelle & Brown (Eds.), *Bosques Nublados del Neotrópico* (p. 623). Inbio.

Bruijnzeel, L.A., Kappelle, M., Mulligan, M. and Scatena, F.N. (2010). Tropical montane cloud forests: state of knowledge and sustainability perspectives in a changing world.

Bush, M. B. (2023). The palaeoecological view from the mountains. In F. O. Sarmiento (Ed.), *Montology palimpsest: A primer of mountain geographies* (pp. 257–273). Springer.

Capello, E. (2018). From imperial pyramids to anticolonial sundials: Commemorating and contesting French geodesy in Ecuador. *Journal of Historical Geography, 62*, 37–50.

Cuatrecasas, J. (1968). Páramo vegetation and its life forms. *Colloquium Geographicum, 9*, 163–186.

Dangles, O. (2023). Climate Change on Mountains: Reviving Humboldt's Approach to Science. Springer Nature.

Esquivel, H. E. (2012). *Flora Arbórea de la Ciudad de Ibagué*. Universidad del Tolima y CORTOLIMA. Diseño e impresión, Imprenta Tolima. 1° reimpresión. 643 pp.

Esquivel, H. E. (2015). *Arvenses de la Familia Asteraceae o Compositae en la Región Central de los Andes de Colombia* (p. 195). Sello editorial Universidad del Tolima. Imprenta Tolima-Ibagué.

Esquivel, H. E., & Nieto, A. (2003). *Diversidad Florística de la Cuenca Alta del Río Combeima* (p. 190). Programa PEI del Cañón del Combeima. Publicidad & Marketing.

Esquivel, H. E., Tinoco, F., & Torres, A. J. (2016). La Sucesión vegetal en los lodos fluviovolcánicos de Armero-Tolima-Colombia 30 años después de la erupción del volcán Arenas del nevado del Ruiz. *Caldasia, 38*(1), 101–116.

Gentry, A. H. (1988). Changes in plant community diversity and floristic composition on environmental and geographical gradients. *Annals of the Missouri Botanical Garden, 75*, 1–34.

Hall, M. L., & Mothes, P. A. (2008). Volcanic impediments in the progressive development of pre-Columbian civilizations in the Ecuadorian Andes. *Journal of Volcanology and Geothermal Research, 176*(3), 344–355.

Hamilton, L. W., Juvik, J. O., & Scatena, F. N. (Eds.). (1994). *Tropical montane cloud forests* (Ecological Studies 110). Springer.

Holdridge, L. R. (1967). *Life zone ecology*. Tropical Science Center. (Revised Edition from Original 1947).

Kappelle, M., & Brown, A. (Eds.). (1991). *Bosques Nublados del Neotrópico*. Instituto Nacional de Biodiversidad (INBIO).

Kesler, M., & Beck, S. G. (1991). Bolivia. In Kapelle & Brown (Eds.), *Bosques Nublados del Neotrópico* (pp. 581–622). Inbio.

Körner, C., Paulsen, J. and Spehn, E.M. (2011). A definition of mountains and their bioclimatic belts for global comparisons of biodiversity data. Alpine Botany, 121, pp.73–78.

Körner, C., & Hoch, G. (2023). Not every high-latitude or high-elevation forest edge is a treeline. *Journal of Biogeography, 50*, 838–845.

Lippi, R. D. (2001). Yumbo demographics before and after 1532. In *29th annual Midwest conference of Andean and Amazonian Archaeology and Ethnohistory*. University of Michigan.

Malanson, G. P. (2023). Inclusions and exclusions in treeline definitions. *Journal of Biogeography, 51*, 54–56. https://doi.org/10.1111/jbi.14729

Montagnini, F. (Ed.). (2022). *Biodiversity islands: Strategies for conservation in human-dominated landscapes*. Springer.

Mutis, J. C. (1828). *El arcano de la quina: discurso que contiene la parte médica de las cuatro especies de quinas oficiales, sus virtudes eminentes y su legítima preparación*. Impresor de Cámara de Su Majestad.

Myster, R. (Ed.). (2023). *Ecological gradients and their analyses*. Springer. https://doi.org/10.1007/978-3-031-22848-3_16

Paulsch, A. (2002). *Development and application of a classification system for undisturbed and disturbed tropical montane forests based on vegetation structure*. Doctoral dissertation.

Rostain, S., Dorison, A., De Saulieu, G., Prümers, H., Le Pennec, J. L., Mejía-Mejía, F., Freire, A. M., Pagán-Jiménez, J. R., & Descola, P. (2024). Two thousand years of garden urbanism in the Upper Amazon. *Science, 383*(6679), 183–189. https://doi.org/10.1126/science.adi6317

Sarmiento, F. O. (1987). *Antología Ecológica del Ecuador: desde la selva hasta el mar*. Casa de la Cultura Ecuatoriana.

Sarmiento, F. O. (2012). *Contesting Páramo: Critical biogeography of the Northern Andean Highlands* (p. 150). Kona Publishing. Higher Education Division. https://news.uga.edu/book-takes-a-critical-view-on-paramo/

Sarmiento, F. O., & Frolich, L. (2002). Andean cloud forest treelines: Naturalness, agriculture and the human dimension. *Mountain Research and Development, 22*(3), 278–287. https://doi.org/10.1659/0276-4741(2002)022[0278:ACFTL]2.0.CO;2

Sarmiento, F.O., J.A. Gonzalez, E.O. Lavilla, M. Donoso and J.T. Ibarra. (2019.) Onomastic misnomers in the construction of faulty Andeanity and week Andeaness: Biocultural Microrefugia in the Andes. Pirineos, Journal of Mountain Ecology, 174, 1–16.

Sarmiento, F. O., & Sarmiento, E. V. (2021). *Flancos Andinos: Paleoecología, Biogeografía Crítica y Ecología Política en los Climas Cambiantes de los Bosques Neotropicales de Montaña*. Editorial INDES_CES/UNTRM. Universidad Nacional Toribio Rodríguez de Mendoza, Amazonas. https://doi.org/10.5281/zenodo.4746912

Sarmiento, F. O., Bush, M. B., Church, W., VanValkenburgh, P., Oliva, M., Delgado, E., Fernandez, S., & Rojas, N. (2020). Mountain science poised to help ecotourism in Peruvian cloud forests. *PAGES, 28*(1), 22–23. https://doi.org/10.22498/pages.28.1.22

Sarmiento, F. O., Rodríguez, J., & Yépez-Noboa, A. (2022). Forest transformation in the wake of colonization: The Quijos Andean-Amazonian Flank, past and present. *Forests, 13*(11), 1–19. https://doi.org/10.3390/f13010011

Sarmiento, F. O., Haller, A., Branca, D., Aguirre, C., Marchant, C., & Yoshida, M. (2023). Chapter 16: Socioecological gradients: Contesting traditional ecoclines to explain the high biocultural diversity of the Andean verdant. In R. Myster (Ed.), *Ecological gradients and their analyses*. Springer.

Sarmiento, F. O., Bush, M. B., McMichael, C. N. H., Chávez, C. R., Cruz, J. F., Rivas, G., Kavoori, A., Weatherford, J., & Hunt, C. A. (2024). Ecological legacies and ethnotourism: Bridging science and community in Ecuador's Amazonia. *Sustainability, 16*(11), 4664. https://doi.org/10.3390/su16114664

Schiferl, J., Kingston, M., Åkesson, C. M., Valencia, B. G., Rozas-Davila, A., McGee, D., Woods, A., Chen, C. Y., Hatfield, R. G., Rodbell, D. T., & Abbott, M. B. (2023). A neotropical perspective on the uniqueness of the Holocene among interglacials. *Nature Communications, 14*(1), 7404.

Solis, R. S. (2006). America's first city? The case of Late Archaic Caral. In *Andean archaeology III: North and south* (pp. 28–66). Springer.

Valdez, F. (2018). *Primeras Sociedades de la Alta Amazonía: La Cultura Mayo Chinchipe-Marañón*. IRD Éditions.

Vidal, J. P. (2014). *Las ocho regiones naturales del Perú*. Terra Brasilis. Revista da Rede Brasileira de História da Geografia e Geografia Histórica. 3.

Wayusada. (2023). https://wayusada.com. Accessed 28 Jan 2024.

White, S. (2013). Grass páramo as hunter-gatherer landscape. *The Holocene, 23*(6), 898–915.

Yépez, A., & León, E. (2015). *Bajo las cenizas del volcán Sangay: ocupaciones, volcanes y adoratorios*. Editorial Casa de la Cultura Ecuatoriana.

Yépez Noboa, A. M. (2017). *Sobre las nieblas de los volcanes Chimborazo y Sangay: Asentamientos humanos milenarios y paleoambiente*. Pontificia Universidad Católica del Ecuador. pucedspace.puce.edu.ec.

Cloud Forests in the Venezuelan Andes: A Review of Functional Characteristics at Ecosystem and Plant Scale

Carlos García-Núñez, Fermin Rada, and Michele Ataroff

1 Introduction

Worldwide, tropical mountain cloud forests (TMCFs) occur anywhere between 800 and 3500 m a.s.l., in the cloud immersion zone with lower and upper limits varying greatly depending on global, regional and local environmental factors that have an influence on cloud formation (Bruijnzeel et al., 2011; Mulligan, 2010). The global extension of TMCFs is about 215,000 km^2 which represents 1.4% of the total area of all tropical forests (Oliveira et al., 2014). Frequent and persistent cloud immersion and/or fog result in a distinct environment (i.e. cool temperature, high air relative humidity, low radiation input), which influences vegetation characteristics (e.g. canopy cloud water interception, low evapotranspiration). These characteristics are linked to the ecosystem services provided by TMCFs such as water supply, ability to store substantial amounts of carbon in the above and belowground biomass as well as in the form of soil organic matter (SOM), and high levels of biodiversity and endemism (Raich et al., 2006; Bruijnzeel et al., 2010; Aparecido et al., 2018; Duque et al., 2021). Changes in temperature, solar radiation and nutrient availability along

C. García-Núñez (✉)
Instituto de Ciencias Ambientales y Ecológicas (ICAE), Facultad de Ciencias, Universidad de Los Andes, Mérida, Venezuela

Center for Urban and Global Studies (CUGS), Trinity College, Hartford, CT, USA

F. Rada
Instituto de Ciencias Ambientales y Ecológicas (ICAE), Facultad de Ciencias, Universidad de Los Andes, Mérida, Venezuela

Laboratorio de Ecología y Fisiología Vegetal, Universidad de Los Andes, Bogotá, Colombia

M. Ataroff
Instituto de Ciencias Ambientales y Ecológicas (ICAE), Facultad de Ciencias, Universidad de Los Andes, Mérida, Venezuela

N. Clerici (ed.), *Conservation of Andean Forests*,
https://doi.org/10.1007/978-3-031-80805-0_2

elevational gradients are the main drivers determining structural and functional characteristics of TMCFs (Raich et al., 1997, 2006; Wang et al., 2003; Homeier et al., 2010; Moser et al., 2007, 2011; Hager & Dohrenbusch, 2011; Girardin et al., 2014a; Clark et al., 2015).

Overall, ecosystem structural traits (e.g. canopy height, basal area, leaf area index (LAI)) and processes (e.g. carbon, nutrients and water budgets) of TMCFs are characterized by relatively lower: canopy height, above-ground biomass, net primary productivity, and relatively higher: plant water potentials (owing to cool temperatures and extra water input by horizontal precipitation), and soil carbon stocks (low decomposition rates), compared to lowland rainforests (Bruijnzeel et al., 2011; Malhi et al., 2017). The ecological interactions that determine the distinctive montane cloud forests characteristics have been well documented and summarized (Grubb, 1977; Cavelier, 1996; Bruijnzeel et al., 2010; Fahey et al., 2016; Aparecido et al., 2018). In particular, Fahey et al. (2016) proposed a hypothetical framework in which the combination of environmental factors (e.g. low temperatures, cloud immersion, high relative humidity, high precipitation, low light levels, nitrogen limited soil) in TMCFs led to a stunted structure and low above-ground production.

Ecosystem processes such as forest productivity and nutrient cycling have great significance in the context of a warming world (Pan et al., 2013). Scientists are facing the enormous challenge of predicting how climate change will affect species distribution and species assemblages (Lavergne et al., 2010; Fadrique et al., 2018; Tito et al., 2020). How tropical forests will respond to this change, dampening or accelerating its ripple effect, requires mechanistic answers (Goldstein & Santiago, 2016; Malhi et al., 2020). These topics have particular importance in cloud forests, for which their persistence depends on narrow bands of environmental conditions (Oliveira et al., 2014). Variations in extreme events such as El Niño-Southern Oscillation, a major driver of seasonal and climatic variability across the tropics producing severe drought-inducing meteorological anomalies, may also have a significant effect (Roman-Cuesta et al., 2014; Oliveras et al., 2017; Rifai et al., 2019). Different models on the future of tropical Andean biomes have been proposed (Tovar et al., 2013; Helmer et al., 2019). Conservative models point to a relative stability, between 74.8% and 83.1% of the tropical Andes will remain stable, while rather catastrophic ones suggest generalized declines in surface area due to the reduction, dryness or disappearance of the habitat. In any case, the environmental variability of the Andean region complicates the predictions that can be attained (Killeen et al., 2007). In view of their constrained elevation range, many TMCFs species present high extinction risks (Still et al., 1999; Pounds et al., 1999; Bush et al., 2004; Helmer et al., 2019).

In the last decade, a great deal of information has been produced on the structure and functioning of Andean TMCFs. At the ecosystem level, studies have focused on the measurement of primary productivity and the factors and processes that control it (Leuschner et al., 2013; Girardin et al. 2014a, b, c; van de Weg et al., 2014; Malhi et al., 2017; Vilanova et al., 2018; Phillips et al., 2019; de la Cruz-Amo et al., 2020). At a plant scale, the emphasis has been placed on functional traits associated to carbon and water relations (Choat et al., 2012; Goldsmith et al., 2013; Atkin et al., 2015; Guan et al., 2015; Asner et al., 2016; Bahar et al., 2016; Blonder et al., 2017; Doughty et al., 2017; Enquist et al., 2017; Fyllas et al., 2017; Bruelheide et al., 2018; Feeley et al., 2020; Martin et al., 2020; Oliveras et al., 2020; Llerena-Zambrano et al., 2021). The data derived from intensive study sites, and the whole carbon budget approach used in some of these investigations, allows further understanding of productivity variation and its allocation along a tropical elevation gradient, as well as the parameterization of process-based models.

In the case of Venezuela, there is an extensive set of data on the structure and floristic composition of cloud forests. These works cover: distribution (forests above 800 m in altitude), and detailed floristic lists with structural characterizations along with environmental variables (Lamprecht, 1954; Veillon, 1965, 1985, 1989; Huber, 1986a, b; Medina, 1986; Vareschi, 1992; Bono, 1996; Michelangeli, 1998; Ataroff, 2001, 2003; Ataroff & García-Núñez, 2013). Conversely, studies on ecosystem processes and plant functional traits are relatively limited and concentrated mainly in two localities of cloud forests in the Venezuelan Andes (La Carbonera (LC) and La Mucuy (LM)), and two localities in the Cordillera de la Costa (Rancho Grande and Altos de Pipe) (Huber, 1986a, b; Medina, 1986; Ataroff & García-Núñez, 2013). Additionally, most of the information on these ecosystem processes and functional traits are of limited access to readers as they are dispersed in theses, book chapters and reports. A compilation and evaluation of this information will help understanding Venezuelan Andean cloud forests and how their characteristics compare to other TMCFs.

Beginning in the 1970s, intensive ecosystem level studies on forest productivity, nutrients and water budgets were performed in LC and LM. These studies benefited from the fact that the sites constitute large extensions of continuous forest in protected areas with different environmental conditions (although the sites are only 30 km apart, climatic, geologic and topographic differences are remarkable) (Steinhardt, 1979; Fassbender & Grimm, 1981a, b; Grimm & Fassbender, 1981a, b; Delaney et al., 1997; Ataroff & Rada, 2000). This information is essential for the comprehension of the relationship between the environmental features of cloud forests and forest productivity, and to understand its resilience to human-driven perturbations (Berendse, 1994; Silver et al., 1996). These ecosystem studies acquired great relevance since Andean cloud forests in Venezuela have undergone a strong deforestation process (1–2.4% year^{-1}, assessment between 1952 and 1997) (Rodríguez-Morales et al., 2009; Ataroff & García-Núñez, 2013). For example, an evaluation of the Capaz River watershed (where the LC protected area is located) showed that since the 1950s the landscape has been fragmented into small forest patches (fragments ranging from 356 ha in 1952 to 80 ha in 1997) surrounded by a matrix of grasslands and crops (Rodríguez-Morales et al., 2009).

This chapter offers a review of the functions of TMCFs at the ecosystem (i.e., forest productivity) and plant (i.e., water and carbon relations) levels, with an emphasis on Venezuelan forests, and particularly on the work carried out at LC and

LM in the Merida Cordillera, two of the most intensively studied cases in the Venezuelan Andes. We gather, apart from material published in specialized journals, valuable literature of limited circulation found in books, book chapters and graduate and undergraduate theses. We analyze and summarize this scattered information covering two different ecological scales to have a synthetic and updated view and to highlight important aspects of current research related to TMCFs. Our objective is to contrast the findings on function in Venezuelan Andean cloud forests with the current state of knowledge for TMCFs, to reveal its status, particularly in the context of the Andean forests. In some instances, we make comparisons with lowland tropical moist forests to place the functioning of TMCFs into perspective.

We specifically address the following topics:

1. Ecosystem functional aspects: Forest productivity (aboveground wood productivity, biomass stocks, litterfall production, decomposition rates and nutrient budgets).
2. Plant scale aspects: Water and carbon relations (stomatal conductance, water potentials and photosynthetic responses).

Regarding the general environmental and vegetation characteristics of the Venezuelan Andean cloud forests, particularly for LC and LM, see Chapter "A multiscale ecological approach for the conservation and restoration of the Venezuelan Andean cloud forests" in this book.

2 Ecosystem Functional Aspects: Forest Productivity

2.1 Aboveground Wood Productivity

The influence of climatic drivers on forest dynamics and aboveground wood productivity (AGWP) has been analyzed in comparative studies of lowland and montane forests in Venezuela (Vilanova et al., 2018) and along the Andes (Duque et al., 2021). In general, the study by Vilanova et al. (2018) indicated that stands located at higher elevations (LC, 2312–2452 m a.s.l.) exhibited high aboveground biomass (AGB) while having lower AGWP and turnover rates. AGWP of high Andean forests showed intermediate values (2.39 Mg C ha^{-1} $year^{-1}$), among the lowland wet forests (3.26 Mg C ha^{-1} $year^{-1}$) and coastal dry forests (1.76 Mg C ha^{-1} $year^{-1}$) (Vilanova et al., 2018). This trend in AGWP could be related, as shown by other studies, to responses to cooler temperatures, frequent cloud cover or mist which reduce light incidence and influence nutrient and water soil conditions (Grubb, 1977; Aylett, 1985; Cavelier, 1996; Jarvis & Mulligan, 2010; Girardin et al., 2014a). Recent findings on the patterns and processes of forest productivity along altitudinal gradients in the Peruvian Andes (Malhi et al., 2017) show that low growth rates at high elevations were primarily driven by low gross primary productivity (GPP), possibly caused by cloud immersion effects within the cloud forest zone. These authors also describe small altitudinal shifts in carbon use efficiency, allocation of net primary productivity (NPP) between wood, fine roots and canopy, mortality rate/residence time and LAI, and invariant leaf photosynthetic rates. Nevertheless, in the Andean Cordillera, forest productivity may be highly variable. For an elevation gradient in Ecuador, Leuschner et al. (2013) found a decline in GPP associated with a strong decline in LAI, and an increased allocation of NPP towards roots at high elevations. On the other hand, for Andean forests, studies of aboveground carbon dynamics in a broad elevation range across the tropical and subtropical Andes (Duque et al., 2021), showed no significant associations between AGWP and elevation (ranging from approximately 0.5 to 3.8 Mg C ha^{-1} year^{-1}), being highest near the equator. Besides the climatic factors, other biotic factors were important in determining AGWP, e.g., size-dependent mortality of trees, arbuscular mycorrhizal fungal associations, and the strong influence of large-statured clades (Duque et al., 2021).

2.2 *Biomass Stocks in Cloud Forests*

Research on organic and mineral reserves in the Andean cloud forests of Venezuela was directed towards biomass inventories using destructive (Grimm & Fassbender, 1981a) and biometric methods (Delaney et al., 1997; Vilanova et al., 2018) and litterfall production and decomposition studies (Fassbender & Grimm, 1981a; Ramírez, 1999). The amount of AGB reported for LC by Grimm and Fassbender (1981a) was 347.7 Mg ha^{-1} of standing biomass. Other compartments amounted to 701.3 Mg ha^{-1} for a total of 1049 Mg ha^{-1} of organic matter in the ecosystem (See Fig. 1 for details on all compartments). Delaney et al. (1997) reported 346 and 314 Mg ha^{-1} in AGB with DBH $\geq$ 10 cm, and 506 and 514 Mg ha^{-1} in the soil at 0 to 100 cm in depth, for LC (2310–2450 m a.s.l.) and LM (2640–3000 m a.s.l.), respectively. More recently, a reassessment of AGB for montane forests in Venezuela (Vilanova et al., 2018) supported the findings of Grimm and Fassbender (1981a) and Delaney et al. (1997) for LC. Vilanova et al. (2018) found that turnover rates (stem dynamics calculated as the average of recruitment and mortality rates) and climate are key drivers of forest biomass. Montane forests with low turnover rates, as a result of low temperatures at higher elevations (e.g. TMCFs in the Andes; 2312–2452 m a.s.l.), tend to have high AGB with stands dominated by medium to high wood density species (Vilanova et al., 2018). One potential explanation for the relatively high AGB in the TMCFs is that despite having a dry season (i.e. less rainfall and shorter periods of clouds and fog on a daily basis), lower temperatures at higher elevations promote lower evapotranspiration and a higher carbon residence time (Vilanova et al., 2018). Furthermore, studies carried out in LC show that topography (e.g., slope) and the physical characteristics of the soil related to water content constitute an important ecological gradient that explains the differences in forest structure, productivity and organic and inorganic soil matter stocks (Hetsch & Hoheisel, 1976; Quiroz, 2010). Organic carbon accumulation in aboveground

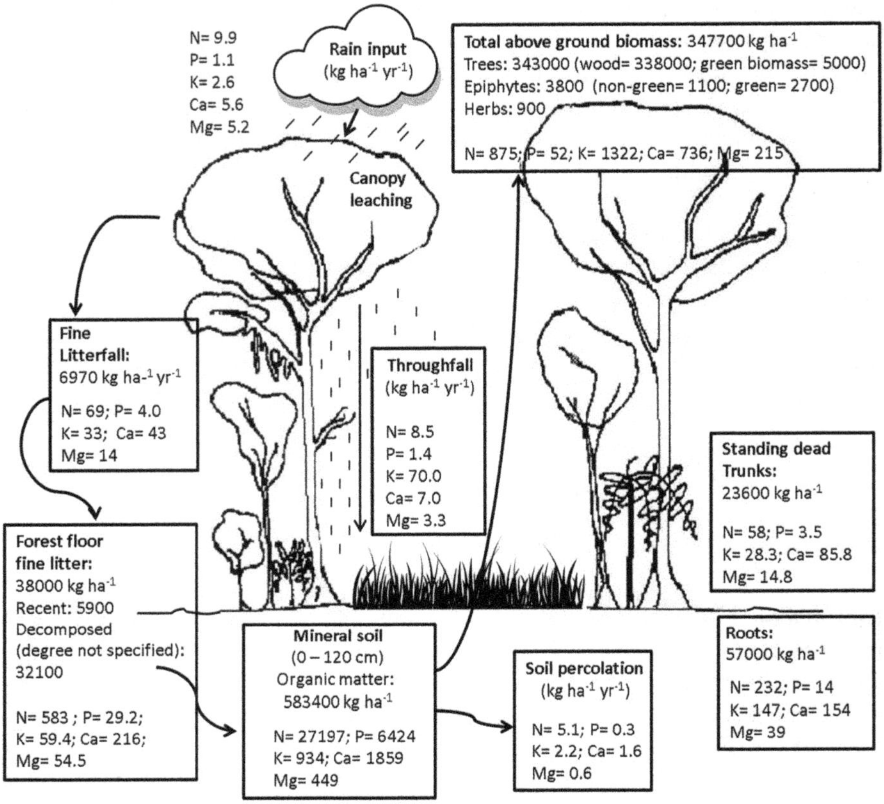

Fig. 1 Simplified diagram of the main carbon and mineral stocks and fluxes for La Carbonera, Venezuela. Values are in kg ha^{-1} or kg ha^{-1} year^{-1} dry weight. Source: Grimm and Fassbender (1981a, b); Fassbender and Grimm (1981a)

biomass (ABG) and soil mineral reserves are strikingly reduced in soils with lower slopes and marked hydromorphism (Hetsch & Hoheisel, 1976; Quiroz, 2010).

The stand structural characteristics of the studied cloud forests (i.e. LC and LM) were typified by a high mean canopy height (20–30 m) with the presence of large-trunk tree species that reach 30–35 m in height, and high basal areas (30–69 m^2 ha^{-1}) (Veillon, 1965; Vareschi, 1992; Quiroz, 2010). For example, in LC, few species (e.g. *Retrophyllum rospiglisii*, *Vochysia meridensis*, *Myrcia acuminata*) comprise almost half (46%) of the basal area (24 m^2 ha^{-1}) in a forest stand (Quiroz, 2010). Similar AGB trends in a few large-trunk tree species has been described in lowland and montane tropical moist forests (Slik et al., 2013; Duque et al., 2021; Maza et al., 2022). Comparatively, canopy height, basal area and AGB, exhibited lower values in a network of six Andean elevation transects from lowland Amazonian rain forests to high-elevation cloud forests in Ecuador, Peru, and Bolivia (Girardin et al., 2014a), with mean values ranging between 15–25 m for canopy height, 20–40 m^2 ha^{-1} for basal area, and 80–210 Mg ha^{-1} for AGB, at elevations between 2000–2500 m a.s.l.

Similarly, Duque et al. (2021) found lower AGB values (ranging between 104.32–148.18 Mg ha^{-1}) at elevations between 2000–2800 m a.s.l. for the whole subtropical and tropical Andean region. An extensive study on AGB of tropical montane forests (TMFs) worldwide showed that it varied from 77 to 785 Mg ha^{-1} with a mean storage of 271 Mg ha^{-1} (Spracklen & Righelato, 2014). This survey shows that AGB declines moderately with both elevation and slope angle but that TMFs store substantial amounts of biomass, both at high elevations (up to 3500 m) and on steep slopes (slope angles of up to 40°).

The AGB reported for the Cordillera de Merida (Grimm & Fassbender, 1981a; Delaney et al., 1997, 1998; Vilanova et al., 2018) were high, and fall within the range for lowland tropical forests (233–435 Mg ha^{-1}) (Golley, 1971; Grubb, 1977; Kira, 1977; Klinge, 1977; Pan et al., 2013). Vitousek and Sanford Jr. (1986) reported this pattern for AGB (including our focal area, LC), showing that montane forests could have similar values to lowland forests and that there is no strong link between biomass accumulation and nutrient availability for lowland (on fertile and infertile soils) and montane tropical forests. Total AGB is strongly dependent on stand age (successional status) as well as climate and soil conditions (Vitousek & Sanford Jr., 1986). Nevertheless, a more recent analysis (i.e. 92 forests in different climate zones) revealed that nutrient availability indeed plays a crucial role in determining global forest carbon balance and particularly on the capacity of forests to sequester carbon (i.e. as more nutrients are available, more carbon is sequestered by forests) (Fernández-Martínez et al., 2014). Besides, for LC and LM, SOM represents 55% of the total reserves (Grimm & Fassbender, 1981a). This result agrees with studies performed in Ecuador and Peru, which found that cloud forest carbon pools were equivalent or even higher than in neotropical lowland forests, mainly due to the relatively high soil carbon stock (Leuschner et al., 2013; de la Cruz-Amo et al., 2020).

2.3 *Litterfall Production and Decomposition Rates*

Studies on the productivity of cloud forests in Venezuela have been mostly directed to the measurement of litter production. The rates of litterfall for LC and LM ranged between 6.9–7.7 Mg ha^{-1} year^{-1} (Fassbender & Grimm, 1981a; Ramírez, 1999), close to the average (6.4 Mg ha^{-1} year^{-1}) reported for montane forests (Vitousek & Sanford Jr., 1986). The rates reported for montane and lowland tropical moist forests ranged between 2.4–12.4 Mg ha^{-1} year^{-1} (Cuevas & Medina, 1986; Vitousek & Sanford Jr., 1986; Scott et al., 1992; Lisanework & Michelsen, 1994; Pendry & Proctor, 1996; Chave et al., 2010). Arends et al. (1991) found similar litter production in a 10 years old secondary vegetation from cut down abandoned plots in LC compared to the original neighboring forest (7.1 vs. 7.4 Mg ha^{-1} year^{-1}), even though biomass and number of species only achieved a small fraction (12% and 14%, respectively) of the original forest. Leuschner et al. (2013) described a strong decline in litter production from 1890 to 2380 m a.s.l. (i.e. 8.3 and 3.7 Mg ha^{-1} year^{-1}, respectively) for a TMCF in Ecuador. This result may be related to a decline in canopy photosynthesis due to the reduction of LAI as a response to nutrient supply limitations with increasing elevation (Weaver & Murphy, 1990; Kitayama & Aiba, 2002; Moser et al., 2007).

The amounts of N transferred annually through litterfall for LC and LM were 69 and 104 kg ha^{-1} year^{-1}, respectively, comparable to the average values reported for tropical moist montane and lowland (on infertile soils) forests (64 and 108 kg ha^{-1} year^{-1}, respectively) (Vitousek & Sanford Jr., 1986). The higher N transferred in LM could be due to the higher soil fertility compared to LC (i.e. Soil CIC of 15 and 45 meq/100 g for LC and LM, respectively) (Vareschi, 1992; Ramírez, 1999). Many species show significant rates of retranslocation before leaf abscission as part of nutrient economics strategies. Retranslocation rates for montane tropical moist forests worldwide range from 16 to 48% of N and 25 to 69% of P (Cavelier, 1996). N retranslocation reported for LC and LM (40 and 44%, respectively) are similar to the average reported for montane forests, while P retranslocation reported for LC (32%) was lower (Fassbender & Grimm, 1981a).

Litter restores a fraction of the assimilated nutrients by plants to the soil surface and depends on decomposition processes (Swift et al., 1979; Salinas et al., 2011; Esquivel et al., 2020; Ostertag et al., 2022). Mulch disappearance rates of 72 and 75% per year have been measured for LC and LM, respectively (Fassbender & Grimm, 1981a; Ramírez, 1999). In LC, total litter decomposition occurred after a 20-month exposure, branches required 12.5 years to disappear (annual rate = 8%) (Fassbender & Grimm, 1981a). For a wide range of montane forests around the world, the rate of litter disappearance was approximately 60% (7 months) (Ostertag et al., 2022).

2.4 Nutrient Budgets

Biogeochemical studies in LC determined the stocks of different nutrient elements in the vegetation and soil (Grimm & Fassbender, 1981a). The element with the highest content in vegetation compartments was N, followed by K, Ca, Mg and P; in the mineral soil (0–120 cm) plus mulch, the highest concentration was N followed by P, Ca, K, and Mg. The mineral reserves, N (Kjeldahl method) and P (digestion with a mixture of acids; forms of P after fractionation with NH_4F (Al-P), with NaOH (Fe-P) and H_2SO_4 (Ca-P)) are total, while K, Ca, and Mg represent only the exchangeable fraction (atomic absorption spectrophotometry) (Fig. 1). In addition to litter production, there is an important movement of nutrients from the AGB to the soil that depends on water flows. Steinhardt (1979) determined that inputs of nutrients by precipitation are higher than the percolation outputs, resulting in a positive balance for the forest (Fig. 1).

For LC, the amount of N accumulated in the vegetation (875 kg ha^{-1}) was higher than the average reported for montane forests (671 kg ha^{-1}) (Vitousek & Sanford Jr., 1986). The range for montane and lowland tropical moist forests is between 32 and

2430 kg ha^{-1} (Vitousek & Sanford Jr., 1986). The soil interchangeable reservoirs of K, Ca, Mg were relatively low due to the low soil pH (Fassbender & Grimm, 1981b; Grimm & Fassbender, 1981a). The N accumulated in the superficial soil (3223 kg ha^{-1} at 10 cm in depth) was close to the average reported for different montane and lowland tropical moist forests (range 1000–6048 kg ha^{-1}) (Grimm & Fassbender, 1981a; Soethe et al., 2008). Analysis of nutrient cycling in tropical moist forests reveals that tropical montane forests in general seem to be low in nitrogen, and upper montane forests cycle substantially less nitrogen than do lower montane forests (Vitousek & Sanford Jr., 1986). Slow nitrogen cycling rates, rather than low nitrogen inputs, appear to be responsible for the decreased nitrogen supply, and the high soil water content of many cloud-immersed montane forests is likely to be a major ultimate cause of declining nitrogen cycling rates (Benner et al., 2010). Vegetation P reserves (52 kg ha^{-1}) were higher than the average reported for montane forests (39 kg ha^{-1}). The range for montane and lowland tropical moist forests is between 2 and 290 kg ha^{-1} (Vitousek & Sanford Jr., 1986). The highest accumulation of P was in the soil (509 kg ha^{-1} at 10 cm in depth), a relatively high value compared to the average for montane tropical moist forests reviewed by Grimm and Fassbender (1981a) (279 kg ha^{-1}) and the values reported along an elevation gradient (1900–3000 m a.s.l.) in a montane forest in Ecuador (200–270 kg ha^{-1}) (Soethe et al., 2008).

In order to test the hypothesis of limited growth due to nutrient deficiencies (Grubb, 1977), fertilization experiments with N and P in an Andean cloud forest site close to LM (2450–2650 m a.s.l.), showed higher tree growth and greater litter production with respect to controls, concluding that N is more limiting than P in this forest (Tanner et al., 1992). Similarly, large-scale nutrient fertilization experiments along a 3000 m elevation transect in the Peruvian Andes and adjacent lowland Amazonia (Fisher et al., 2013) found increased N limitation and decreased P limitation with increasing elevation. Plots fertilized with N at the highest elevation sites (1500 m, 3000 m) showed greater relative growth rates in DBH than did the control plots, again suggesting an increased N constraint with elevation (Fisher et al., 2013).

Overall, the data analyzed on the forest productivity of Andean forests in Venezuela (i.e., LC and LM) do not coincide with the predictions proposed by Fahey et al. (2016) on the stunted nature and low productivity of TMCF. Compared with other studies in Andean cloud forests (Girardin et al., 2014a; Malhi et al., 2017; Duque et al., 2021), LC and LM are characterized by relatively high canopy height, basal area and AGB (Fig. 2). Climatic factors, such as mean annual temperature, rainfall and insolation hours do not seem to differ from other Andean locations in the altitudinal belt between 2200 and 2800 m a.s.l. (Jarvis & Mulligan, 2010). The relatively high P content in the vegetation and soil in LC highlights the importance of nutrient status in determining the capacity of forests to sequester carbon, as suggested by Fernández-Martínez et al. (2014) and Santiago and Goldstein (2016), who reported that the limitation of P is even more critical than that of N for photosynthetic processes.

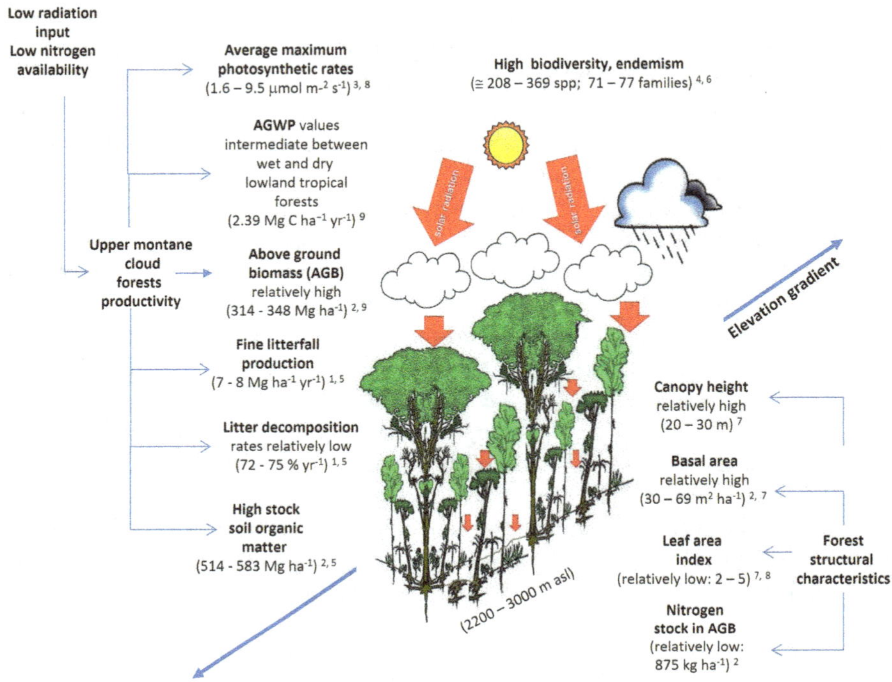

Fig. 2 Diagram illustrating the productivity and structural characteristics of upper montane cloud forests in the Venezuelan Andes. Key features include low radiation and nitrogen input, leading to average maximum photosynthetic rates of 1.6–9.5 μmol m² s^{-1}. Aboveground biomass is relatively high at 314–348 Mg ha^{-1}, with fine litterfall production of 7–8 Mg ha^{-1} yr^{-1} and low litter decomposition rates of 72–75% yr^{-1}. High soil organic matter stock ranges from 514–583 Mg ha^{-1}. Forest structural characteristics include canopy height of 20–30 m, basal area of 30–69 m² ha^{-1}, and a leaf area index of 2–5. Nitrogen stock in above-ground biomass is relatively low at 875 kg ha^{-1}. The forest supports high biodiversity, with 208–369 species across 71–77 families, influenced by solar radiation and elevation gradients from 2200–3000 m above sea level. Source: Fassbender and Grimm (1981a); Grimm and Fassbender (1981a, b); García-Núñez et al. (1995); Engwald (1999); Ramírez (1999); Linares (2008); Schwarzkopf et al. (2011); Quevedo et al. (2016, 2018); Vilanova et al. (2018)

3 Plant Scale Studies in Venezuelan Cloud Forests: Water and Carbon Relations

3.1 *Stomatal Conductance and Leaf Water Potential*

Even though TMCFs are characterized by very humid environmental conditions, plants of different growth-forms (canopy trees, shrubs, subshrubs, palms, ferns and epiphytes) in Andean cloud forests transpire at rates comparable to those recorded in lowland humid forests (Jordan & Kline, 1976; Rada & Jaimez, 1992; García-Núñez et al., 1995; Alvarez, 1996; Vilanova, 1996; Cabrera, 1999; Dulhoste, 2002; Dávila, 2009; Rada et al., 2009; Aparecido et al., 2016; Andriyas et al., 2021).

A significant decrease in stomatal conductance (g_s) may also occur during periods of direct sunlight and higher vapor pressure differences (VPD) (Meinzer et al., 1984; García-Núñez et al., 1995; Rada et al., 2009). Additionally, seasonal precipitation variations may induce short periods of water stress (Bruijnzeel & Veneklass, 1998) with adverse effects on carbon gain. Stomatal closure due to an increase in VPD or dry season conditions constitutes a conservative strategy to avoid greater water plant deficits and has been described for different species of the Venezuelan Andean cloud forest (Meinzer et al., 1984; García-Núñez et al., 1995, Rada et al., 2009). Canopy species respond differently to increases in VPD due to direct sunlight or to the dry season. Species sensitive to increases in VPD such as *Clusia multiflora* Kunth showed a 46% reduction in g_s between wet and dry seasons, affecting maximum assimilation rates (A_{max}), while *Miconia resimoides* Cogn. increased g_s and A_{max} during the dry season (Rada et al., 2009).

In relation to leaf water status, in general, tree species of Venezuelan Andean cloud forests show more positive leaf water potentials and lower fluctuations throughout the year due to rainfall seasonality (Cavelier, 1986, 1996; Rada & Jaimez, 1992; García-Núñez et al., 1995; Vilanova, 1996; Cabrera, 1999; Dulhoste, 2002; Dávila, 2009; Rada et al., 2009) compared to tropical humid lowland forest tree species (Robichaux et al., 1984; Mulkey & Wright, 1996; Santiago et al., 2016). Cloud forest adult tree species show lower g_s, minimum leaf water potentials (Ψ_{min}) and osmotic potentials at turgor loss ($\Psi_\pi°$) during the wet period than the understory growth-forms (subshrubs, shrubs, palms, ferns and epiphytes) and understory tree saplings (Table 1). Trees also display a more drought tolerant character in response to the drier environment at the canopy, and higher hydraulic resistances to reach leaves at heights of ≅ 15–25 m. The avoidance character of the understory growth-forms in response to rainfall seasonality is evidenced by the drastic reduction in g_s between seasons (Table 1). These water relations parameters (i.e. g_s, Ψ_{min}, $\Psi_\pi°$) are comparable to those reported for other TMCFs (Kapos & Tanner, 1985; Aylett, 1985; Cavelier, 1986, 1990, 1996; Sobrado, 2003; Motzer et al., 2005).

Table 1 Mean stomatal conductance (g_s, mmol m^{-2} s^{-1}), minimum leaf water potential (Ψ_{min}, MPa) and osmotic potential at the turgor loss point ($\Psi_\pi°$, MPa) during wet (WS) and dry (DS) seasons for different plant growth-forms of the Andean cloud forests in Venezuela (number of species studied in brackets)

	g_s		Ψ_{min}		$\Psi_\pi°$	
Growth-form	WS	DS	WS	DS	WS	DS
Trees[a] (canopy individuals) [7]	101.5 ± 15.6	104.8 ± 26.4	−1.1 ± 0.1	−1.2 ± 0.1	−1.8 ± 0.2	−1.8 ± 0.2
Trees[b] (understory saplings) [3]	174.2 ± 69.1	159.9 ± 78.5	−0.7 ± 0.2	−1.1 ± 0.4	−1.5 ± 0.1	−1.7 ± 0.3
Understory[c] growth-forms [5]	139.9 ± 26.1	70.7 ± 17.0	−0.5 ± 0.1	−0.8 ± 0.1	−1.4 ± 0.1	−1.6 ± 0.1

Source: [a]Cavelier (1996), Dulhoste (2002), Rada et al., 2009; [b]García-Núñez et al., 1995, Dávila (2009); [c]Rada and Jaimez (1992), Vilanova (1996)
Understory growth-forms (i.e. subshrubs, shrub, palm, fern and epiphyte; 1 sp per growth-form)

Despite the moist nature of Andean cloud forests, plants could be exposed to water stress conditions due to a small hydraulic safety margin (Choat et al., 2012). Yet, the presence of fog and foliar water uptake (Eller et al., 2013; Goldsmith et al., 2013) could minimize the occurrence of tree water deficits and thus favor the survival of TMCF trees (Oliveira et al., 2014). Edge effects as a consequence of forest fragmentation could produce distinct changes in biotic and microclimatic conditions between forest borders and the interior, being these changes more severe as the forest fragments become smaller (Hansen et al., 2020). For example, in *Alchornea triplinervia* Spren. M. Arg., an early succession tree species, g_s did not show significant seasonal changes in juveniles located in small clearings (<100 m^2) within the forest (LC, at 2200 m a.s.l.), while individuals located in the interior and on the edges of a small forest fragment (<10 ha) surrounded by grassland showed significant decreases in g_s and survival during drought (García-Núñez et al., 1995; Dávila, 2009).

3.2 Photosynthetic Responses

Even though brief periods of water deficit that occur in cloud forests could restrict CO_2 assimilation, the light limited environment of the forest understory controls photosynthetic processes. Under these shaded conditions, high light capture efficiency, low dark respiration rates and low light compensation points are essential to maintain a positive carbon balance in understory species (García-Núñez et al., 1995). Different authors agree that many tree species comprise a broad functional group of shade-tolerant generalists, characterized by a limited capacity of photosynthetic acclimation to high radiation conditions (Wright et al., 2003; Quevedo et al., 2018), and specialists, able to acclimate to a sudden radiation increase in treefall gaps (Chazdon et al., 1996; Barker et al., 1997; Krause et al., 2001; Goldstein et al., 2016; Quevedo et al., 2018).

For tree saplings from LC and LM with different shade tolerance and different growth forms of understory species (palms, ferns, shrubs, subshrubs, epiphytes), light compensation points ranged between 1.7 and 7.5 μmol m^{-2} s^{-1}, dark respiration from 0.1 to 0.7 μmol m^{-2} s^{-1}, chlorophyll/nitrogen ratio between 0.39 and 0.85, electron transport rates ranged from 40 to 80 μmol m^{-2} s^{-1}; leaf nitrogen and chlorophyll content values between 0.011–0.043 kg N/kg and 6.98–9.56 mg Chl/g; leaf mass per unit area (kg m^{-2}) ranged between 0.036–0.16 (García-Núñez et al., 1995; Alvarez, 1996; Vilanova, 1996; Cabrera, 1999; Bracho, 2014; Jaimes, 2015; Quevedo et al., 2018). Optimum temperatures for photosynthesis varied between 18 and 21 °C for understory and gap colonizer species with minimum and maximum temperature compensation points between 0 and 40 °C, respectively (García-Núñez et al., 1995; Cavieres et al., 2000).

Due to the high correlation between the rate of photosynthesis (A) and leaf N (Field & Mooney, 1986; Evans, 1989) and considering that nitrogen availability could be limiting in cloud forests (Tanner et al., 1992), plants are expected to have low photosynthetic rates (Aylett, 1985; Cavelier, 1996). Maximum assimilation rates (A_{max}) are relatively low in shade tolerant plants and in mature canopy trees from LC (García-Núñez et al., 1995; Quevedo et al., 2018), comparable to the range reported for various late succession species (sun leaves) of tropical humid forests (1.4–9.4 μmol m^{-2} s^{-1}) (Strauss-Debenedetti & Bazzaz, 1996). However, comparatively high maximum assimilation rates (11.6–20.6 μmol m^{-2} s^{-1}) have been reported for shrubs and pioneer woody species like *Montanoa quadrangularis* Sch. Bip., *Heliocarpus americanus* L., and *Oreopanax reticulatus* (Linden ex B.S. Williams) L.H. Bailey, in disturbed areas and large canopy gaps, of cloud forest sites (LM) (Cabrera, 1999; Ataroff & García-Núñez, 2013). These rates are similar to those of sun leaves of early successional lowland humid forest species (Strauss-Debenedetti & Bazzaz, 1996). Comparisons of functional responses of tree species along an elevation gradient in the Peruvian Andes (Bahar et al., 2016) showed similar average photosynthetic rates for lowland (8.2 μmol m^{-2} s^{-1}) and upland (7.6 μmol m^{-2} s^{-1}) species.

Different studies in Venezuelan cloud forest plants (García-Núñez et al., 1995; Hernández & Medina, 1995; Anten et al., 1996; Quilici & Medina, 1998; Rada et al., 2009) showed a high correlation between A_{max} and N, and a higher nitrogen use efficiency (PNUE = A_{max}/N) in sun plants, characteristic of early succession (Añez, 1987; García-Núñez et al., 1995; Quilici & Medina, 1998). This could be related to a greater carbon investment in leaf formation (high specific leaf area) and nitrogen in the carboxylation and electron transport capacity (Field & Mooney, 1986; Evans, 1989). On the other hand, in shaded environments, plants are more limited by the ability to capture light (greater investment of nitrogen in pigments). Changes (plasticity) in leaf morphology towards greater leaf expansion (greater specific leaf area) are important for survival in the understory (García-Núñez et al., 1995; Hernández & Medina, 1995; Anten et al., 1996; Evans & Poorter, 2001). Anten et al. (1996) reported that growth in *Tetrorchidium rubrivenium* Poepp & Endl., an understory tree species of the cloud forest, is mostly limited by light rather than N availability for photosynthesis. On the contrary, in evergreen species of the cloud forest, most displaying sclerophyllous leaves (Grubb, 1977), nitrogen limitations occur as a result of its allocation to increase longevity (e.g. anti-herbivore compounds such as alkaloids) and not to the photosynthetic apparatus (Hernández & Medina, 1995).

Studies on photosynthetic acclimation responses to light in juvenile LC tree species representative of contrasting shade tolerance groups showed that all were susceptible to photoinhibition. Shade-intolerant species showed dynamic photoinhibition whereas shade-tolerant species presented chronic photoinhibition. The partially shade-tolerant species showed mixed responses; nonetheless, they exhibited larger adjustments in morpho-anatomical and optical properties. Thus, the acclimation responses of these species when subjected to contrasting light conditions may determine their distribution along the light gradient in the understory (Quevedo et al., 2015, 2018). These results agree with other studies in tropical and subtropical wet forests (Mulkey & Pearcy, 1992; Chazdon et al., 1996; Barker et al., 1997; Yamashita et al., 2000; Krause et al., 2001; Zhang et al., 2015; Goldstein et al., 2016).

4 Future Issues

1. Even though most of the remaining Andean cloud forests in Venezuela are located in protected areas, and deforestation rates have been reduced in the last twenty years (0.1% between 2000 and 2022; see Chapter "A multiscale ecological approach for the conservation and restoration of the Venezuelan Andean cloud forests" in this book), forest fragmentation and degradation may still be considered a major threat to the cloud forest. Deforestation processes disrupt natural forest regeneration and successional cycles. Besides, deforestation could disrupt biogeochemical cycles, since part of nutrient reservoirs are within the standing biomass and the soil organic matter, determining low resilience (long time to recover from massive disturbances) of cloud forest communities regarding forest productivity. Forest fragmentation processes in the Andean basins of Venezuela result in a mosaic of small fragments (<10 ha) with strong edge effects from the surrounding grassland matrix (Dávila, 2009; Rodríguez-Morales et al., 2009). The limited capacity for acclimation to light in most understory tree juvenile species (Quevedo et al., 2018) determines a limited capacity to invade the grassland matrix that surrounds the fragments, changing species composition towards gap-specialist tree species that are more resistant to brighter and drier conditions of the forest edges (Ataroff & García-Núñez, 2013). Hence it is likely that in the long run late successional tree species will be endangered. In this context, a better understanding of the consequences of deforestation and forest fragmentation processes at the ecosystem (e.g. carbon, water and nutrient economy of undisturbed and forests fragments) and plant species levels (e.g. responses to disturbance and changes in micro-environmental conditions) is required. Aide et al. (2019), in a recent assessment in the tropical and subtropical Andes including Venezuelan cloud forests (from 2001 to 2014), describe an overall trend of increasing woody vegetation in abandoned pastures and cropland in mid-elevation areas, and shrub encroachment into highland grasslands (at the cloud forest/paramo ecotone). These authors point out an urgent need for a regional real-time land-use, biodiversity, and ecosystem services monitoring network. This information is essential for sustainable forest management and the success of biodiversity conservation and ecological restoration programs.
2. Additional impacts on these systems seem to be in progress as a result of climate change (Fig. 3). The most widespread scenarios on climate change trends are too general to forecast changes above certain elevations (Oliveira et al., 2014). However, a predicted shift upwards of the orographic condensation belt would reduce the amount and frequency of fog interception and increase evapotranspi-

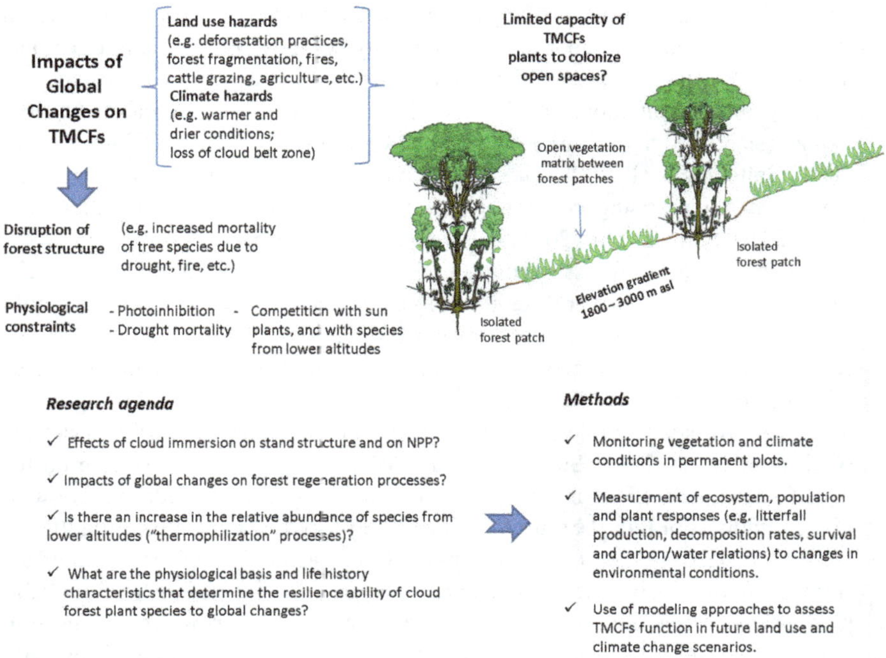

Fig. 3 Summary characterization of the main impacts of climate change on TMCFs, and a research agenda (major questions; gaps in knowledge) for future studies, particularly in Venezuela

ration at the forest's present elevation, eventually giving rise to drought stress (Oliveira et al., 2014). The effects of topographic position on forest structure, productivity and species composition (e.g. hydro-sequence and/or nutrient gradients), underline the vulnerability of ecosystem functional and structural responses in face of the changing climate (Bellingham & Tanner, 2000; Bruijnzeel et al., 2010; Homeier et al., 2010; Werner & Homeier, 2015). A hypothesis to be tested is, for example, if a warmer/drier scenario could lead to better growing conditions on less steep sites (i.e. low slope) in LC due to higher mineralization rates and improved aeration of the soil profile, while communities on medium slopes could be exposed to drought conditions. The marked species-specific gas exchange responses found in canopy species (Rada et al., 2009), emphasizes the possible outcome of the TMCFs under a warmer/drier scenario (e.g. drought resistant species will become dominant), suggesting that changes in forest structure could occur (i.e., species composition; forest dynamics) and thus in functioning (e.g. carbon and nutrient cycling, NPP, etc.). Moreover, fertilization experiments performed in order to test the hypothesis of limited growth due to nutrient deficiencies (Tanner et al., 1992; Fisher et al., 2013), point out the ecosystem responsiveness to moderate variations in nutritional conditions (Homeier et al., 2012) producing significant changes in ecosystem processes (e.g. forest productivity, storage and cycling of carbon and nutrients, etc.) as well as on structural characteristics, e.g. plant community composition, biotic interactions. These results suggest a high vulnerability of tropical cloud forests to the expected increase in nutrient inputs in face of global changes, e.g. increased nutrient deposition related to industrial activities and agricultural practices (Tilman et al., 2001; Galloway et al., 2008) and increased decomposition and nutrient mineralization rates due to rising temperatures (Davidson & Janssens, 2006; Zimmermann et al., 2010; Salinas et al., 2011; Nottingham et al., 2015).

Disruption of the forest structure is likely to occur as a consequence of climate changes (e.g. warmer/drier conditions, increased forest fires), which could produce drought-induced forest diebacks (Allen et al., 2010). In the context of forest successional dynamics treefall gaps could increase due to increased rates of adult tree mortality (Allen et al., 2010; Breshears et al., 2013), shifting dominance towards shade intolerant/drought resistant secondary successional species (Goldstein et al., 2016; Santiago et al., 2016). Climate change scenarios for TMCFs include from warmer and drier conditions to warmer conditions but with higher amounts of rain (Urrutia & Vuille, 2009). Both scenarios agreeing that an increase in seedling and tree mortality could occur, in view of the direct effect from decreased water availability and/or increased competition from lowland plant species (Oliveira et al., 2014). In warmer/drier scenarios, lower elevation plant species seem to be more competitive and could displace cloud forest species from its current altitudinal belt (Foster, 2001; Willis & MacDonald, 2011; Oliveira et al., 2014). While the consequences of global warming on tropical trees have been poorly studied, recent studies of 200 Andean Forest plot inventories spread across wide latitudinal (26.8° S – 7.1° N) and altitudinal (360–3360 m a.s.l.) gradients, report that tropical and subtropical tree communities are undergoing directional changes in composition showing greater relative abundances of species from lower, warmer elevations, a phenomenon known as "thermophilization" (Fadrique et al., 2018). Moreover, the epiphytes would possibly be the first group of cloud forest plants to be affected, so they have been proposed as biological indicators of climate change (Nadkarni & Solano, 2002). We have recently (year 2015) established permanent plots in LM (Red de Bosques Andinos, CONDESAN) to monitor changes in vegetation structure, composition and plant functional traits. If "thermophilization" processes are occurring, it is important to identify the physiological basis and life history characteristics of lower elevation and cloud forest species to understand the greater competitiveness of the former and the resilience ability of the latter which will undoubtedly determine shifts in community composition.

3. Finally, it is important to direct the Venezuelan research agenda towards the construction and implementation of mechanistic and process-based models to improve our understanding and prediction of TMCFs function under future scenarios of land use and climate change.

5 Conclusions

1. In relation to biomass stocks and forest productivity, cloud forests are viewed as more limited than lowland wet forests. Nevertheless, standing biomass and litterfall production reported in this review compare to lowland tropical humid forests, which denotes a high efficiency in organic matter production in spite of its radiation limitations, relatively lower LAI values and restrictions in nitrogen availability. The relatively high canopy height (particularly in LC), basal areas, wood density, and low turnover rates, seem to be the main factors determining high AGB values compared to other TMCFs in the Andean region.
2. Nutrient economy studies at LC showed that mineral reservoirs in both vegetation and mineral soil are comparable (N) or even higher (P) to those reported for other TMFs locations. The highest ecosystem reserves of N and P were found in the mineral soil, while the vegetation content was relatively low. The vegetation and soil reservoirs of K, Ca, Mg were fairly low due to the low soil pH. Nutrient fertilization experiments revealed that N is more limiting than P in this forest. Overall, LC appears as a relatively nitrogen limited forest with low litter decomposition rates, a general pattern for TMCFs.
3. Even though TMCFs are characterized by very humid environmental conditions, plants of different growth-forms in Andean cloud forests transpire at rates comparable to those recorded in lowland humid forests, and a significant decrease in g_s may occur during periods of direct sunlight and higher VPD. In relation to leaf water status, in general, tree species of the studied Andean cloud forests show more positive leaf water potentials and lower fluctuations throughout the year due to rainfall seasonality than tropical humid lowland forest tree species.
4. Brief periods of water deficit that occur in cloud forests may generate restrictions to CO_2 assimilation. However, low radiation inputs due to frequent and persistent clouds could be the main limitation to the photosynthetic processes in the forest understory and differentially affect leaf gas exchange of canopy species. Maximum assimilation rates of tree species (sun and shade conditions) were comparable to those reported for lowland wet forest tree species. For LC, many tree species at sapling stages comprise a broad functional group of shade-tolerant generalists, characterized by a limited capacity of photosynthetic acclimation to high radiation conditions, while a much smaller group of tree species behave as specialists, able to acclimate to sudden radiation increases in treefall gaps.

Acknowledgments A first version of this review was part of the workshop "South American Montane Forests in a Warming World Nerc/Fapesp", UNICAMP, Campinas, Brasil (July, 2019). We thank the organizers for sponsoring the assistance of CG-N to this meeting. CG-N thanks the Institute of International Education's Scholar Rescue Fund (IIE-SRF) and CUGS, Trinity College, for their academic and financial support. FR thanks the Open Society University Network (OSUN) and UniAndes for their academic and financial support.

Disclosure Statement The authors declare that the research was conducted in the absence of any commercial or financial relationships that could be construed as a potential conflict of interest.

Funding This work was supported by the Instituto de Ciencias Ambientales y Ecológicas (ICAE), Facultad de Ciencias, Universidad de Los Andes, Venezuela.

References

Aide, T. M., Grau, H. R., Graesser, J., Andrade-Nuñez, M. J., Aráoz, E., Barros, A. P., Campos-Cerqueira, M., Chacon-Moreno, E., Cuesta, F., Espinoza, R., et al. (2019). Woody vegetation dynamics in the tropical and subtropical Andes from 2001 to 2014: Satellite image interpretation and expert validation. *Global Change Biology, 25*(6), 2112–2126. https://doi.org/10.1111/gcb.14618

Allen, C. D., Macalady, A. K., Chenchouni, H., Bachelet, D., McDowell, N., Vennetier, M., Kitzberger, T., Rigling, A., Breshears, D. D., Hogg, E. H., et al. (2010). A global overview of drought and heat-induced tree mortality reveals emerging climate change risks for forests. *Forest Ecology and Management, 259*(4), 660–684. https://doi.org/10.1016/j.foreco.2009.09.001

Alvarez, L. (1996). *Relaciones hídricas y de carbono en plantas epífitas de una selva nublada tropical, La Mucuy, Estado Mérida*. Undergraduate thesis. Facultad de Ciencias, Universidad de Los Andes, Mérida (Venezuela).

Andriyas, T., Leksungnoen, N., & Tor-ngern, P. (2021). Comparison of water-use characteristics of tropical tree saplings with implications for forest restoration. *Scientific Reports, 11*, 1745. https://doi.org/10.1038/s41598-021-81334-0

Añez, M. A. (1987). *Relaciones hídricas y nutricionales en especies arbóreas de un bosque nublado a lo largo de un gradiente sucesional*. Masters thesis. Instituto Venezolano de Investigaciones Científicas, Caracas (Venezuela), p. 205.

Anten, N. P. R., Hernández, R., & Medina, E. (1996). The photosynthetic capacity and leaf nitrogen concentration as related to light regime in shade leaves of a montane tropical forest tree, *Tetrorchidium rubrivenium*. *Functional Ecology, 10*, 491–500. https://doi.org/10.2307/2389942

Aparecido, L. M. T., Miller, G. R., Cahill, A. T., & Moore, G. W. (2016). Comparison of tree transpiration under wet and dry canopy conditions in a Costa Rican premontane tropical forest. *Hydrological Processes, 30*(26), 5000–5011. https://doi.org/10.1002/hyp.10960

Aparecido, L. M. T., Gs, T., Mosquera, G., Brum, M., Barros, F. D. V., Pompeu, P. V., & Oliveira, R. S. (2018). Ecohydrological drivers of Neotropical vegetation in montane ecosystems. *Ecohydrology, 11*(3), e1932. https://doi.org/10.1002/eco.1932

Arends, E., Villaquiran, A., & Calderon, O. (1991). Caracterización de la vegetación secundaria en un área talada de la selva nublada, Estado Mérida. *Rev Forest Venez, 15*, 13–22.

Asner, G. P., Martin, R. E., Anderson, C. B., Kryston, K., Vaughn, N., Knapp, D. E., Bentley, L. P., Shenkin, A., Salinas, N., Sinca, F., et al. (2016). Scale dependence of canopy trait distributions along a tropical forest elevation gradient. *The New Phytologist, 214*(3), 973–988. https://doi.org/10.1111/nph.14068

Ataroff, M. (2001). Venezuela. In M. Kappelle & A. D. Brown (Eds.), *Bosques Nublados del Neotrópico* (pp. 397–442). Editorial IMBIO.

Ataroff, M. (2003). Selvas y bosques de montaña. In M. Aguilera, A. Azócar, & E. González-Jiménez (Eds.), *Biodiversidad en Venezuela* (pp. 762–810). Ediciones CONICIT-Fundación Polar.

Ataroff, M., & García-Núñez, C. (2013). Selvas y bosques nublados de Venezuela. In E. Medina, O. Huber, J. Nassar, & P. Navarro (Eds.), *Recorriendo el Paisaje Vegetal de Venezuela* (pp. 125–155). Ediciones Instituto Venezolano de Investigaciones Científicas (IVIC).

Ataroff, M., & Rada, F. (2000). Deforestation impact on water dynamics in a Venezuelan Andean cloud forest. *Ambio, 29*(7), 438–442. https://doi.org/10.1579/0044-7447-29.7.440

Atkin, O. K., Bloomfield, K. J., Reich, P. R., Tjoelker, M. G., Asner, G. P., Bonal, D., Bönisch, G., Bradford, M. G., Cernusak, L. A., Cosio, E. G., et al. (2015). Global variability in leaf respiration in relation to climate, plant functional types and leaf traits. *The New Phytologist, 206*, 614–636. https://doi.org/10.1111/nph.13253

Aylett, G. P. (1985). Irradiance interception, leaf conductance and photosynthesis in Jamaican upper montane rainforest trees. *Photosynthetica, 19*, 323–327.

Bahar, N. H. A., Ishida, F. Y., Weerasinghe, L. K., Guerrieri, R., O'Sullivan, O. S., Bloomfield, K. J., Asner, G. P., Martin, R. E., Lloyd, J., Malhi, Y., et al. (2016). Leaf-level photosynthetic capacity in lowland Amazonian and high-elevation Andean tropical moist forests of Peru. *The New Phytologist, 214*(3), 1002–1018. https://doi.org/10.1111/nph.14079

Barker, M. G., Press, M. C., & Brown, N. D. (1997). Photosynthetic characteristics of dipterocarp seedlings in three tropical rain forest light environments: A basis for niche partitioning? *Oecologia, 112*, 453–463. https://doi.org/10.1007/s004420050332

Bellingham, P. J., & Tanner, E. V. J. (2000). The influence of topography on tree growth, mortality, and recruitment in a tropical montane forest. *Biotropica, 32*, 378–384. https://doi.org/10.1111/j.1744-7429.2000.tb00484.x

Benner, J., Vitousek, P. M., & Ostertag, R. (2010). Nutrient cycling and nutrient limitation in tropical montane cloud forests. In L. A. Bruijnzeel, F. N. Scatena, & L. S. Hamilton (Eds.), *Tropical montane cloud forests: Science for conservation and management* (pp. 90–100). Cambridge University Press. https://doi.org/10.1017/CBO9780511778384

Berendse, F. (1994). Ecosystem stability, competition and nutrient cycling. In E. D. Schulze & H. A. Mooney (Eds.), *Biodiversity and ecosystem function* (pp. 409–431). Springer.

Blonder, B., Salinas, N., Bentley, L. P., Shenkin, A., Chambi Porroa, P. O., Valdez Tejeira, Y., Violle, C., Fyllas, N. M., Goldsmith, G. R., Martin, R. E., et al. (2017). Predicting trait environment relationships for venation networks along an Andes Amazon elevation gradient. *Ecology, 98*(5), 1239–1255. https://doi.org/10.1002/ecy.1747/suppinfo

Bono, G. (1996). *Flora y Vegetación del Estado Táchira* (p. 951). Museo Reg. Scienze Naturali, Torino. Monografie XX.

Bracho, J. (2014). *Respuestas ecofisiológicas de tres especies arbóreas a diferentes regímenes de luz en el bosque nublado San Eusebio*. Undergraduate thesis. Facultad de Ciencias, Universidad de Los Andes, Mérida (Venezuela). p. 52.

Breshears, D. D., Adams, H. D., Eamus, D., McDowell, N. G., Law, D. J., Will, R. E., Williams, A. P., & Zou, C. B. (2013). The critical amplifying role of increasing atmospheric moisture demand on tree mortality and associated regional die-off. *Frontiers in Plant Science, 4*(266), 1–4. https://doi.org/10.3389/fpls.2013.00266

Bruelheide, H., Dengler, J., Purschke, O., Lenoir, J., Jiménez-Alfaro, B., Hennekens, S. M., Botta-Dukát, Z., Chytrý, M., Field, R., Jansen, F., et al. (2018). Global trait–environment relationships of plant communities. *Nature Ecology & Evolution, 2*(12), 1906–1917. https://doi.org/10.1038/s41559-018-0699-8

Bruijnzeel, L. A., & Veneklass, E. J. (1998). Climatic conditions and tropical montane forest productivity: The fog has not lifted yet. *Ecology, 79*(1), 3–9. https://doi.org/10.1890/0012-9658(1998)079[0003:CCATMF]2.0.CO;2

Bruijnzeel, L. A., Scatena, F. N., & Hamilton, L. S. (Eds.). (2010). *Tropical montane cloud forests. Science for conservation and management*. Cambridge University Press.

Bruijnzeel, L. A., Mulligan, M., & Scatena, F. N. (2011). Hydrometeorology of tropical montane cloud forests: Emerging patterns. *Hydrological Processes, 25*(3), 465–498. https://doi.org/10.1002/hyp.7974

Bush, M. B., Silman, M. R., & Urrego, D. H. (2004). 48,000 years of climate and forest change in a biodiversity hot spot. *Science, 303*(5659), 827–829. https://doi.org/10.1126/science.1090795

Cabrera, H. M. (1999). *Mecanismos de aclimatación a la luz en especies de la sucesión de bosques tropicales montanos*. Doctorate thesis. Facultad de Ciencias, Universidad de Los Andes, Mérida (Venezuela). p. 171.

Cavelier, J. (1986). *Relaciones Hídricas y de Nutrientes en Bosques Enanos Nublados*. Masters thesis. Facultad de Ciencias, Universidad de Los Andes, Mérida (Venezuela). p. 273.

Cavelier, J. (1990). Tissue water relations in elfin cloud forest tree species of Serrania de Macuira, Guajira, Colombia. *Trees, 4*, 155–163. https://doi.org/10.1007/BF00225780

Cavelier, J. (1996). Environmental factors and ecophysiological processes along altitudinal gradients in wet tropical mountains. In S. S. Mulkey, R. L. Chazdon, & A. P. Smith (Eds.), *Tropical forest plant ecophysiology* (pp. 399–439). Chapman & Hall.

Cavieres, L. A., Rada, F., Azócar, A., García-Núñez, C., & Cabrera, H. M. (2000). Gas exchange and low temperature resistance in two tropical high mountain tree species from the Venezuelan Andes. *Acta Oecologica, 21*, 203–211. https://doi.org/10.1016/S1146-609X(00)01077-8

Chave, J., Navarrete, D., Almeida, S., Alvarez, E., Aragão, L. E. O. C., Bonal, D., Châtelet, P., Silva-Espejo, J. E., Goret, J. Y., von Hildebrand, P., et al. (2010). Regional and seasonal patterns of litterfall in tropical South America. *Biogeosciences, 7*, 43–55. https://doi.org/10.5194/bg-7-43-2010

Chazdon, R., Pearcy, R., Lee, D., & Fetcher, N. (1996). Photosynthetic responses of tropical forest plants to contrasting light environments. In S. S. Mulkey, R. L. Chazdon, & A. P. Smith (Eds.), *Tropical forest plant ecophysiology* (pp. 5–55). Chapman & Hall.

Choat, B., Jansen, S., Brodribb, T. J., Cochard, H., Delzon, S., Bhaskar, R., Bucci, S. J., Field, T. S., Gleason, S. M., Hacke, U. G., et al. (2012). Global convergence in the vulnerability of forests to drought. *Nature, 491*, 752–755. https://doi.org/10.1038/nature11688

Clark, D. B., Hurtado, J., & Saatchi, S. S. (2015). Tropical rain forest structure, tree growth and dynamics along a 2700-m elevational transect in Costa Rica. *PLoS One, 10*(4), 1–18. https://doi.org/10.1371/journal.pone.0122905

Cuevas, E., & Medina, E. (1986). Nutrient dynamics within Amazonian forest: II. Fine root growth, nutrient availability and leaf litter decomposition. *Oecologia, 76*, 222–235. https://doi.org/10.1007/BF00379956

Davidson, E. A., & Janssens, I. A. (2006). Temperature sensitivity of soil carbon decomposition and feedbacks to climate change. *Nature, 440*, 165–173. https://doi.org/10.1038/nature04514

Dávila, Y. (2009). Efecto de borde y respuestas funcionales durante el establecimiento de *Alchornea triplinervia* y *Myrcia acuminata* en un fragmento de selva nublada andina. Masters thesis. Facultad de Ciencias, Universidad de Los Andes, Mérida (Venezuela). p. 80.

de la Cruz-Amo, L., Bañares-de-Dios, G., Cala, V., Granzow-de la Cerda, I., Espinosa, C. I., Ledo, A., Salinas, N., Macía, M. J., Cayuela, L., et al. (2020). Trade-offs among aboveground, belowground, and soil organic carbon stocks along altitudinal gradients in andean tropical montane forests. *Frontiers in Plant Science, 11*, 106. https://doi.org/10.3389/fpls.2020.00106

Delaney, M., Brown, S., Lugo, A. E., Torres-Lezama, A., & Bello, N. (1997). The distribution of organic carbon in major components of forests located in five life zones of Venezuela. *Journal of Tropical Ecology, 13*(5), 697–708. https://doi.org/10.1017/S0266467400010877

Delaney, M., Brown, S., Lugo, A.E., Torrez-Lezama, A., & Bello, N. (1998). The quantity and turnover of dead wood in permanent forest plots in six life zones of Venezuela. *Biotropica, 30*(1), 2–11. https://doi.org/10.1111/j.1744-7429.1998.tb00364.x

Doughty, C. E., Santos-Andrade, P. E., Goldsmith, G. R., Blonder, B., Shenkin, A., Bentley, L., Chavana-Bryant, C., Huaraca-Huasco, W., Díaz, S., Salinas, N., et al. (2017). Can leaf spectroscopy predict leaf and forest traits along a Peruvian tropical forest elevation gradient? *Journal of Geophysical Research: Biogeosciences, 122*, 2952–2965. https://doi.org/10.1002/2017JG003883

Dulhoste, R. (2002). *Caracterización florístico-estructural y ecofisiologia de las especies dominantes de un bosque secundario en la Selva Nublada*. Masters thesis. Facultad de Ciencias, Universidad de Los Andes, Mérida (Venezuela). p. 87.

Duque, A., Peña, M. A., Cuesta, F., González-Caro, S., Kennedy, P., Phillips, O. L., Calderón-Loor, M., Blundo, C., Carilla, J., Cayola, L., et al. (2021). Mature Andean forests as globally important carbon sinks and future carbon refuges. *Nature Communications, 12*, 2138. https://doi.org/10.1038/s41467-021-22459-8

Eller, C. B., Lima, A. L., & Oliveira, R. S. (2013). Foliar uptake of fog water and transport belowground alleviates drought effects in the cloud forest tree species, *Drimys brasiliensis* (Winteraceae). *The New Phytologist, 199*(1), 151–162. https://doi.org/10.1111/nph.12248

Engwald, S. (1999). *Diversitat und Okologie der vaskularen Epiphyten in einem Berg-und einem Tieflandergenwald in Venezuela*. Doctorate thesis. Universität Bonn, Bonn (Deutschland). p. 390.

Enquist, B. J., Bentley, L. P., Shenkin, A., Maitner, B., Savage, V., Michaletz, S., Blonder, B., Buzzard, V., Boza Espinoza, T. E., Farfan-Rios, W., et al. (2017). Assessing trait-based scaling theory in tropical forests spanning a broad temperature gradient. *Global Ecology and Biogeography, 26*(12), 1357–1373. https://doi.org/10.1111/geb.12645

Esquivel, J., Park, B. B., Casanoves, F., Delgado, D., Park, G.-E., & Finegan, B. (2020). Altitude and species identity drive leaf litter decomposition rates of ten species on a 2950 m altitudinal gradient in Neotropical rain forests. *Biotropica, 52*(1), 11–21. https://doi.org/10.1111/btp.12730

Evans, J. R. (1989). Phosynthesis and nitrogen relationships in leaves of C3 plants. *Oecologia, 78*, 9–19. https://doi.org/10.1007/BF00377192

Evans, J. R., & Poorter, H. (2001). Photosynthetic acclimation of plants to growth irradiance: The relative importance of specific leaf area and nitrogen partitioning in maximizing carbon gain. *Plant, Cell & Environment, 24*(8), 755–767. https://doi.org/10.1046/j.1365-3040.2001.00724.x

Fadrique, B., Báez, S., Duque, A., Malizia, A., Blundo, C., Carilla, J., Osinaga-Acosta, O., Malizia, L., Silman, M., Farfán-Ríos, W., et al. (2018). Widespread but heterogeneous responses of Andean forests to climate change. *Nature, 564*, 207–212. https://doi.org/10.1038/s41586-018-0715-9

Fahey, T., Sherman, R., & Tanner, E. (2016). Tropical montane cloud forest: Environmental drivers of vegetation structure and ecosystem function. *Journal of Tropical Ecology, 32*(5), 355–367. https://doi.org/10.1017/S0266467415000176

Fassbender, H., & Grimm, U. (1981a). Ciclos biogeoquímicos en un ecosistema forestal de Los Andes Occidentales de Venezuela. II. Producción y descomposición de los residuos vegetales. *Turrialba, 31*, 39–47.

Fassbender, H., & Grimm, U. (1981b). Ciclos biogeoquímicos en un ecosistema forestal de Los Andes Occidentales de Venezuela. IV. Modelos y conclusiones. *Turrialba, 31*, 101–108.

Feeley, K., Martinez-Villa, J., Perez, T., Silva Duque, A., Gonzalez, D. T., & Duque, A. (2020). The thermal tolerances, distributions, and performances of tropical montane tree species. *Frontiers in Forests and Global Change, 3*, 25. https://doi.org/10.3389/ffgc.2020.00025

Fernández-Martínez, M., Vicca, S., Janssens, I. A., Sardans, J., Luyssaert, S., Campioli, M., Chapin, F. S., III, Ciais, P., Malhi, Y., & Obersteiner, M. (2014). Nutrient availability as the key regulator of global forest carbon balance. *Nature Climate Change, 4*, 471–476. https://doi.org/10.1038/NCLIMATE2177

Field, C., & Mooney, H. A. (1986). The phosynthesis-nitrogen relationship in wild plants. In T. J. Givnish (Ed.), *On the economy of plant form and function* (pp. 25–49). Cambridge University Press.

Fisher, J. B., Malhi, Y., Cuba Torres, I., Metcalfe, D. B., van de Weg, M. J., Meir, P., Silva-Espejo, J. E., & Huasco, W. H. (2013). Nutrient limitation in rainforests and cloud forests along a 3,000-m elevation gradient in the Peruvian Andes. *Oecologia, 172*, 889–902. https://doi.org/10.1007/s00442-012-2522-6

Foster, P. (2001). The potential negative impacts of global climate change on tropical montane cloud forests. *Earth Science Reviews, 55*, 73–106. https://doi.org/10.1016/s0012-8252(01)00056-3

Fyllas, N. M., Bentley, L. P., Shenkin, A., Asner, G. P., Atkin, O. K., Díaz, S., Enquist, B. J., Farfan-Rios, W., Gloor, E., Guerrieri, R., et al. (2017). Solar radiation and functional traits explain the decline of forest primary productivity along a tropical elevation gradient. *Ecology Letters, 20*(6), 730–740. https://doi.org/10.1111/ele.12771

Galloway, J. N., Townsend, A. R., Erisman, J. W., Bekunda, M., Cai, Z., Freney, J. R., Martinelli, L. A., Seitzinger, S. P., Sutton, M. A., et al. (2008). Transformation of the nitrogen cycle: Recent trends, questions, and potential solutions. *Science, 320*, 889–892. https://doi.org/10.1126/science.1136674

García-Núñez, C., Azócar, A., & Rada, F. (1995). Photosynthetic acclimation to light in juveniles of two cloud forest tree species. *Trees, 10*, 114–124. https://doi.org/10.1007/bf00192192

Girardin, C. A. J., Farfan-Rios, W., Garcia, K., Feeley, K. J., Jørgensen, P. M., Murakami, A. A., Cayola Pérez, L., Seidel, R., Paniagua, N., Fuentes Claros, A. F., et al. (2014a). Spatial patterns of above-ground structure, biomass and composition in a network of six Andean elevation transects. *Plant Ecology and Diversity, 7*, 161–171. https://doi.org/10.1080/17550874.2013.820806

Girardin, C. A. J., Malhi, Y., Feeley, K. J., Rapp, J. M., Silman, M. R., Meir, P., Huasco, W. H., Salinas, N., Mamani, M., Silva-Espejo, J. E., et al. (2014b). Seasonality of above-ground net primary productivity along an Andean altitudinal transect in Peru. *Journal of Tropical Ecology, 30*(6), 503–519. https://doi.org/10.1017/S0266467414000443

Girardin, C. A. J., Silva Espejo, J. E. S., Doughty, C. E., Huasco, W. H., Metcalfe, D. B., Durand-Baca, L., Marthews, T. R., Aragaõ, L. E. O. C., Farfán-Rios, W., García-Cabrera, K., et al. (2014c). Productivity and carbon allocation in a tropical montane cloud forest in the Peruvian Andes. *Plant Ecology and Diversity, 7*, 107–123. https://doi.org/10.1080/17550874.2013.820222

Goldsmith, G. R., Matzke, N. J., & Dawson, T. E. (2013). The incidence and implications of clouds for cloud forest plant water relations. *Ecology Letters, 16*, 307–314. https://doi.org/10.1111/ele.12039

Goldstein, G., & Santiago, L. S. (2016). *Tropical tree physiology: Adaptations and responses in a changing environment*. Springer.

Goldstein, G., Santiago, L. S., Campanello, P. I., Avalos, G., Zhang, Y. J., & Villagra, M. (2016). Facing shortage or excessive light: How tropical and subtropical trees adjust their photosynthetic behavior and life history traits to a dynamic forest environment. In G. Goldstein & L. S. Santiago (Eds.), *Tropical tree physiology: Adaptations and responses in a changing environment* (pp. 319–336). Springer. https://doi.org/10.1007/978-3-319-27422-5_15

Golley, F. B. (1971). La biomasa y la estructura mineral de algunos bosques de Darién, Panamá. *Turrialba, 21*, 186–196.

Grimm, U., & Fassbender, H. (1981a). Ciclos biogeoquímicos en un ecosistema forestal de Los Andes Occidentales de Venezuela. I. Inventario de las reservas orgánicas y minerales (N, P, K, Ca, Mg, Mn, Fe, Al, Na). *Turrialba, 31*, 27–37.

Grimm, U., & Fassbender, H. (1981b). Ciclos biogeoquímicos en un ecosistema forestal de Los Andes Occidentales de Venezuela. III. Ciclo hidrológico y translocación de elementos químicos con el agua. *Turrialba, 31*, 89–99.

Grubb, P. J. (1977). Control of forest growth and distribution on wet tropical mountains: With special reference to mineral nutrition. *Annual Review of Ecology and Systematics, 8*, 83–107. https://doi.org/10.1146/annurev.es.08.110177.000503

Guan, K., Pan, M., Li, H., Wolf, A., Wu, J., Medvigy, D., Caylor, K. K., Sheffield, J., Wood, E. F., Malhi, Y., et al. (2015). Photosynthetic seasonality of global tropical forests constrained by hydroclimate. *Nature Geoscience, 8*, 284–289. https://doi.org/10.1038/ngeo2382

Hager, A., & Dohrenbusch, A. (2011). Hydrometeorology and structure of tropical montane cloud forests under contrasting biophysical conditions in North-Western Costa Rica. *Hydrological Processes, 25*, 392–401. https://doi.org/10.1002/hyp.7726

Hansen, M. C., Wang, L., Song, X.-P., Tyukavina, A., Turubanova, S., Potapov, P. V., & Stehman, S. V. (2020). The fate of tropical forest fragments. *Science Advances, 6*(11), eaax8574. https://doi.org/10.1126/sciadv.aax857

Helmer, E. H., Gerson, E. A., Baggett, L. S., Bird, B. J., Ruzycki, T. S., & Voggesser, S. M. (2019). Neotropical cloud forests and páramo to contract and dry from declines in cloud immersion and frost. *PLoS One, 14*(4), e0213155. https://doi.org/10.1371/journal.pone.0213155

Hernández, R., & Medina, E. (1995). Nitrogen content and photosyhnthesis of trees in a tropical cloud forest as affected by light climate. In P. Mathis (Ed.), *Photosynthesis: From light to biosphere* (pp. 671–674). Springer Netherlands: Kluwer Academic Publ.

Hetsch, W., & Hoheisel, H. (1976). Standorts –und Vegetationsgliederung in einem tropischen Nebelwald. *Allgemeine. Forst-und. Jagd-Zeitung, 147*, 200–209.

Homeier, J., Breckle, S. W., Gunter, S., Rollenbeck, R. T., & Leuschner, C. (2010). Tree diversity, forest structure and productivity along altitudinal and topographical gradients in a species rich Ecuadorian montane rain Forest. *Biotropica, 42*(2), 140–148. https://doi.org/10.1111/j.1744-7429.2009.00547.x

Homeier, J., Hertel, D., Camenzind, T., Cumbicus, N. L., Maraun, M., Martinson, G. O., Poma, L. N., Rillig, M. C., Sandmann, D., Scheu, S., et al. (2012). Tropical Andean forests are highly susceptible to nutrient inputs – Rapid effects of experimental N and P addition to an Ecuadorian montane Forest. *PLoS One, 7*(10), e47128. https://doi.org/10.1371/journal.pone.0047128

Huber, O. (1986a). El clima. In O. Huber (Ed.), *La Selva Nublada de Rancho Grande, Parque Nacional Henri Pittier* (pp. 17–30). Fondo Editorial Acta Científica Venezolana.

Huber, O. (1986b). Las selvas nubladas de Rancho Grande: observaciones sobre su fisionomía, estructura y fenología. In O. Huber (Ed.), *La Selva Nublada de Rancho Grande, Parque Nacional Henri Pittier* (pp. 131–170). Fondo Editorial Acta Científica Venezolana.

Jaimes, J. C. (2015). *Dinámica de la aclimatación del aparato fotosintético a cambios en el ambiente de luz en individuos juveniles de especies arbóreas de la sucesión temprana en la Selva Nublada Andina*. Undergraduate thesis. Facultad de Ciencias, Universidad de Los Andes, Mérida (Venezuela). p. 54.

Jarvis, A., & Mulligan, M. (2010). The climate of cloud forests. In L. A. Bruijnzeel, F. N. Scatena, & L. S. Hamilton (Eds.), *Tropical montane cloud forests: Science for conservation and management* (pp. 39–56). Cambridge University Press. https://doi.org/10.1017/CBO9780511778384

Jordan, C. F., & Kline, J. R. (1976). Transpiration of trees in a tropical rainforest. *Journal of Applied Ecology, 14*(3), 853–860. https://doi.org/10.2307/2402816

Kapos, V., & Tanner, E. V. J. (1985). Water relations of Jamaica upper montane rain forests trees. *Ecology, 66*, 241–250. https://doi.org/10.2307/1941324

Killeen, T., Douglas, M., Consiglio, T., Jørgensen, P. M., & Mejia, J. (2007). Dry spots and wet spots in the Andean hotspot. *Journal of Biogeography, 34*(8), 1357–1373. https://doi.org/10.1111/j.1365-2699.2006.01682.x

Kira, T. (1977). Community architecture and organic matter dynamics in tropical lowland rain forest of Southwest Asia, with special reference to Pasoh Forest, West Malaysia. In P. B. Tomlinson & M. H. Zimmermann (Eds.), *Tropical trees as living systems, 4th Cabot symposium, Harvard* (pp. 561–590). Cambridge University Press.

Kitayama, K., & Aiba, S.-I. (2002). Ecosystem structure and productivity of tropical rain forests along altitudinal gradients with contrasting soil phosphorus pools on Mount Kinabalu, Borneo. *Journal of Ecology, 90*, 37–51.

Klinge, H. (1977). Bilanzierung von Hauptnährstoffen im Ökosystem tropisher Regenwald (manaus) – vorläufige Daten. *Biogeographica Den Haag, 7*, 59–77.

Krause, G. H., Koroleva, O. Y., Dalling, J. W., & Winter, K. (2001). Acclimation of tropical tree seedlings to excessive light in simulated tree-fall gaps. *Plant, Cell & Environment, 24*(12), 1345–1352. https://doi.org/10.1046/j.0016-8025.2001.00786.x

Lamprecht, H. (1954). *Estudios Selviculturales en los Bosques del Valle de La Mucuy, Cerca de Mérida*. Facultad de Ciencias Forestales, Universidad de Los Andes.

Lavergne, S., Mouquet, N., Thuiller, W., & Ronce, O. (2010). Biodiversity and climate change: Integrating evolutionary and ecological responses of species and communities. *Annual Review of Ecology, Evolution, and Systematics, 41*, 321–350. https://doi.org/10.1146/annurev-ecolsys-102209-144628

Leuschner, C., Zach, A., Moser, G., Homeier, J., Graefe, S., Hertel, D., Wittich, B., Soethe, N., Iost, S., Röderstein, M., et al. (2013). The carbon balance of tropical mountain forests along an altitudinal transect. In J. Bendix, E. Beck, A. Bräuning, F. Makeschin, R. Mosandl, S. Scheu, & W. Wilcke (Eds.), *Ecosystem services, biodiversity and environmental change in a tropical mountain ecosystem of South Ecuador* (Vol. 22, pp. 117–139). Springer.

Linares, A. (2008). *Análisis florístico y estructural de la vegetación de una Selva Nublada Andina en un gradiente altitudinal en La Mucuy, Estado Mérida*. Doctorate thesis. Facultad de Ciencias, Universidad de Los Andes, Mérida (Venezuela). p. 139.

Lisanework, N., & Michelsen, A. (1994). Litterfall and nutrient release by decomposition in three plantations compared with a natural forest in the Ethiopian highland. *Forest Ecology and Management, 65*, 149–164. https://doi.org/10.1016/0378-1127(94)90166-X

Llerena-Zambrano, M., Ordoñez, J. C., Llambí, L. D., van der Sande, M., Pinto, E., Salazar, L., & Cuesta, F. (2021). Minimum temperature drives community leaf trait variation in secondary montane forests along a 3000-m elevation gradient in the tropical Andes. *Plant Ecology and Diversity, 14*(1–2), 47–63. https://doi.org/10.1080/17550874.2021.190360

Malhi, Y., Girardin, C. A., Goldsmith, G. R., Doughty, C. E., Salinas, N., Metcalfe, D. B., Huasco, W. H., Silva-Espejo, J. E., Aguilla-Pasquell, J., Amézquita, F. F., et al. (2017). The variation of productivity and its allocation along a tropical elevation gradient: A whole carbon budget perspective. *The New Phytologist, 214*(3), 1019–1032. https://doi.org/10.1111/nph.14189

Malhi, Y., Franklin, J., Seddon, N., Solan, M., Turner, M. G., Field, C. B., & Knowlton, N. (2020). Climate change and ecosystems: Threats, opportunities and solutions. *Philosophical Transactions of the Royal Society B, 375*(1794), 20190104. https://doi.org/10.1098/rstb.2019.0104

Martin, R. E., Asner, G. P., Bentley, L. P., Shenkin, A., Salinas, N., Huaypar, K. Q., Pillco, M. M., Ccori Álvarez, F. D., Enquist, B. J., Diaz, S., et al. (2020). Covariance of sun and shade leaf traits along a tropical forest elevation gradient. *Frontiers in Plant Science, 10*, 1810. https://doi.org/10.3389/fpls.2019.01810

Maza, B., Rodes-Blanco, M., & Rojas, E. (2022). Aboveground biomass along an elevation gradient in an evergreen Andean–Amazonian forest in Ecuador. *Frontiers in Forests and Global Change, 5*, 738585. https://doi.org/10.3389/ffgc.2022.738585

Medina, E. (1986). Aspectos ecofisiológicos de plantas de bosques nublados tropicales: el bosque nublado de Rancho Grande. In O. Huber (Ed.), *La Selva Nublada de Rancho Grande, Parque Nacional Henri Pittier* (pp. 189–196). Fondo Editorial Acta Científica Venezolana.

Meinzer, F., Goldstein, G., & Jaimes, M. (1984). The effect of atmospheric humidity on stomatal control of gas exchange in two tropical coniferous species. *Canadian Journal of Botany, 62*(3), 591–595. https://doi.org/10.1139/b84-089

Michelangeli, F. (1998). *La Selva Nublada: Rancho Grande*. Armitano Editores.

Moser, G., Hertel, D., & Leuschner, C. (2007). Altitudinal change in LAI and stand leaf biomass in tropical montane forests: A transect study in Ecuador and a pantropical meta-analysis. *Ecosystems, 10*, 924–935. https://doi.org/10.1007/s10021-007-9063-6

Moser, G., Leuschner, C., Hertel, D., Graefe, S., Soethe, N., & Iost, S. (2011). Elevation effects on the carbon budget of tropical mountain forests S Ecuador: The role of the belowground compartment. *Global Change Biology, 17*, 2211–2226. https://doi.org/10.1111/j.1365-2486.2010.02367.x

Motzer, T., Munz, N., Küppers, M., Schmitt, D., & Anhuf, D. (2005). Stomatal conductance, transpiration and sap flow of tropical montane rain forest trees in the southern Ecuadorian Andes. *Tree Physiology, 25*, 1283–1293. https://doi.org/10.1093/treephys/25.10.1283

Mulkey, S. S., & Pearcy, R. W. (1992). Interactions between acclimation and photoinhibition of photosynthesis of a tropical forest understory herb, *Alocasia macrorrhiza*, during simulated canopy gap formation. *Functional Ecology, 6*(6), 719–729. https://doi.org/10.2307/2389969

Mulkey, S. S., & Wright, S. J. (1996). Influence of seasonal drought on the carbon balance of tropical forest plants. In S. S. Mulkey, R. L. Chazdon, & A. P. Smith (Eds.), *Tropical forest plant ecophysiology* (pp. 187–216). Chapman & Hall.

Mulligan, M. (2010). Modeling the tropics-wide extent and distribution of cloud forest and cloud forest loss, with implications for conservation priority. *Tropical Montane Cloud Forests: Science for Conservation and Management, 740*, 16–38. https://doi.org/10.1017/CBO9780511778384.004

Nadkarni, N. M., & Solano, R. (2002). Potential effects of climate change on canopy communities in a tropical cloud forest: An experimental approach. *Oecologia, 131*(4), 580–586. https://doi.org/10.1007/s00442-002-0899-3

Nottingham, A. T., Whitaker, J., Turner, B. L., Salinas, N., Zimmermann, M., Malhi, Y., & Meir, P. (2015). Climate warming and soil carbon in tropical forests: Insights from an elevation gradient in the Peruvian Andes. *Bioscience, 65*(9), 906–921. https://doi.org/10.1093/biosci/biv109

Oliveira, R. S., Eller, C. B., Bittencourt, P. R. L., & Mulligan, M. (2014). The hydroclimatic and ecophysiological basis of cloud forest distributions under current and projected climates. *Annals of Botany, 113*(6), 909–920. https://doi.org/10.1093/aob/mcu060

Oliveras, I., Roman-Cuesta, R. M., Urquiaga-Flores, E., Quintano-Loayza, J. A., Kala, J., Huaman, V., Lizarraga, N., Sans, G., Quispe, K., Lopez, E., et al. (2017). Fire effects and ecological recovery pathways of tropical montane cloud forests along a time chronosequence. *Global Change Biology, 24*(2), 758–772. https://doi.org/10.1111/gcb.13951

Oliveras, I., Bentley, L., Fyllas, N. M., Gvozdevaite, A., Shenkin, A. F., Prepah, T., Morandi, P., Peixoto, K. S., Boakye, M., Adu-Bredu, S., et al. (2020). The influence of taxonomy and environment on leaf trait variation along tropical abiotic gradients. *Frontiers in Forests and Global Change, 3*, 18. https://doi.org/10.3389/ffgc.2020.00018

Ostertag, R., Restrepo, C., Dalling, J. W., Martin, P. H., Abiem, I., Aiba, S.-I., Alvarez-Dávila, E., Aragón, R., Ataroff, M., Chapman, H., et al. (2022). Litter decomposition rates across tropical montane and lowland forests are controlled foremost by climate. *Biotropica, 54*(2), 309–326. https://doi.org/10.1111/btp.13044

Pan, Y., Birdsey, R. A., Phillips, O. L., & Jackson, R. B. (2013). The structure, distribution and biomass of the world's forests. *Annual Review of Ecology, Evolution, and Systematics, 44*, 593–622. https://doi.org/10.1146/annurev-ecolsys-110512-135914

Pendry, C., & Proctor, J. (1996). The causes of altitudinal zonation of rain forest on Bukit Belalong, Brunei. *Journal of Ecology, 84*(3), 407–418. https://doi.org/10.2307/2261202

Phillips, J., Ramirez, S., Wayson, C., & Duque, A. (2019). Differences in carbon stocks along an elevational gradient in tropical mountain forests of Colombia. *Biotropica, 51*(4), 490–499. https://doi.org/10.1111/btp.12675

Pounds, J. A., Fogden, M. P. L., & Campbell, J. H. (1999). Biological response to climate change on a tropical mountain. *Nature, 398*, 611–615. https://doi.org/10.1038/19297

Quevedo, A. M., Jerez, M., Schwarzkopf, T., & García-Núñez, C. (2015). Distribution of juveniles of tree species along a canopy closure gradient in a tropical cloud forest of the Venezuelan Andes. *iForest, 9*, 363–369. https://doi.org/10.3832/ifor1568-008

Quevedo, A. M., Schwarzkopf, T., García, C., & Jerez, M. (2016). Ambiente de luz del sotobosque de una selva nublada andina: estructura del dosel y estacionalidad climática. *Revista de Biología Tropical, 64*(4), 1699–1707.

Quevedo, A. M., García-Núñez, C., Jerez, M., Jaimez, R., & Schwarzkopf, T. (2018). Leaf acclimation strategies to contrasting light conditions in saplings of different shade tolerance in a tropical cloud forest. *Functional Plant Biology, 45*(9), 968–982. https://doi.org/10.1071/FP17308

Quilici, A., & Medina, E. (1998). Photosynthesis-nitrogen relationships in pioneer plants of disturbed tropical montane forest sites. *Photosynthetica, 35*(4), 525–534. https://doi.org/10.1023/a:1006974806455

Quiroz, L. (2010). *Definición de comunidades arbóreas asociadas a sitios específicos en una selva nublada de Los Andes venezolanos*. Masters thesis. Facultad de Ciencias, Universidad de Los Andes, Mérida (Venezuela). p. 144.

Rada, F., & Jaimez, R. (1992). Comparative ecophysiology and anatomy of terrestrial and epiphytic *Anthurium bredemeyeri* Schott in a tropical Andean cloud forest. *Journal of Experimental Botany, 43*, 723–727. https://doi.org/10.1093/jxb/43.5.72

Rada, F., García-Núñez, C., & Ataroff, M. (2009). Leaf gas exchange in canopy species of a Venezuelan Cloud Forest. *Biotropica, 41*(6), 679–778. https://doi.org/10.1111/j.1744-7429.2009.00537.x

Raich, J. W., Russell, A. E., & Vitousek, P. M. (1997). Primary productivity and ecosystem development along an elevational gradient on Mauna Loa, Hawai'i. *Ecology, 78*, 707–721. https://doi.org/10.1890/0012-9658(1997)078[0707:PPAEDA]2.0.CO;2

Raich, J. W., Russell, A. E., Kitayama, K., Parton, W. J., & Vitousek, P. M. (2006). Temperature influences carbon accumulation in moist tropical forests. *Ecology, 87*(1), 76–87. https://doi.org/10.1890/05-0023

Ramírez, M. E. (1999). *Producción y descomposición de hojarasca en la selva nublada de La Mucuy, Parque Nacional Sierra Nevada, Estado Mérida*. Masters thesis. Facultad de Ciencias, Universidad de Los Andes, Mérida (Venezuela). p. 86.

Rifai, S. W., Li, S., & Malhi, Y. (2019). Coupling of El Niño events and long-term warming leads to pervasive climate extremes in the terrestrial tropics. *Environmental Research Letters, 14*(10), 105002. https://doi.org/10.1088/1748-9326/ab402f

Robichaux, R. H., Rundel, P. W., Steemmerman, L., Canfield, J. E., Morse, S. R., & Friedman, W. E. (1984). Tissue water deficits and plant growth in wet tropical environments. In E. Medina, H. A. Mooney, & C. Vazquez-Yanes (Eds.), *Physiological ecology of plants of the wet tropics* (pp. 99–112). Lancaster.

Rodríguez-Morales, M., Chacón-Moreno, E., & Ataroff, M. (2009). Transformación del paisaje de selvas de montaña en la Cuenca del río Capaz, Andes Venezolanos. *Ecotropicos, 22*, 64–82.

Roman-Cuesta, R. M., Carmona-Moreno, C., Lizcano, G., New, M., Silman, M., Knoke, T., Malhi, Y., Oliveras, I., Asbjornsen, H., & Vuille, M. (2014). Synchronous fire activity in the tropical high Andes: An indication of regional climate forcing. *Global Change Biology, 20*(6), 1929–1942. https://doi.org/10.1111/gcb.12538

Salinas, N., Malhi, Y., Meir, P., Silman, M., Roman-Cuesta, R., Huaman, J., Salinas, D., Huaman, V., Gibaja, A., Mamani, M., et al. (2011). The sensitivity of tropical leaf litter decomposition to temperature: Results from a large-scale leaf translocation experiment along an elevation gradient in Peruvian forests. *The New Phytologist, 189*, 967–977. https://doi.org/10.1111/j.1469-8137.2010.03521.x

Santiago, L. S., & Goldstein, G. (2016). Is photosynthesis nutrient limited in tropical trees? In G. Goldstein & L. S. Santiago (Eds.), *Tropical tree physiology: Adaptations and responses in a changing environment* (pp. 243–258). Springer. https://doi.org/10.1007/978-3-319-27422-5_14

Santiago, L. S., Damien, B., De Guzman, M. E., & Ávila-Lovera, E. (2016). Drought survival strategies of tropical trees. In G. Goldstein & L. S. Santiago (Eds.), *Tropical tree physiology: Adaptations and responses in a changing environment* (pp. 299–315). Springer. https://doi.org/10.1007/978-3-319-27422-5_11

Schwarzkopf, T., Susan, J. R., Fahey, T. J., & Degloria, S. (2011). Are cloud forest tree structure and environment related in the Venezuelan Andes? *Austral Ecology, 36*(3), 280–289. https://doi.org/10.1111/j.1442-9993.2010.02160.x

Scott, D. A., Proctor, J., & Thompson, J. (1992). Ecological studies on a lowland evergreen rain forest on Maraca Island, Roraima, Brazil. II. Litter and nutrient cycling. *Journal of Ecology, 80*(4), 705–717. https://doi.org/10.2307/2260861

Silver, W. L., Brown, S., & Lugo, A. (1996). Biodiversity and biogeochemical cycles. In G. Orians, R. Dirzo, & J. H. Cushman (Eds.), *Biodiversity and ecosystem processes in tropical forests, ecological studies* (Vol. 122, pp. 49–67). Springer.

Slik, J. W. F., Paoli, G., McGuire, K., Amaral, I., Barroso, J., Bastian, M., Blanc, L., Bongers, F., Boundja, P., Clark, C., et al. (2013). Large trees drive forest aboveground biomass variation in moist lowland forests across the tropics. *Global Ecology and Biogeography, 22*(12), 1261–1271. https://doi.org/10.1111/geb.12092

Sobrado, M. (2003). Hydraulic characteristics and leaf water use efficiency in trees from tropical montane habitats. *Trees, 17*, 400–406. https://doi.org/10.1007/s00468-003-0251-5

Soethe, N., Lehmann, J., & Engels, C. (2008). Nutrient availability at different altitudes in a tropical montane forest in Ecuador. *Journal of Tropical Ecology, 24*, 397–406. https://doi.org/10.1017/S026646740800504X

Spracklen, D. V., & Righelato, R. (2014). Tropical montane forests are a larger than expected global carbon store. *Biogeosciences, 11*, 2741–2754. https://doi.org/10.5194/bg-11-2741-2014

Steinhardt, U. (1979). Untersuchungen uber den Wasser- und Nahrstoffhaushalt eines Andinen Wolkenwaldes in Venezuela. *Göttinger Bodenkundliche Berichte, 56*, 1–185.

Still, C. J., Foster, P. N., & Schneider, S. H. (1999). Simulating the effects of climate change on tropical montane cloud forests. *Nature, 398*, 608–610. https://doi.org/10.1038/19293

Strauss-Debenedetti, S., & Bazzaz, F. (1996). Photosynthetic characteristics of tropical trees along successional gradients. In S. S. Mulkey, R. L. Chazdon, & A. P. Smith (Eds.), *Tropical forest plant ecophysiology* (pp. 162–186). Chapman & Hall.

Swift, M. J., Heal, O. W., & Anderson, J. M. (1979). *Decomposition in terrestrial ecosystems*. Blackwell.

Tanner, E. V., Kapos, V., & Franco, W. (1992). Nitrogen and phosphorus fertilization effects on Venezuelan montane forest trunk growth and litterfall. *Ecology, 73*(1), 78–86. https://doi.org/10.2307/1938722

Tilman, D., Fargione, J., Wolff, B., D'Antonio, C., Dobson, A., Howarth, R., et al. (2001). Forecasting agriculturally driven global environmental change. *Science, 292*, 281–284. https://doi.org/10.1126/science.1057544

Tito, R., Vasconcelos, H. L., & Feeley, K. J. (2020). Mountain ecosystems as natural laboratories for climate change experiments. *Frontiers in Forests and Global Change, 3*, 38. https://doi.org/10.3389/ffgc.2020.00038

Tovar, C., Arnillas, C. A., Cuesta, F., & Buytaert, W. (2013). Diverging responses of tropical Andean biomes under future climate conditions. *PLoS One, 8*(5), e63634. https://doi.org/10.1371/journal.pone.0063634

Urrutia, R., & Vuille, M. (2009). Climate change projections for the tropical Andes using a regional climate model: Temperature and precipitation simulations for the end of the 21st century. *Journal of Geophysical Research, 114*, D02108. https://doi.org/10.1029/2008JD011021

van de Weg, M., Meir, P., Williams, M., Girardin, C., Malhi, Y., Silva-Espejo, J., & Grace, J. (2014). Gross primary productivity of a high elevation tropical montane cloud forest. *Ecosystems, 17*, 751–764. https://doi.org/10.1007/s10021-014-9758-4

Vareschi, V. (1992). *Ecología de la Vegetación Tropical*. Sociedad Venezolana de Ciencias Naturales.

Veillon, J. P. (1965). Variación altitudinal de la masa forestal de los bosques primarios en la vertiente nor-occidental de la Cordillera de Los Andes, Venezuela. *Turrialba, 15*, 216–224.

Veillon, J. P. (1985). El crecimiento de algunos bosques naturales de Venezuela en relación con los parámetros del medio ambiente. *Rev Forest Venez, 29*, 1–122.

Veillon, J. P. (1989). *Los Bosques Naturales de Venezuela. Parte I: el Medio Ambiente*. Instituto de Silvicultura, Universidad de Los Andes.

Vilanova, I. (1996). *Intercambio de gases y relaciones hídricas en plantas del sotobosque de la selva nublada de La Mucuy, Estado Mérida, Venezuela*. Undergraduate thesis. Facultad de Ciencias, Universidad de Los Andes, Mérida (Venezuela). p. 117.

Vilanova, E., Ramírez-Angulo, H., Torres-Lezama, A., Aymard, G., Gámez, L., Durán, C., Hernández, L., Herrera, R., van der Heijden, G., Phillips, O. L., et al. (2018). Environmental drivers of forest structure and stem turnover across Venezuelan tropical forests. *PLoS One, 13*(6), e0198489. https://doi.org/10.1371/journal.pone.0198489

Vitousek, P. M., & Sanford, R. L., Jr. (1986). Nutrient cycling in moist tropical forest. *Annual Review of Ecology, Evolution, and Systematics, 17*(1), 137–167.

Wang, H., Hall, C. A. S., Scatena, F. N., Fetcher, N., & Wu, W. (2003). Modeling the spatial and temporal variability in climate and primary productivity across the Luquillo Mountains, Puerto Rico. *Forest Ecology and Management, 179*, 69–94. https://doi.org/10.1016/S0378-1127(02)00489-9

Weaver, P. L., & Murphy, P. G. (1990). Forest structure and productivity in Puerto Rico's Luquillo Mountains. *Biotropica, 22*, 69–82.

Werner, F. A., & Homeier, J. (2015). Is tropical montane forest heterogeneity promoted by a resource-driven feedback cycle? Evidence from nutrient relations, herbivory and litter decomposition along a topographical gradient. *Functional Ecology, 29*(3), 430–440. https://doi.org/10.1111/1365-2435.12351

Willis, K. J., & MacDonald, G. M. (2011). Long-term ecological records and their relevance to climate change predictions. *Annual Review of Ecology, Evolution, and Systematics, 42*, 267–287. https://doi.org/10.1146/annurev-ecolsys-102209-144704

Wright, S. J., Muller-Landau, H. C., Condit, R., & Hubbell, S. P. (2003). Gap-dependent recruitment, realized vital rates, and size distributions of tropical trees. *Ecology, 84*(12), 3174–3185. https://doi.org/10.1890/02-0038

Yamashita, N., Ishida, A., Kushima, H., & Tanaka, N. (2000). Acclimation to sudden increase in light favoring an invasive over native trees in subtropical islands, Japan. *Oecologia, 125*, 412–419. https://doi.org/10.1007/s004420000475

Zhang, Q., Zhang, T. J., Chow, W. S., Xie, X., Chen, Y. J., & Peng, C. L. (2015). Photosynthetic characteristics and light energy conversions under different light environments in five tree species occupying dominant status at different stages of subtropical forest succession. *Functional Plant Biology, 42*(7), 609–619. https://doi.org/10.1071/FP14355

Zimmermann, M., Meir, P., Silman, M. R., Fedders, A., Gibbon, A., Malhi, Y., Urrego, D. H., Bush, M. B., Feeley, K. J., Garcia, K. C., et al. (2010). No differences in soil carbon stocks across the tree line in the Peruvian Andes. *Ecosystems, 13*(1), 62–74. https://doi.org/10.1007/s10021-009-9300-2

The Role of Litterfall in Understanding the Ecological Integrity of Endangered Upper Andean Successional Forests

Dennis Castillo-Figueroa and Juan Manuel Posada

1 Introduction

Litterfall is the main link between above- and belowground carbon dynamics as it constitutes one of the major pathways of organic material and nutrients from the aerial part of plants to the soil (Liu et al., 2003; Zhang et al., 2014). Overall, it has been estimated that about 70% of litterfall is composed of leaves, and the remaining fraction is made up of logs, branches, flowers and fruits (Zhang et al., 2014). Litter production represents, on average, a third of the total net above-ground net primary production (NPP) (Melillo et al., 1993; Chave et al., 2010), from which over half is returned to the soil via litter decomposition (Cebrian, 1999; Wardle et al., 2004). The quantity and quality of litter inputs largely influences the environmental conditions and the biological communities of soils in forests (Zhang et al., 2024; Castillo-Avila et al., 2025). For instance, litter depth can modulate soil humidity (Sotes et al., 2013), temperature (Fekete et al., 2016) and microhabitat heterogeneity (Palace et al., 2007). In addition, litter quality is one of the main drivers of detrital food webs structure, given that the chemical and physical properties of fallen leaves are the cornerstone of soil community composition and trophic interactions (Brousseau et al., 2018; Raymond-Leonard et al., 2019). Although litterfall has been extensively studied across most of the world biomes in the last few decades (Chave et al., 2010; Vendrami et al., 2012; Zhang et al., 2014), our understanding of it is still very limited in tropical mountain ecosystems.

The Northern region of the Andes is a global hotspot of biodiversity (Myers et al., 2000), that hosts not only a high number of plant species, but also the highest levels of vascular plant endemicity in the planet (Myster, 2021). With only 2% of

---

D. Castillo-Figueroa · J. M. Posada (✉)
Biology Department, Faculty of Natural Sciences, Universidad del Rosario,
Bogotá, Colombia
e-mail: juan_posada@me.com

N. Clerici (ed.), *Conservation of Andean Forests*,
https://doi.org/10.1007/978-3-031-80805-0_3

the terrestrial surface of Earth, the Andes encompass 10% of the global plant biodiversity (ca. 30,000 spp.; Mittermeier et al., 2011; Rahbek et al., 2019), from which half are unique to this region (Hoorn et al., 2010). However, the tropical upland Andes have been largely disturbed due to extensive landscape transformation since Spanish colonization (Etter & van Wyngaarden, 2000; Etter et al., 2008), which has resulted in an increase of secondary forest cover (Rubiano et al., 2017; Hurtado-M et al., 2021; Calbi et al., 2020, 2021). These upper Andean forests are, therefore, a conservation priority with considerable levels of irreplaceability and vulnerability, and are included in many regional and global conservation agendas (Bax & Francesconi, 2019; Homeier et al., 2021). For instance, the IUCN Red List of Terrestrial Ecosystems in Colombia indicates that a quarter of the country's ecosystems exhibit critical risk levels (CR), that is 20 out of the 81 identified ecosystem types, many of which belong to the Andean region (Etter et al., 2018, 2021). Moreover, Andean forest communities are highly sensitive to pressures related to global warming and their concomitant effects, like thermophilisation, which consists in the movement of lowland species into higher elevations, causing compositional changes in the native Andean flora (Duque et al., 2015; Fadrique et al., 2018). Considering the high specialization of mountain forest species to cold climates, raising temperatures can increase recruitment and growth rates of thermophilic species, but at the same time, warmer and drier environments can elevate the mortality of native upper Andean species (Duque et al., 2015, 2021). The consequences of these demographic and compositional changes on the functioning and integrity of ecosystems are uncertain and difficult to predict (Homeier et al., 2021), along with shifts in precipitation patterns caused by climate change (Urrutia & Vuille, 2009) and other current human-induced changes, such as increasing nitrogen deposition (Wilcke et al., 2013).

Studying how upper Andean tropical forests regenerate after being disturbed is fundamental to assess their resilience under different historical and ecological scenarios (Hurtado-M et al., 2022). In particular, the evaluation of litterfall along successional gradients provides fundamental information for a better understanding of the recovery of ecosystem function (Souza et al., 2019), which is linked to the provision of ecosystem services such as carbon sequestration and nutrient cycling (Liu et al., 2003; Zhang et al., 2014). Litterfall can be considered one of the main components of ecosystem integrity in tropical secondary forests (Rosenfield et al., 2023), because it largely influences ecosystem dynamics, such as soil food webs and nutrient fluxes during succession (Brown & Lugo, 1990; Krishna & Mohan, 2017). Therefore, litter production rates can serve as suitable indicators of ecosystem conservation status and recovery, and can help inform forest restoration programs (Rosenfield et al., 2023). During succession, aboveground biomass and canopy height increase with time (Poorter et al., 2021, 2023), and drive litter inputs to the soils which, in turn, may determine key factors of the belowground compartment such as fine root production (An et al., 2017), soil organic layer thickness (Wilcke, 2021), and the assembly of animal communities (Castillo-Avila et al., 2025).

Importantly, functional relationships between litterfall and soil attributes contribute to carbon fluxes (Wardle et al., 2004) and can determine the storage of substantial carbon stocks in Andean forests (Yepes et al., 2015, 2016). Recent studies have highlighted the importance of mature Andean forests as global carbon sinks considering that aboveground carbon stocks of the mountain Andes are similar or even higher than that from lowland mature forests of Africa, Amazonia or southeast Asia (Duque et al., 2021). Due to this, Andean montane forests are currently posed as strategic ecosystems not only for biodiversity conservation, but also for climate change mitigation (Myster, 2021; Duque et al., 2021). Yet, despite its importance, there is a lack of research about litterfall in upper Andean tropical forests, and it is not clear to what extent litterfall is related to the edaphic characteristics (Castillo-Figueroa, 2021).

In this study, we quantify litterfall rates along a successional gradient in tropical upper Andean montane forests of Colombia, and established its functional link with soil attributes. Specifically, our objectives were to: (1) analyze the changes of litterfall with succession and, (2) determine the relationships between litterfall and fine root production, soil organic layer thickness, soil respiration, and soil fauna communities. We discuss the implications of these functional relationships for the resilience and conservation of Andean forests, as well as the knowledge gaps that remain to develop a comprehensive understanding of litterfall dynamics in these threatened ecosystems.

2 Materials and Methods

2.1 Study Area

The Colombian Andes covers 24.5% of the country and constitutes the most populated region, hosting almost 70% of the population. (DANE, 2018; Etter & van Wyngaarden, 2000; Etter et al., 2021). Our study area is located in the Eastern branch of the Andes, in the Cundiboyacense high plains, which includes sites in the Capital city of Bogotá (Torca) and three surrounding municipalities (i.e., Guasca, Guatavita and Tabio; Fig. 1 and Table 1). Precipitation in the area varies between 600 mm in the center to 1200 mm in the western part, and displays a bimodal pattern with two rainy seasons (April to June and October to December) and two dry seasons (January to March and July to September) (Mendoza & Etter, 2002; IDEAM, 2024). In these peri-urban sites, cattle, agriculture and mining are the dominant economic activities (Montañez et al., 1994; Antonio-Fragala and Obregón-Neira, 2011). TMS-4 dataloggers (TOMST, Czech Republic) installed in the study sites, indicated that mean atmospheric temperature 15 cm aboveground is 10.68 °C, soil surface temperature is 11.19 °C, and mean volumetric soil water content is 37.91% (Fig. 1). The four locations ranged in elevation from 2685 to 3140 m.

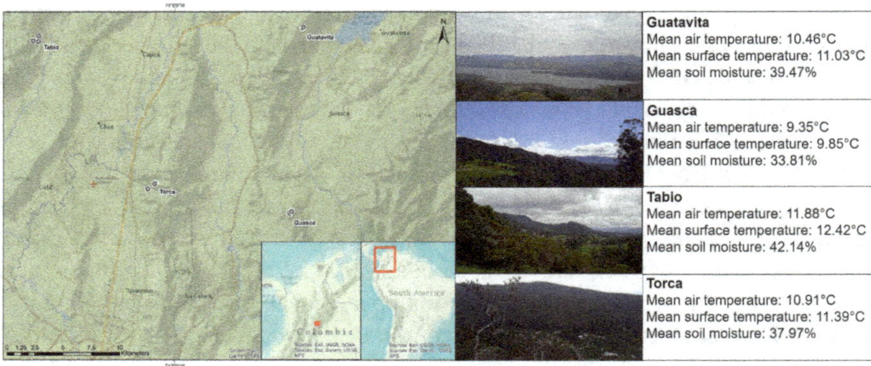

Fig. 1 Study area in upper Andean tropical forest from Colombia. Permanent plots are represented as small grey circles. Additional information about climate is given in the right panel. Air (15 cm above-ground) and soil surface temperature as well as volumetric soil water content were recorded by using TMS-4 dataloggers (TOMST, Czech Republic) installed in each of the 14 permanent plots. The urban section of the city of Bogotá is in the lower left corner of the map

2.2 *Plant Communities*

We established 14 permanent plots (20 m × 20 m) in secondary and old-growth forests (Table 1) where we tagged and identified all individuals with a diameter at 'ankle' height (DAH, at 5 cm aboveground; Hurtado-M et al., 2021) larger or equal to 5 cm. We used this approach because several individuals branched near the ground or had twisted stems, especially in secondary forests. The dominant plant families in terms of abundance were Ericaceae, Melastomataceae, Cunoniaceae, Primulaceae, and Asteraceae, representing 57.7% of individuals, and the most dominant species were *Weinmannia tomentosa* Linnaeus filius 1782, *Cavendishia bracteata* Hoerold 1909, *Miconia squamulosa* Triana 1872, and *Myrcianthes leucoxyla* McVaugh 1963, and *Myrsine latifolia* Spreng. 1824, with 46,16% of individuals (Castillo-Figueroa et al., 2023) (Table 1).

2.3 *Litter Itraps*

To estimate litterfall rate, we randomly placed ten 0.25 m^2 (50 × 50 cm) litter traps in each plot (total = 140), avoiding the edge of the plots and large forest gaps (Fig. 2). The traps were made of PVC tubes (1.27 cm diameter), raised ca. 1 m above-ground, and were covered with 1 mm mesh nylon screen. Litter was collected every 15–20 days to minimize loss by decomposition, and sampling was carried out during a whole year. Litter was oven-dried at 60 °C until constant weight, and then weighed with an analytical scale (precision: 0.0001 g). All branches with a diameter

Table 1 Study sites and plant diversity of our 14 permanent plots in upper Andean tropical forests in Colombia. Dominant plant species show the three most abundant species in each plot

Location	Plot ID	Succession	Northing	Easting	Elevation (m)	Dominant plant species
Guatavita	1	Secondary	4° 56′ 9,716″	73° 53′ 54,237″	3036	*Myrcianthes leucoxyla, Miconia squamulosa, Macleania rupestris,*
	2	Secondary	4° 56′ 12,618″	73° 53′ 51,825″	3028	*Miconia squamulosa, Myrcianthes leucoxyla, Macleania rupestris*
Guasca	3	Old-growth	4° 47′ 20,318″	73° 54′ 31,812″	3140	*Weinmannia tomentosa, Bejaria resinosa, Cavendishia bracteata*
	4	Secondary	4° 47′ 28,667″	73° 54′ 25,886″	3086	*Cavendishia bracteata, Gaiadendron punctatum, Alnus acuminata*
	5	Old-growth	4° 47′ 24,124″	73° 54′ 31,332″	3107	*Weinmannia tomentosa, Viburnum triphyllum, Myrsine guianensis*
	6	Secondary	4° 47′ 26,609″	73° 54′ 25,904″	3095	*Cavendishia bracteata, Gaiadendron punctatum, Alnus acuminata*
Tabio	7	Secondary	4° 55′ 40,858″	74° 6′ 29,194″	2696	*Miconia squamulosa, Daphnopsis caracasana, Myrcianthes leucoxyla*
	8	Secondary	4° 55′ 47,149″	74° 6′ 31,021″	2708	*Myrsine guianensis, Daphnopsis caracasana, Miconia squamulosa*
	9	Old-growth	4° 55′ 33,961″	74° 6′ 47,225″	2821	*Miconia squamulosa, Croton bogotanus, Xylosma spiculifera*
	10	Old-growth	4° 55′ 31,683″	74° 6′ 31,579″	2685	*Viburnum triphyllum, Miconia squamulosa, Daphnopsis caracasana*
Torca	11	Old-growth	4° 48′ 48,674″	74° 0′ 58,527″	2946	*Clusia multiflora, Drimys granadensis, Cavendishia nitida*
	12	Old-growth	4° 48′ 47,937″	74° 0′ 56,997″	2966	*Weinmannia tomentosa, Clusia multiflora, Bejaria resinosa,*
	13	Secondary	4° 48′ 31,216″	74° 0′ 19,178″	2709	*Miconia squamulosa, Viburnum triphyllum, Palicourea lineariflora*
	14	Old-growth	4° 48′ 45,912″	74° 0′ 58,852″	2954	*Clusia multiflora, Weinmannia tomentosa, Bejaria resinosa*

Fig. 2 Litter traps randomly placed in the 14 permanent plots for estimating litterfall rate (10 per plot)

above 1.0 cm were excluded. Branches with a diameter lower than 1.0 cm, but wider than the trap were cut to the length of the traps.

2.4 *Edaphic Attributes*

Edaphic attributes comprise all soil properties that result from the interplay between climate, organisms (including human disturbance), topography, parental rock, and time (Rajakaruna & Boyd, 2008). These attributes include several features of the physical, chemical, and biological properties of soil (Rajakaruna & Boyd, 2008). In this study, we focused on edaphic attributes related to soil carbon, plant roots, and soil fauna. First, we quantified the thickness of the soil organic layer (O horizon) by excavating soil pits next to each plot (Fig. 3). Second, we estimated fine root productivity using the root-ingrowth core technique (Vogt et al., 1998; Steingrobe et al., 2000; Girardin et al., 2013). We installed nine root-in-growth cores (mesh cages 11.5 cm diameter and 30 cm long) 2 m away from each other in a 3 × 3 grid pattern, and 3 m away from the corner points. Thus, we obtained 126 independent sampling units across the 14 plots. To prepare the cores, we sampled soil by first removing the surface soil litter and then used a post-hole digger to excavate a 30 cm depth cylindrical core (Phillips et al., 2001). We extracted all roots from the soil sample and then wrapped this rootless soil core with the mesh cage before replacing back into the ground. We harvested roots after 6 and 12 months; we manually removed the roots over a period of 40 min for each root-ingrowth core (four intervals of 10 min). We separated live fine roots (<2 mm) following standard

Fig. 3 Soil organic layer thickness (cm) measured by digging soil pits in (**a**) Guatavita-plot 1, (**b**) Guasca-plot 6 (**c**) Tabio-plot 10 and (**d**) Torca-plot 12

methodologies (Metcalfe et al., 2007; Brassard et al., 2011; Mathews et al., 2012), and then oven-dried them at 60 °C in the lab until constant weight; dry weight was measured with an analytical scale (0.0001 g). Third, we measured soil respiration during the wet season (May 2018) in the morning and afternoon for one day at each point by placing a Soil CO_2 Flux Chamber (6400-09, LiCor Inc., Lincoln, Nebraska) attached to an infrared gas analyzer (LI-6400XT, LiCor Inc., Lincoln, Nebraska) atop PVC collars in the same points of the 126 root-ingrowth cores; the PVC collars were installed three months before doing the measurements. Fourth, we collected soil fauna from three soil samples in 30 cm × 30 cm × 5 cm depth volumes in each plot (total n = 42) during the dry season (Jan-Feb 2022). The three soil samples had a minimal distance of 5 m between them and were stored in separate plastic bags to distinguish soil fauna from: (1) litter and (2) soil depth at 0–5 cm. In the lab, invertebrate specimens were preserved in 70% alcohol and were sorted into morphospecies and identified to the best resolution possible with the available taxonomical keys and the assistance of experts. We calculated mean soil fauna richness, soil fauna density, and mean soil predator richness per plot.

2.5 Data Analysis

Litterfall rate was expressed in grams per meter square per day (g m^{-2} d^{-1}). We compared litterfall rate between individual sites using a non-parametric analysis of variance (Kruskal-Wallis test, H). Also, we related litterfall rates to forest canopy height, which was used as a surrogate of successional status (i.e., younger forests were shorter than older forests). To compare old-growth and secondary forests per site, we used a U Mann-Whitney-Wilcoxon test. Finally, we did reduced major axis regressions (RMA) between litter rates and each of the soil attributes measured.

3 Results

3.1 Litterfall in Forest Succession

On average litterfall rate was 1.86 ± 1.19 g m^{-2} d^{-1} and ranged between 0.40 and 6.51, being highest in Tabio and then decreasing in the following order: Torca>Guasca>Guatavita (Fig. 4 and Table 1).

Overall, old-growth forests had a 64% higher litterfall rate than secondary ones ($p < 0.001$; U = 3779; rrb = 0.54; n = 140; Fig. 5a and Table 2). This was confirmed by the positive relation between forest canopy height and litterfall rate ($p < 0.0001$; $r^2 = 0.71$; n = 14; Fig. 5b). When comparisons between old-growth and secondary forests were done within sites (excluding Guatavita that only had second-growth forests, Table 1), significant differences were found only for Tabio ($p < 0.001$; U = 398; rrb = 0.99; n = 40, Fig. 5c), being higher in old-growth forests (Table 2); in Torca, old-growth forests also had slightly higher mean litterfall than secondary forest but the difference was not significant (Fig. 5c and Table 2).

3.2 Relationships with Belowground Attributes

Litterfall rates showed strong directly proportional relationships with key edaphic attributes including soil organic layer thickness ($p < 0.0001$; $r^2 = 0.63$; n = 14; CI = 14.66, 31.84; Fig. 6a), soil respiration ($p < 0.0001$; $r^2 = 0.66$; n = 14; CI = 4.74, 13.45; Fig. 6b), annual fine root production ($p < 0.01$; $r^2 = 0.39$; n = 14; CI = 0.69, 4.18; Fig. 6c), mean soil fauna species richness ($p = 0.003$; $r^2 = 0.53$; n = 14; CI = 0.52, 4.09; Fig. 6d), soil fauna density ($p = 0.035$; $r^2 = 0.32$; n = 14; CI = 16.51, 119.8; Fig. 6e), and mean predator species richness ($p < 0.0001$; $r^2 = 0.77$; n = 14; CI = 1.21, 2.13; Fig. 6f).

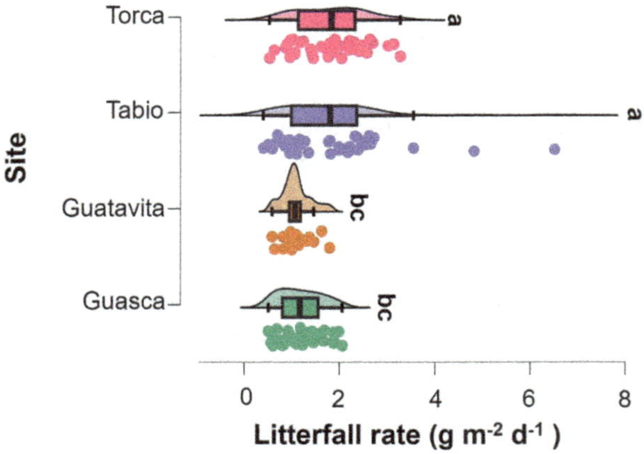

Fig. 4 Rainclouds comparing litterfall rate among sites in upper Andean tropical forests. Different letters represent statistical differences between sites (p < 0.05). Dots depict data from individual litter traps

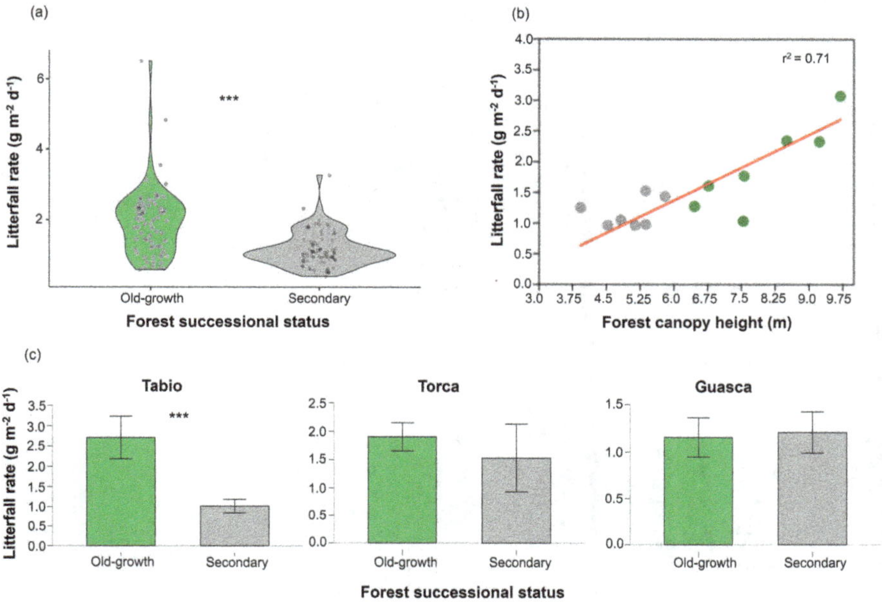

Fig. 5 Litterfall rate and succession. (**a**) violin plots comparing litterfall rate between secondary (grey) and old-growth forest (green) from the four study sites. Each dot in the violin plots represents an individual litter trap (n = 140) (**b**) linear regression between litterfall rate and forest canopy height as a surrogate of succession. Each dot in the linear regression represents a permanent plot (n = 14). Green dots represent old-growth forests and grey dots indicate secondary forests. (**c**) Barplots showing 95% Confidence intervals of Tabio, Torca and Guasca comparing litterfall rate between secondary and old-growth forest. Significant differences ($p < 0.001$) are denoted by ***

Table 2 Mean, range and coefficient of variation in litterfall rate (g m^{-2} d^{-1}) by site and forest successional status, in upper Andean tropical forests in Colombia; n is the number of litter traps

SITE/succession stage	n	Mean ± SD	Range	Coefficient of variation
TABIO	40	1.86 ± 1.19	0.40–6.51	0.64
Old-growth	20	2.71 ± 1.11	1.77–6.51	0.41
Secondary	20	1.01 ± 0.37	0.40–1.81	0.37
TORCA	40	1.81 ± 0.72	0.53–3.27	0.40
Old-growth	30	1.90 ± 0.66	0.63–3.13	0.35
Secondary	10	1.53 ± 0.84	0.53–3.27	0.55
GUASCA	40	1.18 ± 0.45	0.51–2.06	0.38
Old-growth	20	1.15 ± 0.45	0.59–2.06	0.39
Secondary	20	1.21 ± 0.47	0.51–1.99	0.39
GUATAVITA	20	1.10 ± 0.30	0.58–1.80	0.28
Old-growth overall	70	1.92 ± 0.96	0.59–6.51	0.50
Secondary overall	70	1.17 ± 0.49	0.40–3.27	0.42

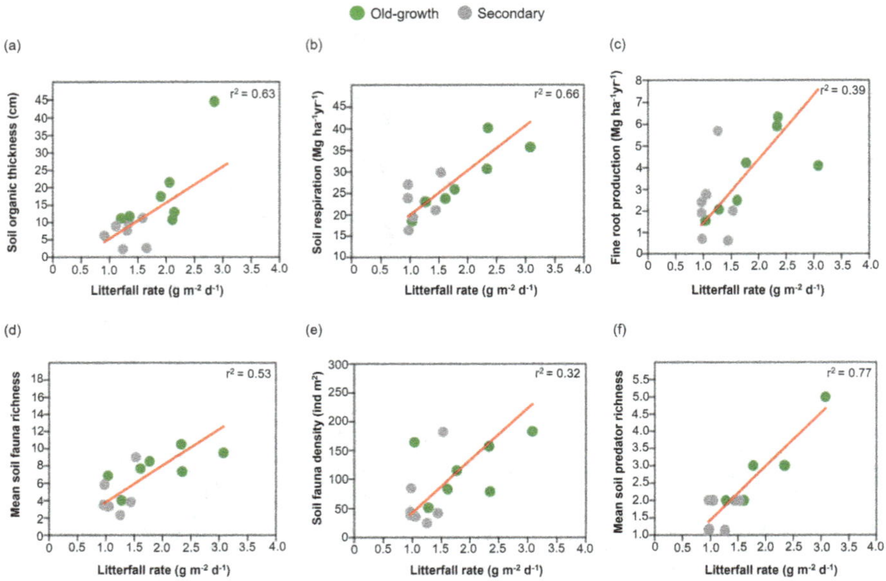

Fig. 6 Relationships between litterfall rate and belowground attributes: (**a**) soil organic layer thickness, (**b**) soil respiration, (**c**) annual fine root production, (**d**) mean soil fauna richness, (**e**) soil fauna density, (**f**) mean soil predator richness. Green dots represent old-growth forests and grey dots indicate secondary forests

4 Discussion

Despite extensive long-term landscape transformation in the central Andes of Colombia (Etter & van Wyngaarden, 2000; Etter et al., 2021), we show that the remaining forests around the capital city of Bogotá produce substantial litterfall, which is fundamental for carbon sequestration and soil functioning (Figs. 4, 5, and 6). We found that, on average, litterfall production rate was about 64% higher in old-growth than in secondary forests (Fig. 5a). Yet, considering the ubiquity of secondary forests in the Colombian Andes, which account for more than 70% of the remaining mountain forest cover (Etter et al., 2021), they provide an important flux of carbon to the soil. Moreover, the positive relationship between litterfall rate and forest canopy height (Fig. 5b) is an indication that litterfall should keep increasing with forests age.

Litterfall rates were found to be strongly related to multiple edaphic components (Fig. 6). First, an approximately threefold increase in litterfall production was related to an approximately tenfold increase in the thickness of the soil organic horizon (Fig. 6a), suggesting a strong causal link between these two variables. Thick organic horizons in montane ecosystems can be related to low air temperature (Pinos et al., 2017), limited nitrogen and phosphorous in soils (Tanner et al., 1998; Wilcke et al., 2008), high acidity, hypoxic soils (Schuur et al., 2001; Wilcke et al., 2002), and reduced microbial diversity (Nottingham et al., 2018). In addition, the production of a less decomposable leaf-litter, with high C:N ratio, leaf mass per area, leaf toughness and thickness (Asner et al., 2016; Homeier et al., 2021) can also slowdown decomposition (Castillo-Figueroa et al., 2025). This phenomenon can be more pronounced in mature forests where conservative species can be dominant in the community (Pinho et al., 2018; Poorter et al., 2021), and produce leaves characterized by high carbon and secondary compounds contents with low water-soluble components (Wright et al., 2004; Reich, 2014; Chua & Potts, 2018). Thus, the high increase in the soil organic horizon with litterfall could be, in part, attributed to changes in species composition with succession that led to slower decomposition in forests with higher litter production.

Second, soil respiration can be affected by multiple variables, such as land cover changes that alter soil temperature, microbial community composition, or plant species composition (Davidson et al., 1998; Yan et al., 2006; Fekete et al., 2014), which could have impacts on the global carbon balance (Han et al., 2015). Here, we found an increase in soil respiration with increasing litter rate production (Fig. 6b). This can be partially explained because litter is the substrate for microorganisms and an increase in leaf litter inputs contributes to more microbial activity and higher fluxes of CO_2 to the atmosphere (Berg & Mcclaugherty, 2020). Litterfall manipulation experiments have shown that soil respiration decreases by 35% when litterfall is removed and increases by 77% when litter is added again (Han et al., 2015). Interestingly, in our study, soil respiration increased by around 100% along the threefold (300%) increase in leaf litter input rate. This supports the interpretation that forests that produce more litter, which are also taller old-growth forests (Fig. 5b), may be producing leaves that are more difficult to decompose. Further studies should investigate whether litter traits change with succession, but also if there are parallel changes in other soil properties, such as microbial composition or nutrient availability.

Third, annual fine root production increased about sixfold with increasing litterfall (Fig. 6c). This increase can be interpreted as an increment in belowground allocation to capture additional nutrients from litter, and an illustration of the overall importance of litter in the transfer of nutrients from the aboveground to the belowground compartment (Liu et al., 2003). Yet, there are other, non-exclusive, mechanisms that may explain this increment such as a higher allocation to root production to compensate for low nutrient supply in more acidic soils (Vitousek & Sanford, 1986; Hendricks et al., 1993). In addition, mineralization rates may be lower in old-growth than in secondary forests due to the dominance of species with conservative leaf traits that produce lower quality litter for decomposers in older forests. Future research in these forests should explore the link between root production and nutrient cycling dynamics (Wardle et al., 2004), through nutrient mineralization rates and nutrient resorption (Vera et al., 1999; Castillo-Figueroa et al., 2025).

Fourth, soil fauna biodiversity and density increased with litterfall rate (Fig. 6d–f). Litter production and accumulation regulate energy flow and carbon cycling (Attiwill & Adams, 1993), and litter is also an important matrix that creates micro-environmental conditions for soil biodiversity, including both microorganisms and edaphic fauna (Lavelle, 1997; Castillo-Avila et al., 2025). The increase in the organic horizon thickness with litterfall production suggest that there is a parallel increase in the upper soil structural complexity created by lumping different shapes and sizes of leaves, logs, twigs, flowers and fruits at different stages of decomposition, resulting in more microhabitat heterogeneity for soil fauna, enhancing the diversity and abundance of the different animal groups (Barberena-Arias & Aide, 2003; Kattan et al., 2006; Castillo-Figueroa et al., 2024; Castillo-Avila et al., 2025). Particularly, many predators including centipedes, pseudoscorpions, and spiders increase in well-conserved habitats where there is a high number of potential preys (Barnes et al., 2014). This is consistent with the increase in soil fauna density with litter that we observed (Fig. 6e), suggesting a higher availability of preys that, in turn, favor soil predator species richness (Fig. 6f). Importantly, soil fauna is still poorly studied and there is a myriad of species for all the biological groups that remain undescribed in Andean forests (Guarderas et al., 2022; Castillo-Avila & Castillo-Figueroa, 2024). We encourage the conduction of further surveys to document the still obscure edaphic animal diversity in these forests. This will help prevent silent centinelan extinctions (Wilson, 1992), considering that undescribed species are at higher risk than known species (Liu et al., 2021), which is even more likely to happen in ecosystems under high pressure such as the Andean montane forests.

5 Concluding Remarks

In conclusion, we showed that litterfall is a suitable indicator of upper Andean forest recovery and resilience given the strong relationships that litterfall has with succession, forest structure, soil organic horizon thickness, soil respiration, root production, and the maintenance of edaphic faunal biodiversity. Our study also highlights the importance of conserving the ecological integrity of tropical upper Andean montane forests in Colombia, as they represent some of the few natural areas that provide a large array of ecosystem services for the wellbeing of urban and peri-urban populations. Indeed, based on the IUCN Red List of Terrestrial Ecosystems in Colombia, highly populated ecosystems such as Sub-Humid Tropical Forests of the Andean Highlands have less than 20% of their original cover and are categorized with a critical risk levels of collapse (Etter et al., 2021). This scenario is similar in other parts of the Tropical Andean region given that some highly populated cities, such as Quito (Ecuador) and Mérida (Venezuela), are also placed in tropical montane forests, and these ecosystems are some of the last natural areas left to the urban populations. In addition, conservation targets, besides the protection of old-growth forest remnants, should prioritize secondary forests, which have experienced an increase in the last decades in the landscape of the Andean montane region, and can develop into old-growth forests with several added benefits in terms of carbon capture and biodiversity conservation.

Acknowledgments This study is the result of the "Rastrojos project", a long-term collaboration between researchers from Universidad del Rosario, Pontificia Universidad Javeriana, and Instituto Humboldt. The permanent plots network is led by Natalia Norden and Juan Manuel Posada, and was initially installed and monitored thanks to the financial support of Universidad del Rosario, Pontificia Universidad Javeriana and Minciencias. This study was part of the project "*Estudio de dinámicas socio-ecológicas ante escenarios de cambio climático en bosques secundarios periurbanos Altoandinos*" funded by Minciencias (No. FP44842-046-2017) and part of the project "*Dinámicas de regeneración y descomposición de hojarasca en un gradiente sucesional de bosque Altoandino*" funded by Universidad del Rosario (No. IV-FPD003). We thank Carolina Álvarez, Sergio Venegas-Carrillo, María Alejandra Molina, Ana Belen Hurtado, Nicole Morales, Luisa Vargas, and Camilo Castillo-Avila for their support in collecting and processing the data at different stages of the process. We are grateful to the owners, residents, managers, and neighbors of the private areas where we conducted this research for their generosity and hospitality, in particular to Martha Giraldo and Estefania Cabo (Torca), Gonzalo Martínez (Tabio), Carlos Castillo (Guasca), the staff of the Encenillo Biological Reserve (Fundación Natura), and Pedro Rodríguez (Guatavita).

References

An, J. Y., Park, B. B., Chun, J. H., & Osawa, A. (2017). Litterfall production and fine root dynamics in cool-temperate forests. *PLoS One, 12*(6), e0180126. https://doi.org/10.1371/journal.pone.0180126

Antonio-Fragala, F., & Obregón-Neira, N. (2011). Recharge estimation in aquifers of the Bogotá savannah. *Ingenieria y Universidad, 15*, 145–169. https://doi.org/10.11144/Javeriana.iyu15-1.erma

Asner, G. P., Knapp, D. E., Anderson, C. B., Martin, R. E., & Vaughn, N. (2016). Large-scale climatic and geophysical controls on the leaf economics spectrum. *Proceedings of the National Academy of Sciences of the United States of America, 113*(28), E4043–E4051. https://doi.org/10.1073/pnas.1604863113

Attiwill, P. M., & Adams, M. A. (1993). Nutrient cycling in forests. *The New Phytologist, 124*, 561–582. https://doi.org/10.1111/j.1469-8137.1993.tb03847.x

Barberena-Arias, M. F., & Aide, T. M. (2003). Species diversity and trophic composition of litter insects during plant secondary succession. *Caribbean Journal of Science, 39*, 161–169.

Barnes, A. D., Jochum, M., Mumme, S., Haneda, N. F., Farajallah, A., Widarto, T. H., & Brose, U. (2014). Consequences of tropical land use for multitrophic biodiversity and ecosystem functioning. *Nature Communications, 5*, 5351. https://doi.org/10.1038/ncomms6351

Bax, V., & Francesconi, W. (2019). Conservation gaps and priorities in the Tropical Andes biodiversity hotspot: Implications for the expansion of protected areas. *Journal of Environmental Management, 232*, 387–396. https://doi.org/10.1016/j.jenvman.2018.11.086

Berg, B., & Mcclaugherty, C. A. (Eds.). (2020). *Plant litter: Decomposition, humus formation, carbon sequestration* (p. 569). Springer Nature.

Brassard, B. W., Chen, H. Y. H., Bergeron, Y., & Paré, D. (2011). Differences in fine root productivity between mixed- and single-species stands. *Functional Ecology, 25*, 238–246. https://doi.org/10.1111/j.1365-2435.2010.01769.x

Brousseau, P. M., Gravel, D., & Handa, I. T. (2018). Trait-matching and phylogeny as predictors of predator-prey interactions involving ground beetles. *Functional Ecology, 32*, 192–202. https://doi.org/10.1111/1365-2435.12943

Brown, S., & Lugo, A. E. (1990). Tropical secondary forest. *Journal of Tropical Ecology, 6*, 1–32. https://doi.org/10.1017/S0266467400003989

Calbi, M., Clerici, N., Borsch, T., & Brokamp, G. (2020). Reconstructing long term high Andean forest dynamics using historical aerial imagery: A case study in Colombia. *Forests, 11*, 788. https://doi.org/10.3390/f11080788

Calbi, M., Fajardo-Gutiérrez, F., Posada, J. M., Lücking, R., Brokamp, G., & Borsch, T. (2021). Seeing the wood despite the trees: Exploring human disturbance impact on plant diversity, community structure, and standing biomass in fragmented high Andean forests. *Ecology and Evolution, 11*, 2110–2172. https://doi.org/10.1002/ece3.7182

Castillo-Figueroa, D. (2021). Carbon cycle in tropical upland ecosystems: A global review. *Web Ecology, 21*, 109–136. https://doi.org/10.5194/we-21-109-2021

Castillo-Avila, C., & Castillo-Figueroa, D. (2024). Notable range extension of the genus Biramus Oswald, 1993 (Neuroptera: Hemerobiidae) and first record in Colombia. *Universitas Scientiarum, 29*(3), 218–227. https://doi.org/10.11144/Javeriana.SC293.nreo

Castillo-Avila, C., Castillo-Figueroa, D., & Posada, J. M. (2025). Drivers of soil fauna communities along a successional gradient in upper andean tropical forests. *Soil Biology and Biochemistry, 202*, 109692. https://doi.org/10.1016/j.soilbio.2024.109692

Castillo-Figueroa, D., Soler-Marín, D. & Posada, J. M. (2025). Functional traits and species identity drive decomposition along a successional gradient in upper Andean tropical forests. *Biotropica, 57*, e13425. https://doi.org/10.1111/btp.13425

Castillo-Figueroa, D., González-Melo, A., & Posada, J. M. (2023). Wood density is related to aboveground biomass and productivity along a successional gradient in upper Andean tropical forests. *Frontiers in Plant Science, 14*, 1276424. https://doi.org/10.3389/fpls.2023.1276424

Castillo-Figueroa, D., Castillo-Avila, C., Moreno-González, J. A., & Posada, J. (2024). Habitat of two threatened short-tailed whip-scorpions (Arachnida: Schizomida) in the tropical Andes of Northern South America. *Journal of Insect Conservation, 28*, 503–509. https://doi.org/10.1007/s10841-024-00565-4

Cebrian, J. (1999). Patterns in the fate of production in plant communities. *The American Naturalist, 154*, 449–468. https://doi.org/10.1086/303244

Chave, J., Navarrete, D., Almeida, S., Álvarez, E., Aragão, L. E., Bonal, D., Châtelet, P., Silva Espejo, J., Goret, J.-Y., von Hildebrand, P., Jiménez, E., Patiño, S., Peñuela, M. C., Phillips, O. L., Stevenson, P., & Malhi, Y. (2010). Regional and temporal patterns of litterfall in tropical South America. *Biogeosciences, 6*, 7565–7597. https://doi.org/10.5194/bg-7-43-2010

Chua, S. C., & Potts, M. D. (2018). The role of plant functional traits in understanding forest recovery in wet tropical secondary forests. *Science of the Total Environment, 642*, 1252–1262. https://doi.org/10.1016/j.scitotenv.2018.05.397

Davidson, E. A., Belk, E., & Boone, R. D. (1998). Soil water content and temperature as independent or confounded factors controlling soil respiration in a temperate mixed hardwood forest. *Global Change Biology, 4*, 217–227. https://doi.org/10.1046/j.1365-2486.1998.00128.x

Departamento Administrativo Nacional de Estadística (DANE). (2018). *Censo Nacional de Población y Vivienda*. Available at: https://www.dane.gov.co/index.php/estadisticas-por-tema/demografia-y-poblacion/censo-nacional-de-poblacion-y-vivenda-2018/donde-estamos (Accessed December 24, 2021).

Duque, A., Stevenson, P. R., & Feeley, K. J. (2015). Thermophilization of adult and juvenile tree communities in the northern tropical Andes. *Proceedings of the National Academy of Sciences of the United States of America, 112*, 10744–10749.

Duque, A., Peña, M. A., Cuesta, F., González-Caro, S., Kennedy, P., Phillips, O. L., Calderón-Loor, M., Blund, C., Carilla, J., Cayola, L., Farfán-Ríos, W., Fuentes, A., Grau, R., Homeier, J., Loza-Rivera, M. I., Malhi, Y., Malizia, A., Malizia, L., Martínez-Villa, J. A., Myers, J. A., Osinaga-Acosta, O., Peralvo, M., Pinto, E., Saatchi, S., Silman, M., Tello, J. S., Terán-Valdez, A., & Feeley, K. J. (2021). Mature Andean forests as globally important carbon sinks and future carbon refuges. *Nature Communications, 12*, 2138. https://doi.org/10.1038/s41467-021-22459-8

Etter, A., & van Wyngaarden, W. (2000). Patterns of landscape transformations in Colombia, whit emphasis in the Andean region. *Ambio, 29*, 432–439. https://doi.org/10.1579/0044-7447-29.7.432

Etter, A., McAlpine, C., & Possingham, H. (2008). A historical analysis of the spatial and temporal drivers of landscape change in Colombia since 1500. *Annals of the American Association of Geographers, 98*, 223. https://doi.org/10.1080/00045600701733911

Etter, A., Andrade, A., Saavedra, K., & Cortés, J. (2018). Actualización de la Lista Roja de los Ecosistemas Terrestres de Colombia: Herramienta para la gestión de los ecosistemas. In L. A. Moreno, C. Rueda, & G. I. Andrade (Eds.), *Biodiversidad 2017: Estado y tendencias de la biodiversidad continental de Colombia*. Instituto de Investigación de Recursos Biológicos Alexander von Humboldt.

Etter, A., Andrade, A., Saavedra, K., Amaya, P., Cortés, J., & Arévalo, P. (2021). *Ecosistemas colombianos: amenazas y riesgos*. Editorial Pontificia Universidad Javeriana.

Fadrique, B., Báez, S., Duque, Á., Malizia, A., Blundo, C., Carilla, J., Osinaga-Acosta, O., Malizia, L., Silman, M., Farfán-Ríos, W., Malhi, Y., Young, K. R., Cuesta, C. F., Homeier, J., Peralvo, M., Pinto, E., Jadan, O., Aguirre, N., Aguirre, Z., & Feeley, K. J. (2018). Widespread but heterogeneous responses of Andean forests to climate change. *Nature, 564*(7735), 207–212. https://doi.org/10.1038/s41586-018-0715-9

Fekete, I., Kotroczó, Z., Varga, C., Nagy, P. T., Várbíró, G., Bowden, R. D., Tóth, J. A., & Lajtha, K. (2014). Alterations in forest detritus inputs influence soil carbon concentration and soil respiration in a Central-European deciduous forest. *Soil Biology and Biochemistry, 74*, 106–114. https://doi.org/10.1016/j.soilbio.2014.03.00

Fekete, I., Varga, C., Biró, B., Tóth, J. A., Várbíró, G., Lajtha, K., et al. (2016). The effects of litter production and litter depth on soil microclimate in a central European deciduous forest. *Plant and Soil, 398*, 291–300. https://doi.org/10.1007/s11104-015-2664-5

Girardin, C. A. J., Aragao, L., Malhi, Y., Huaraca Huasco, W., Metcalfe, D., Durand, L., Mamani, M., Silva-Espejo, J. E., & Whittaker, R. J. (2013). Fine root dynamics along an elevational gradient in tropical Amazonian and Andean forests. *Global Biogeochemical Cycles, 27*, 252–264. https://doi.org/10.1029/2011GB004082

Guarderas, P., Trávez, K., Boeraeve, F., Cornelis, J.-T., & Dufrêne, M. (2022). Native forest conversion alters soil macroinvertebrate diversity and soil quality in tropical mountain landscapes of northern Ecuador. *Frontiers in Forests and Global Change, 5*, 959799. https://doi.org/10.3389/gc.2022.9597

Han, T., Huang, W., Liu, J., et al. (2015). Different soil respiration responses to litter manipulation in three subtropical successional forests. *Scientific Reports, 5*, 18166. https://doi.org/10.1038/srep18166

Hendricks, J. J., Nadelhoffer, K. J., & Aber, J. D. (1993). Assessing the role of fine roots in carbon and nutrient cycling. *Trends in Ecology & Evolution, 8*, 174–178. https://doi.org/10.1016/0169-5347(93)90143-D

Homeier, J., Seeler, T., Pierick, K., & Leuschner, C. (2021). Leaf trait variation in species-rich tropical Andean forests. *Scientific Reports, 11*, 9993. https://doi.org/10.1038/s41598-021-89190-8

Hoorn, C., Wesselingh, F. P., ter Steege, H., Bermudez, M. A., Mora, A., Sevink, J., Sanmartín, I., Sanchez-Meseguer, A., Anderson, C. L., Figueiredo, J. P., Jaramillo, C., Riff, D., Negri, F., Hooghiemstra, H., Lundberg, J., Stadler, T., Särkinen, T., & Antonelli, A. (2010). Amazonia through time: Andean uplift, climate change, landscape evolution, and biodiversity. *Science, 330*, 927–931.

Hurtado-M, A. B., Echeverry-Galvis, M. A., Salgado-Negret, B., Muñoz, J. C., Posada, J. M., & Norden, N. (2021). Little trace of floristic homogenization in peri-urban Andean secondary forests despite high anthropogenic transformation. *Journal of Ecology, 109*, 1468–1478. https://doi.org/10.1111/1365-2745.13570

Hurtado-M, A. B., Muñoz, J. C., Echeverry-Galvis, M. A., & Norden, N. (2022). Bosques sucesionales en Colombia: una oportunidad para la recuperación de paisajes transformados. *Caldasia, 44*, 332–344. https://doi.org/10.15446/caldasia.v44n2.82255

Instituto de Hidrología Meteorología y Estudios Ambientales (IDEAM). (2024). *Características climatológicas de 594 ciudades principales y municipios turísticos*. http://www.ideam.gov.co/documents/21021/418894/Caracter%C3%ADsticas+de+Ciudades+Principales+y597%20

+Municipios+Tur%C3%ADsticos.pdf/c3ca90c8-1072-434a-a235-91baee8c73fc (Accesed 25 Febrary 2024).

Kattan, G. H., Correa, D., Escobar, F., & Medina, C. (2006). Leaf-litter arthropods in restored forests in the Colombian Andes: A comparison between secondary forest and tree plantations. *Restoration Ecology, 14*, 95–102. https://doi.org/10.1111/j.1526-100X.2006.00109.x

Krishna, M. P., & Mohan, M. (2017). Litter decomposition in forest ecosystems: A review. *Energy, Ecology and Environment, 2*, 236–249. https://doi.org/10.1007/s40974-017-00649

Lavelle, E. P. (1997). Faunal activities and soil processes: Adaptative strategies that determine ecosystem function. *Advances in Environmental Research, 27*, 93–102. https://doi.org/10.1016/S0065-2504(08)60007-0

Liu, W., Fox, J. E. D., & Xu, Z. (2003). Litterfall and nutrient dynamics in a montane moist evergreen broad-leaved forest in Ailao Mountains, SW China. *Plant Ecology, 164*, 157–170. http://www.jstor.org/stable/20146352

Liu, J., Slik, F., Zheng, S., & Lindenmayer, D. B. (2021). Undescribed species have higher extinction rates than known species. *Conservation Letters, 15*, e12876. https://doi.org/10.1111/conl.12876

Mathews, T. R., et al. (2012). *Measuring tropical forest carbon allocation and cycling: A RAINFOR-GEM Field Manual for Intensive Census Plots*, ver. 2.2, Global Ecosystems Monitoring network. Available at http://gem.tropicalforests.ox.ac.uk.

Melillo, J. M., McGuire, A. D., Kicklighter, D. W., Moore, B., Vorosmarty, C., & Schloss, A. L. (1993). Global climate change and terrestrial net primary production. *Nature, 363*, 234–240. https://doi.org/10.1038/363234a0

Mendoza, J. E., & Etter, A. (2002). Multitemporal analysis, (1940–1996) of land cover changes in the southwestern Bogotá Highplain. *Landscape and Urban Planning, 59*, 147–158. https://doi.org/10.1016/S0169-2046(02)00012-9

Metcalfe, D. B., Meir, P., & Williams, M. (2007). A comparison of methods for converting rhizotron root length measurements in to estimates of root mass production per unit ground area. *Plant and Soil, 301*, 279–288. https://doi.org/10.1007/s11104-007-9447-6

Montañez, G., Arcila, O., & Pacheco, J. C. (1994). *Hacia Dónde va la Sabana de Bogotá: Modernización, Conflicto, Ambiente y Sociedad*. Universidad Nacional de Colombia.

Mittermeier, R. A., Turner, W. R., Larsen, F. W., Brooks, T. M., & Gascon, C. (2011). Global biodiversity conservation: The critical role of hotspots. In F. E. Zachos & J. C. Habel (Eds.), *Biodiversity hotspots* (pp. 3–22). Springer–Verlag Berlin.

Myers, N., Mittermeier, R. A., Mittermeier, C. G., da Fonseca, G. A. B., & Kent, J. (2000). Biodiversity hotspots for conservation priorities. *Nature, 403*, 853–885. https://doi.org/10.1038/35002501

Myster, R. W. (2021). Introduction. In R. W. Myster (Ed.), *The Andean cloud forest* (pp. 1–23). Springer.

Nottingham, A. T., Fierer, N., Turner, B. L., Whitaker, J., Ostle, N. J., McNamara, N. P., Bardgett, R. D., Leff, J. W., Salinas, N., Silman, M. R., Kruuk, L. E. B., & Meir, P. (2018). Microbes follow Humboldt: Temperature drives plant and soil microbial diversity patterns from the Amazon to the Andes. *Ecology, 99*(11), 2455–2466. https://doi.org/10.1002/ecy.2482

Palace, M., Keller, M., Asner, G. P., Silva, J. N. M., & Passos, C. (2007). Necromass in undisturbed and logged forests in the Brazilian Amazon. *Forest Ecology and Management, 238*, 309–318. https://doi.org/10.1016/j.foreco.2006.10.026

Phillips, O., Baker, T., Feldpausch, T., & Brienen, R. (2001). RAINFOR field manual for plot establishment and remeasurement. In *Pan-amazonia* (1st ed., pp. 1–30). The Royal Society.

Pinho, B. X., Melo, F. P. L., Arroyo-Rodríguez, V., Pierce, S., Lohbeck, M., & Tabarelli, M. (2018). Soil-mediated filtering organizes plant assemblages in regenerating tropical forests. *Journal of Ecology, 106*, 137–147. https://doi.org/10.1111/1365-2745.12843

Pinos, J., Studholme, A., Carabajo, A., & García, C. (2017). Leaf Litterfall and decomposition of Polylepis reticulata in the Treeline of the Ecuadorian Andes. *MRD, 37*(1), 87–96. https://doi.org/10.1659/MRD-JOURNAL-D-16-00004.1

Poorter, L., Craven, D., Jakovac, C. C., van der Sande, M. T., Amissah, L., Bongers, F., Chazdon, R. L., Farrior, C. E., Kambach, S., Meave, J. A., Muñoz, R., Norden, N., Rüger, N., van Breugel, M., Almeyda Zambrano, A. M., Amani, B., Andrade, J. L., Brancalion, P. H. S., Broadbent, E. N., de Foresta, H., Dent, D. H., Derroire, G., DeWalt, S. J., Dupuy, J. M., Durán, S. M., Fantini, A. C., Finegan, B., Hernández-Jaramillo, A., Hernández-Stefanoni, J. L., Hietz, P., Junqueira, A. B., N'dja, J. K., Letcher, S. G., Lohbeck, M., López-Camacho, R., Martínez-Ramos, M., Melo, F. P. L., Mora, F., Müller, S. C., N'guessan, A. E., Oberleitner, F., Ortiz-Malavassi, E., Pérez-García, E. A., Pinho, B. X., Piotto, D., Powers, J. S., Rodríguez-Buriticá, S., Rozendaal, D. M. A., Ruíz, J., Tabarelli, M., Teixeira, H. M., Valadares de Sá Barretto Sampaio, E, van der Wal, H., Villa, P. M., Fernandes, G. W., Santos, B. A., Aguilar-Cano, J., de Almeida-Cortez, J. S., Alvarez-Davila, E., Arreola-Villa, F., Balvanera, P., Becknell, J. M., GAL, C., Castellanos-Castro, C., de Jong, B. H. J., Nieto, J. E., Espírito-Santo, M. M., Fandino, M. C., García, H., García-Villalobos, D., Hall, J. S., Idárraga, A., Jiménez-Montoya, J., Kennard, D., Marín-Spiotta, E., Mesquita, R., Nunes, Y. R. F., Ochoa-Gaona, S., Peña-Claros, M., Pérez-Cárdenas, N., Rodríguez-Velázquez, J., Villanueva, L. S., Schwartz, N. B., Steininger, M. K., Veloso, M. D. M., Vester, H. F. M., Vieira, I. C. G., Williamson, G. B., Zanini, K., & Hérault, B. (2021). Multidimensional tropical forest recovery. *Science, 374*(6573), 1370.

Poorter, L., Amissah, L., Bongers, F., Hordijk, I., Kok, J., Laurance, S. G. W., Lohbeck, M., Martínez-Ramos, M., Matsuo, T., Meave, J. A., Muñoz, R., Peña-Claros, M., & van der Sande, M. T. (2023). Successional theories. *Biological Reviews, 98*, 2049–2077. https://doi.org/10.1111/brv.12995

Rahbek, C., Borregaard, M. K., Colwell, R. K., Dalsgaard, B., Holt, B. G., Morueta-Holme, N., Nogues-Bravo, D., Whittaker, R. J., & Fjeldså, J. (2019). Humboldt's enigma: What causes global patterns of mountain biodiversity? *Science, 365*, 1108. https://doi.org/10.1126/science.aax0149

Rajakaruna, N., & Boyd, R. S. (2008). Edaphic factor. In S. E. Jørgensen & B. D. Fath (Eds.), *Encyclopedia of ecology*. Academic. https://doi.org/10.1016/B978-008045405-4.00484-5

Raymond-Leonard, L. J., Gravel, D., & Handa, I. T. (2019). A novel set of traits to describe Collembola mouthparts: Taking a bite out of the broad chewing mandible classification. *Soil Biology and Biochemistry, 138*, 107608. https://doi.org/10.1016/j.soilbio.2019.107608

Reich, P. B. (2014). The world-wide 'fast–slow' plant economics spectrum: A traits manifesto. *Journal of Ecology, 102*, 275–301. https://doi.org/10.1111/1365-2745.12211

Rosenfield, M. F., Jakovac, C. C., Vieira, D. L. M., Poorter, L., Brancalion, P. H. S., Vieira, I. C. G., de Almeida, D. R. A., Massoca, P., Schietti, J., Albernaz, A. L. M., Ferreira, M. J., & Mesquita, R. C. G. (2023). Ecological integrity of tropical secondary forests: Concepts and indicators. *Biological Reviews of the Cambridge Philosophical Society, 98*(2), 662–676. https://doi.org/10.1111/brv.12924

Rubiano, K., Clerici, N., Norden, N., & Etter, A. (2017). Secondary forest and shrubland dynamics in a highly transformed landscape in the Northern Andes of Colombia (1985–2015). *Forests, 8*, 1–17. https://doi.org/10.3390/f8060216

Schuur, E. A., Chadwick, O. A., & Matson, P. A. (2001). Carbon cycling and soil carbon storage in mesic to wet Hawaiian montane forests. *Ecology, 82*, 3182–3196. https://doi.org/10.2307/2679843

Sotes, G. J., Bustamante, R. O., & Henríquez, C. A. (2013). Distribución de plántulas y germinación de semillas del lúcumo chileno (*Pouteria splendens*) en Los Molles, Chile. *Revista Chilena de Historia Natural, 86*, 337–344. https://doi.org/10.4067/S0716-078X2013000300010

Souza, S. R., Veloso, M. D. M., Espírito-Santo, M. M., Silva, J. O., Sánchez-Azofeifa, A., Souzae Brito, B., & Fernandes, G. W. (2019). Litterfall dynamics along a successional gradient in a Brazilian tropical dry forest. *Forest Ecosystems, 6*, 35. https://doi.org/10.1186/s40663-019-0194-y

Steingrobe, B., Schmid, H., & Claassen, N. (2000). The use of the ingrowth core method for measuring root production of arable crops: Influence of soil conditionsinside theingrowth coreon rootgrowth. *Journal of Plant Nutrition and Soil Science, 163*, 617–622. https://doi.org/10.1002/1522-2624(200012)163:63.0.CO;2-0

Tanner, E. V. J., Vitousek, P. M., & Cuevas. (1998). Experimental investigation of nutrient limitation of forest growth on wet tropical mountains. *Ecology, 79*, 10–22. https://doi.org/10.2307/176860

Urrutia, R., & Vuille, M. (2009). Climate change projections for the tropical Andes using a regional climate model: Temperature and precipitation simulations for the end of the 21st century. *Journal of Geophysical Research, 114*, D02108. https://doi.org/10.1029/2008JD011021

Vera, M., Cavelier, J., & Santamaría, J. (1999). Reabsorción de nitrógeno y fósforo foliar en árboles de bosques montanos en los Andes centrales de Colombia. *Revista de Biología Tropical, 47*(1–2), 33–43.

Vendrami, J. L., Jurinitz, C. F., Castanho, C. T., Lorenzo, L., & Oliveira, A. A. (2012). Litterfall and leaf decomposition in forest fragments under different successional phases on the Atlantic Plateau of the state of Sao Paulo, Brazil. *Biota Neotropical, 12*(3), 136–143. https://doi.org/10.1590/S1676-06032012000300016

Vitousek, P. M., & Sanford, R. L. (1986). Nutrient cycling in moist tropical forest. *Annual Review of Ecology, Evolution, and Systematics, 17*, 137–167.

Vogt, K. A., Vogt, D. J., & Bloomfield, J. (1998). Analysis of some direct and indirect methods for estimating root biomass and production of forests at an ecosystem level. *Plant and Soil, 200*, 71–89. https://doi.org/10.1023/A:1004313515294

Wardle, D. A., Bardgett, R. D., Klironomos, J. N., Setälä, H., van der Puttern, W. H., & Wall, D. H. (2004). Ecological linkages between aboveground and belowground biota. *Science, 304*, 1629–1633. https://doi.org/10.1126/science.1094875

Wilcke, W. (2021). Litterfall in Andean forests: Quantity, composition, and environmental drivers. In R. W. Myster (Ed.), *The Andean cloud forest*. Springer. https://doi.org/10.1007/978-3-030-57344-7_5

Wilcke, W., Yasin, S., Abramowski, U., Valarezo, C., & Zech, W. (2002). Nutrient storage and turnover in organic layers under tropical montane rain forest in Ecuador. *European Journal of Soil Science, 53*, 15–27. https://doi.org/10.1046/j.1365-2389.2002.00411.x

Wilcke, W., Oelmann, Y., Schmitt, A., Valarezo, C., Zech, W., & Homeier, J. (2008). Soil properties and tree growth along an altitudinal transect in Ecuadorian tropical montane forest. *Journal of Plant Nutrition and Soil Science, 171*, 220–230. https://doi.org/10.1002/jpln.200625210

Wilcke, W., Leimer, S., Peters, T., Emck, P., Rollenbeck, R., Trachte, K., Valarezo, C., & Bendix, J. (2013). The nitrogen cycle of tropical montane forest in Ecuador turns inorganic under environmental change. *Global Biogeochemical Cycles, 27*, 1194–1120.

Wilson, E. O. (1992). *The diversity of life*. Belknap Press.

Wright, I. J., Reich, P. B., Westoby, M., Ackerly, D. D., Baruch, Z., Bongers, F., Cavender-Bares, J., Chapin, T., Cornelissen, J. H., Diemer, M., Flexas, J., Garnier, E., Groom, P. K., Gulias, J., Hikosaka, K., Lamont, B. B., Lee, T., Lee, W., Lusk, C., Midgley, J. J., Navas, M. L., Niinemets, U., Oleksyn, J., Osada, N., Poorter, H., Poot, P., Prior, L., Pyankov, V. I., Roumet, C., Thomas, S. C., Tjoelker, M. G., Veneklaas, E. J., & Villar, R. (2004). The worldwide leaf economics spectrum. *Nature, 428*(6985), 821–827. https://doi.org/10.1038/nature02403

Yan, J. H., Wang, Y. P., Zhou, G. Y., & Zhang, D. Q. (2006). Estimates of soil respiration and net primary production of three forests at different succession stages in South China. *Global Change Biology, 12*, 810–821. https://doi.org/10.1111/j.1365-2486.2006.01141.x

Yepes, A., Herrera, J., Phillips, J., Cabrera, E., Galindo, G., Granados, E., Duque, A., Barbosa, A., Olarte, C., & Cardona, M. (2015). Contribución de los bosques tropicales de montaña en el almacenamiento de carbono en Colombia. *Revista de Biología Tropical, 63*, 69–82. https://doi.org/10.15517/rbt.v63i1.14679

Yepes, A., Sierra, A., Niño, L. M., López, M., Garay, C., Vargas, D., Cabrera, E., & Barbosa, A. (2016). Biomasa y carbono total almacenado en robledales del sur de los Andes Colombianos: aportes para el enfoque REDD+ a escala de proyectos. *Revista de Biología Tropical, 64*, 399–412. https://doi.org/10.15517/rbt.v64i1.18221

Zhang, H., Yuan, W., Dong, W., & Liu, S. (2014). Seasonal patterns of litterfall in forest ecosystem worldwide. *Ecological Complexity, 20*, 240–247.

Zhang, P., Zhou, Z., Liu, W., Wu, D., & Scheu, S. (2024). Detritivores maintain stoichiometric homeostasis, but alter body size and population density in response to altitude induced stoichiometric mismatches. *Geoderma, 446*, 116897. https://doi.org/10.1016/j.geoderma.2024.116897

The Significance of Successional Forests in Human-Modified Andean Forests Landscape

Carolina Alvarez-Garzón, Ana Belén Hurtado-M, Jaime Burbano-Girón, Gabriela del Mar Castaño, Juan Camilo Muñoz, María Ángela Echeverry, Juan Posada, and Natalia Norden

1 Introduction

The Andes is one of the world's most extended and biodiverse mountain ranges. It is found across Colombia, Ecuador, Perú, Bolivia, Argentina, and Chile and covers an approximate area of 1.543.000 km^2 (Cuesta et al., 2009; Tejedor-Garavito et al., 2012). Due to a wide variation in altitude, climate, and geographical features, this mountain range harbors a diverse mosaic of ecosystems characterized by different types of vegetation cover, such as páramos and punas at higher elevations, montane, and cloud forests at the mid-range and tropical wet and dry forests at lower altitudes (Josse et al., 2009; Tejedor-Garavito et al., 2012). Andean montane forests, in turn, also exhibit a wide variety of physiognomies arising from climatic and latitudinal variations (Tejedor-Garavito et al., 2012). These features result in a rich tapestry of plant and animal species, many of which are endemic or show narrow ranges (Myers et al., 2000; Cuesta et al., 2009). Despite the critical importance of conserving this unique biodiversity, deforestation in Andean forests has historically been a common practice (Christmann et al., 2023) due to the expansion of the agricultural frontier,

C. Alvarez-Garzón (✉) · G. del Mar Castaño · J. Posada
Facultad de Ciencias Naturales, Universidad del Rosario, Bogotá, Colombia
e-mail: carolina.alvarezg@urosario.edu.co

A. B. Hurtado-M · N. Norden
Instituto de Investigación de Recursos Biológicos Alexander von Humboldt, Bogotá, Colombia

J. Burbano-Girón
The Nature Conservancy, Bogotá, Colombia

J. C. Muñoz
Fundación Humedales; Bosques de Canoas SCA, Bogotá, Colombia

M. Á. Echeverry
Pontificia Universidad Javeriana, Bogotá, Colombia

N. Clerici (ed.), *Conservation of Andean Forests*,
https://doi.org/10.1007/978-3-031-80805-0_4

increases in the extension of cattle ranching areas for pastures, and other economic activities such as mining and road constructions, among others (Malizia et al., 2020). Such a long history of occupation reflects the legacy of pre-Hispanic indigenous populations and the Spanish colonization, during which heavy anthropogenic disturbances in Andean ecosystems took place y (Mathez-Stiefel et al., 2017; Duque et al., 2021). The combination of its extraordinary species richness and broad landscape transformation makes Andean montane forests a global biodiversity hotspot (Myers et al., 2000; Hughes, 2016).

Nowadays, Andean landscapes are mostly transformed, resulting in a heterogeneous mosaic of forests with different levels of disturbance (Asner et al., 2013). Many studies have found that Andean forests are recovering after land abandonment and rural-urban migration (Sánchez-Cuervo et al., 2012; Sanchez-Cuervo & Aide, 2013; Chazdon et al., 2016), yet this process is highly variable in time and space (Christmann et al., 2023). For instance, between 2001 and 2014, almost 50,000 km^2 of woody vegetation in the tropical and subtropical Andes was cleared in the 1000–1499 m.a.s.l., while an impressive net gain in vegetation was observed in the 1500–2499 m.a.s.l (Aide et al., 2019). Thus, whereas some regions are undergoing continuous deforestation, in others, land abandonment or land-use dis-intensification is common (Mathez-Stiefel et al., 2017; Aide et al., 2019), and in others, forest recovery and landscape degradation co-occur (Rubiano et al., 2017; Aide et al., 2019; Duque, 2021). Drivers of recovery are highly context-dependent. Forest regrowth usually occurs near remnants and far from human settlements and infrastructure to understand the dynamics of increases and decreases in forest cover in Andean landscapes (Camelo et al., 2017; Rubiano et al., 2017). Such patterns show that Andean forests are complex socio-ecosystems where successional forests are integral to the landscape.

Successional tropical forests support biodiversity and ecosystem functioning (Chazdon et al., 2016; Rozendaal et al., 2019). They contribute to the recovery of critical habitats and offer refuge to a diverse array of plants (Hurtado-M et al., 2021) and animal species (Davies et al., 2020), thereby enhancing taxonomic and functional diversity (Davies et al., 2020). Likewise, they serve as valuable carbon sinks, mitigating climate change (Rozendaal et al., 2019; Lennox et al., 2018). Despite the multifaceted ecological functions of successional forests, they have yet to receive the attention they deserve, and we know little about their contributions to sustaining the health and balance of the Andean ecosystems and human populations. Recognizing successional forests' trends and diversity patterns as fundamental components of the Andean landscape is crucial for developing effective conservation strategies. Therefore, this chapter aims to (1) examine the trends in successional forest cover change in the Colombian Andes and (2) evaluate patterns in diversity recovery in successional forests occurring in a heavily transformed Andean landscape in Colombia. This emphasis underscores the importance of analyzing the recovery trajectories of these successional forests.

2 The Significance of Successional Forests in the Colombian Andes

Colombia is one of the five megadiverse nations worldwide (Rodríguez et al., 2006), and this is partly because the Andean Mountain range crosses the country from South to North, spanning approximately 25% of the country's area (around 28 million hectares; Etter & van Wyngaarden, 2000; Rodríguez Eraso et al., 2012). The Colombian Andean region exhibits considerable biological, cultural, and socioeconomic complexity. It is part of a global biodiversity hotspot (Myers et al., 2000), comprising the Sierra Nevada de Santa Marta in the north, the Colombian Massif in the south, and the Western, Central, and Eastern Cordilleras originating from the "Nudo de los Pastos" in the south (Etter & van Wyngaarden, 2000). This region has experienced the most significant deforestation processes in Colombia, with 28,984 ha of forest deforested by 2020 (Castellanos et al., 2021). Despite historical land use transformation (Etter et al., 2008), forest cover has presented net gains in several regions of the country (Sánchez-Cuervo et al., 2012; Aide et al., 2013). To examine the trends in successional forest cover change in the Colombian Andes, we used land cover maps from four time periods: 2000–2002, 2005–2009, 2010–2012, and 2018. We followed the cover classification adapted from the Corine Land Cover methodology (IDEAM, 2010).

We used the cover units from the latest official national cover map to determine the extent of successional forests in the Colombian Andes (Castellanos et al., 2021). We classified them into four categories while retaining the names as presented in the official map: "secondary vegetation," "mature forests," "other natural covers," and "transformed covers." Andean forests were montane forests located between 500 m above sea level (m.a.s.l.) and the tree line (Cuesta et al., 2009). We differentiated high from low mountains using a threshold of 1500 m.a.s.l. where the lowest montane forests were the dry valleys of Cauca and Magdalena. Considering the land cover units (IDEAM, 2010), we first calculated the area covered by successional vegetation for each period as the sum of the polygons included in the categories: "fragmented forests with secondary vegetation" and "secondary or transition vegetation." Then, we calculated the vegetation surface occupied by successional forests in the most recent period (see Hurtado-M et al., 2022). To understand changes in successional forest cover in the Andes in a larger context, we compared the observed patterns in the Colombian Andes with those observed in the other biogeographical regions of the country: Amazonia, Caribbean, Orinoquia, and Pacific.

The results show that the area covered by successional forests in the country increased by 55% during the last 18 years (from 52 239 km2 in 2000 to 80 793 km2 in 2018) (Table 1). Overall, almost all regions presented an increase in the coverage of successional vegetation over time, but the Pacific and Orinoco regions decreased in the last analysis period (2010–2012 to 2018) (Fig. 1). The Andean highlands were the region that exhibited the highest fraction of "secondary" vegetation, increasing its coverage by 48% from 2000 to 2018. Only the Amazon region showed a comparable pattern, with an increase of 37% during this period (Table 1). This pattern was

Table 1 Area (km^2) of the successional forest in Colombia by regions for the analysis periods

Region	2000–2002	2005–2009	2010–2012	2018
Andean highlands	19251.81 (37%)	19570.09 (35%)	25157.35 (37%)	28578.03 (35%)
Andean lowlands	4681.7 (9%)	4849.49 (9%)	7984.76 (12%)	10620.11 (13%)
Amazon	12481.37 (24%)	11704.69 (21%)	13040.1 (19%)	17160.48 (21%)
Caribbean	6638.11 (13%)	7184.72 (13%)	9180.05 (13%)	13954.3 (17%)
Orinoco	2966.61 (6%)	3775.29 (7%)	4073.68 (6%)	3373.97 (4%)
Pacific	6220.36 (12%)	8636.03 (15%)	9276.05 (13%)	7106.88 (9%)
Total	52239.96 (100%)	55720.31 (100%)	68711.99 (100%)	80793.77 (100%)

The values correspond to the proportion of successional forest in each region considering the country's total area of successional forest (100%)

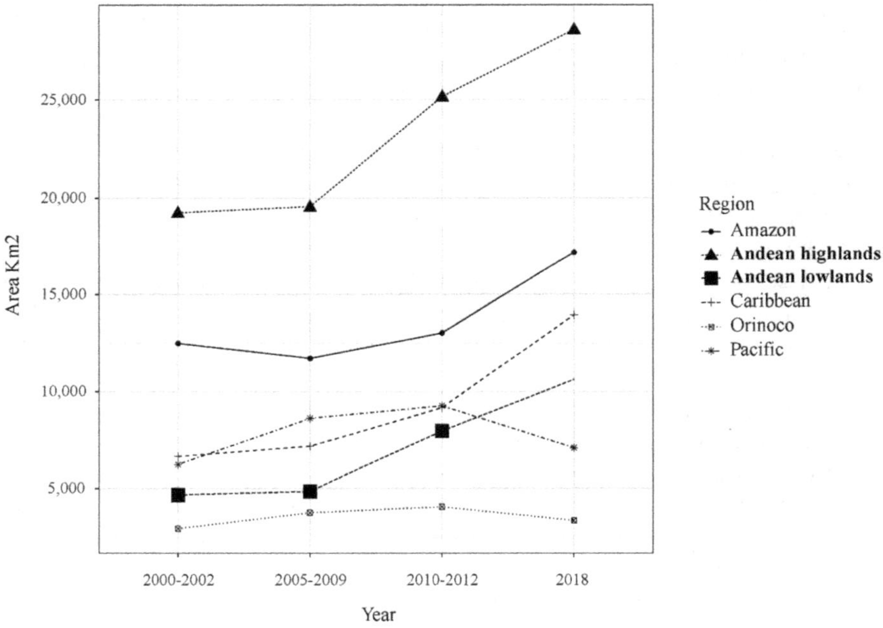

Fig. 1 Patterns in the change of successional vegetation coverage across different regions during the analysis periods

consistent over time; both regions accounted for about 60% of the country's successional forests during all the analysis periods. Compared to the Andean highlands, the Caribbean, the Andean lowlands, and the Pacific showed successively lower percentages of successional vegetation coverage (Table 1). Finally, the Orinoquia consistently showed a tiny fraction of regrowing vegetation, which is unsurprising as this region mainly comprises savannas and gallery forests, which are only present along rivers and have been less disturbed.

Increases in the area covered by successional forests in Colombia reflect the complex socioecological relationships that determine patterns of deforestation, land use, and land abandonment and are the result of factors that are taking place at different spatial scales in widely different contexts (Sánchez-Cuervo et al., 2012). Indeed, many studies have shown an increase in successional forest cover during the early twenty-first century attributed to various factors (Sánchez-Cuervo et al., 2012; Aide et al., 2019; Fagua & Ramsey, 2019). For instance, Aide et al. (2019) associated the secondary vegetation gain in the Colombian Andes with a shift in the distribution of coffee cultivations to regions that produce higher quality and eco-friendly coffee as well as with the expansion of silvopastoral projects that promoted the introduction of foraging tree species into cattle pastures. The Andean highlands have also undergone decreased soil productivity due to intensive land use, global market influences, and rural-to-urban migration (Sánchez-Cuervo et al., 2012; Rubiano et al., 2017), all of which likely contributed to the increase in successional forest cover. In contrast, land abandonment might have been less critical in the Andean lowlands as soils are more productive, and land abandonment has not been so frequent. Government policies promoting the transition of areas previously used for coca crops to legal productive uses might have created opportunities for forest regeneration (Sánchez-Cuervo et al., 2012), particularly in the Amazon.

Increases in successional forest cover over the last few decades have led these forests to become important landscape components. To quantify this, we assessed the "secondary vegetation" ratio to other land covers in 2018, the last year of analysis. (Fig. 2). Low and high-elevation successional forests in the Andean region cover approximately 10–15% of the area. However, mature forests in the highlands

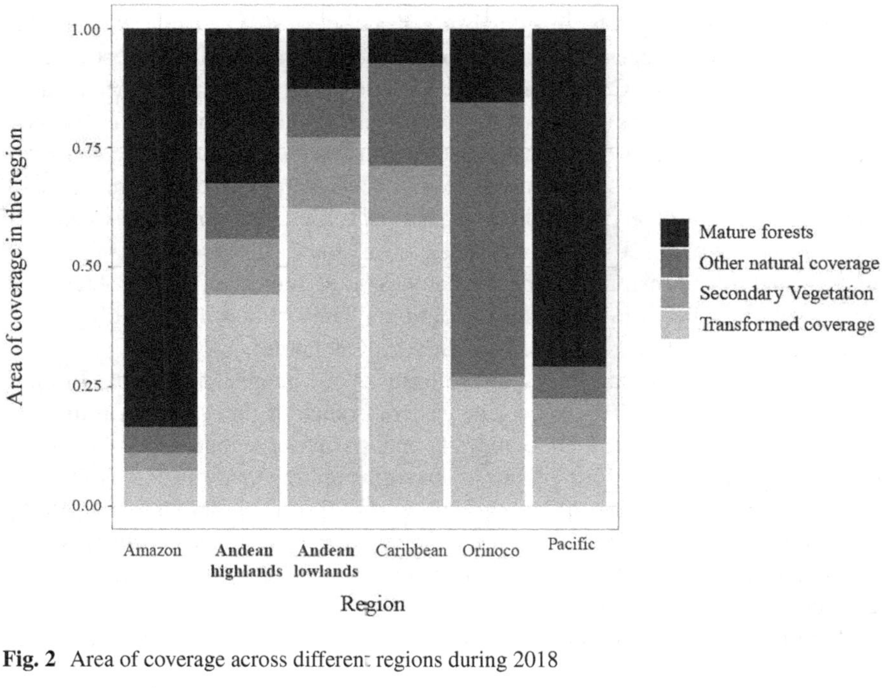

Fig. 2 Area of coverage across different regions during 2018

represent a significant fraction of the landscape (Fig. 2), probably due to the steeper slopes, which restrict human access and reduce the extent of forest use and transformation.

Conversely, slopes are gentler at low elevations and transformed landscapes cover more land (Armenteras et al., 2011). Compared to the Andean region, the other regions shared some characteristics. In the Caribbean region, the most deforested region of the country, together with the Andes, successional forests occupied a significant fraction of the landscape. In contrast, in the Pacific and the Amazon, characterized by low human densities and large extents of mature tropical humid forests, successional forests do not constitute a substantial portion of the landscape—despite the large area in the Amazon where forests are regenerating (Table 1) (Fig. 3).

3 Peri-Urban Forests: Flora Reservoirs and Their Importance for the Conservation of Successional Andean Mountain Forests

In Colombia, the Andean region harbors seven of the country's ten major cities (Etter & Van Wyngaarden, 2000). Peri-urban forests may, therefore, play a critical role in the connectivity and integrity of urban-regional landscapes. Despite its potential importance, it is unclear whether these regrowing forests will be able to recover a structure and composition similar to those observed previous to disturbance or, because they are embedded in a heavily transformed matrix, they will exhibit a truncated succession. The region surrounding one of the largest cities in Latin America, Bogotá, Colombia's capital, is one of the most transformed areas of the North-Eastern Andes, home to over 7.5 approx., million inhabitants (DANE-Departamento et al., 2019). Nevertheless, few studies have evaluated the conservation status of these forests (Hurtado-M et al., 2021). Here, we address this issue by evaluating floristic diversity at three stages of the life cycle and some aspects of ecosystem functioning. To do so, we use data from the Rastrojos Project, a research initiative led by the Instituto de Investigaciones de Recursos Biológicos Alexander von Humboldt (Bogotá, Colombia), Pontificia Universidad Javeriana (Bogotá, Colombia), and the Rosario University (Bogotá, Colombia).

The study area is in the intermontane basin of the Sabana de Bogotá. In these high plains, Bogotá lies (Fig. 4), in the eastern branch of the Andean Cordilleras, has an average altitude of 2600 m.a.s.l. and an average temperature of about 14 °C. Original forests of the plains were dominated by species such as *Quercus humboldtii*, *Weinmania tomentosa,* and *Vallea stipularis* (Van der Hammen, 1986), and mature forests hosted many species belonging to the Lauraceae family. However, only small patches of intact forest remain, where the most common and species-rich families are Melastomataceae and Ericaceae. We established a monitoring network of 20 plots located in four sites in the suburban municipalities of Guatavita (Gva),

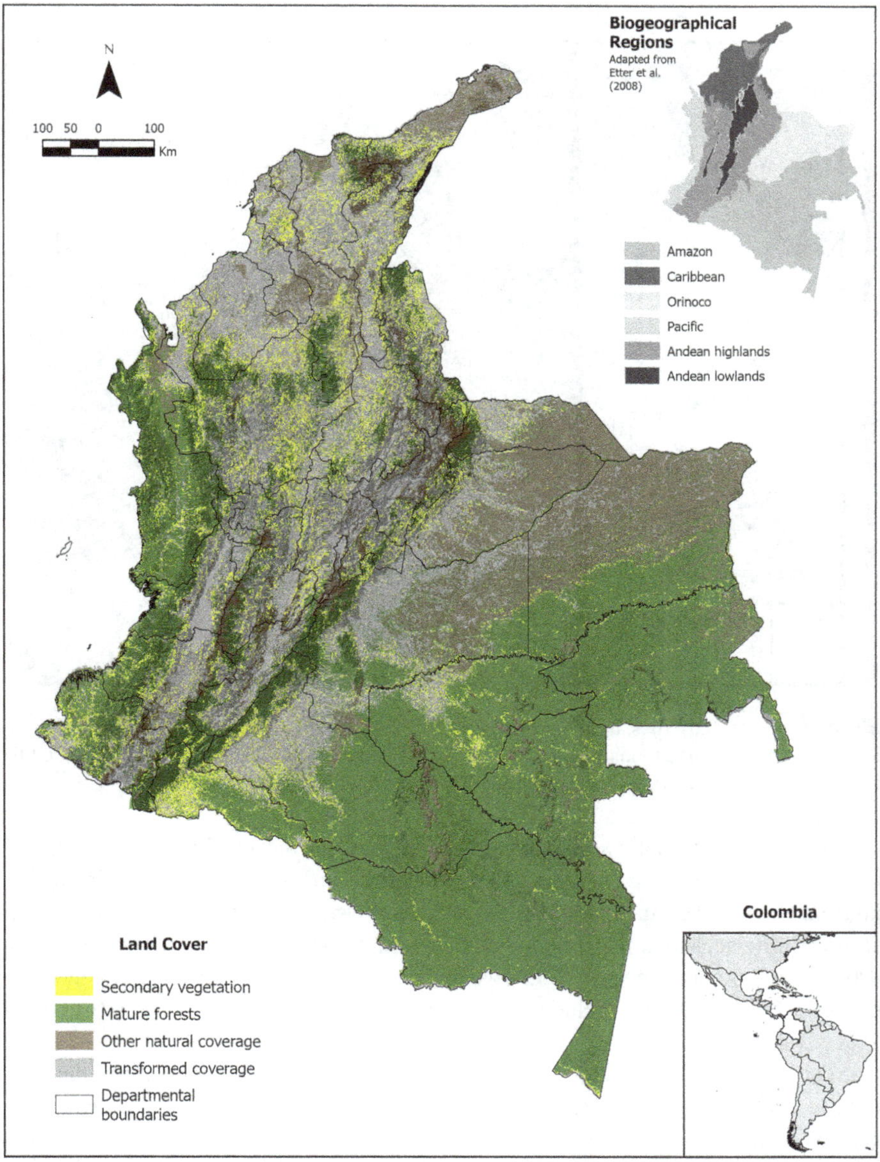

Fig. 3 Geographical distribution of successional forests in Colombia, Map made from the Land Cover Map, CORINE Land Cover Methodology Adapted for Colombia. 2018 period (Castellanos et al., 2021)

Guasca (Gu), Tabio (Ta), and the last one, Torca (To), in a large forest fragment at the periphery of Bogotá. Six of these sites are on private properties, many of which are part of the Colombian Network of Natural Reserves of Civil Society (Resnatur), and one on Encenillo, a biological station of the Natura Foundation. The sites

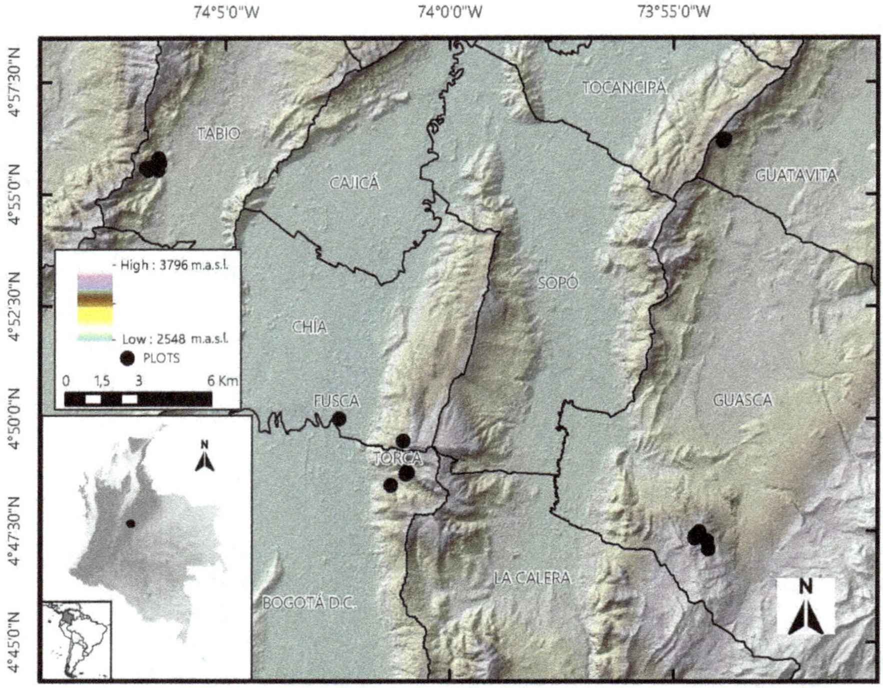

Fig. 4 The four study sites are in remnant forests in Bogotá city's high plain. The study sites are located in the suburban municipalities of Guatavita (Gva), Guasca (Gu), and Tabio (Ta), and the last one, Torca (To), is located in a large forest fragment at the periphery of Bogotá

exhibit different climatic and edaphic characteristics (Hurtado-M et al., 2021), with Guasca, Tabio, and Torca showing a marked dry season and Inceptisols (Fig. 4). The landscapes in Guatavita and Tabio are primarily pastures. Guasca experienced limestone mining activities from the late 1990s, while Torca's forests are surrounded by urban development. The selection of sites aimed to encompass early and late stages of forest succession and included old-growth areas whenever possible (Hurtado-M et al., 2021).

Successional stages were categorized as either early or late based on the minimum estimated age provided by landowners on physiognomic and structural data and the analysis of historical aerial imagery (Calbi et al., 2020). Specifically, early stands were characterized by high stem density and low canopy height (age: 10–30 years), while older stands exhibited a multi-layered canopy with sparse stem density and a varied canopy exceeding 12 meters in height (age: 60+ years; Hurtado-M et al., 2021).

We evaluated diversity in trees, shrubs, herbs, lianas, ferns, and palms in three stages of the life cycle: adults, seedlings, and soil seed banks. Despite the high level of transformation in the Sabana de Bogotá, we recorded high levels plant diversity, with nineteen endemic species to Colombia and four species categorized as vulnerable or endangered recorded either as trees, seedlings, or seeds in our study sites (e.g., *Prunus buxifolia, Solanum humboldtianum, Meriania aguaditensis,* and *Q. humboldtii;* Bernal et al., 2019). Indeed, the Bogotá High Plain is home to a diverse array of plant species, each contributing to the unique character of this landscape. Among the notable species recorded is the Bogotá tea species, *Simplocos theiformis*, with which José Celestino Mutis experimented as a potential commercial beverage. Another species of interest is the mulato, *Ilex kunthiana*, which, although once a dominant species in the region, is now showing signs of population decline (Manrique et al., 2008). Other species likely found in old-growth forests include cedar (*Cedrela montana*), oak (*Quercus humboldtii*), Colombian pine tree (*Prumnopitys montana*), "uché" (*Prunus buxifolia*), and various members of the Lauraceae family, typical of more preserved forests. Together, these species showcase the richness and diversity of the Bogotá high plains flora.

Another significant finding highlighting the relevance of conserving and managing peri-urban landscapes is the uniqueness of each site's floristic composition (Fig. 5). Over 50% of the adult and seedling species were recorded in only one of the four sample sites, underscoring the high beta diversity found in the area and, therefore, the importance of preserving each forest patch to maintain the full range of biodiversity (Hurtado-M et al., 2021). In contrast, only four species were found to have a wide distribution, occurring at all four sites: *Macleania rupestris*, *Oreopanax incisus*, *Vallea stipularis*, and *Myrsine coriacea*. These widely distributed species may be crucial to the ecosystem's functioning and resilience.

Because seeds and seedlings set the stage for future forests, we evaluated changes in diversity across ontogenetic stages. We did not find differences in diversity between the three stages of adults, seedlings, and seeds (Fig. 5a), nor between early and late-successional stages (Fig. 5b). These results align with previous studies from lowland tropical forests, showing that tree species diversity recovers within the first decades of succession (Rozendaal et al., 2019; Prach & Walker, 2019). Here, we further show that this pattern is maintained in early life stages, indicating that seed arrival into these regrowing forest patches is diverse despite intense fragmentation and may promote the recovery of tree species richness. Further research is needed to evaluate the recovery in terms of floristic composition, which may take centuries (Rozendaal et al., 2019).

To assess the recovery of ecosystem functioning in peri-urban forests, it is crucial to evaluate changes in biomass accumulation through stem growth, leaf litter production, and net primary productivity along succession. These indicators are valuable for understanding ecosystem functioning as they offer insights into the ecosystem's productivity, nutrient cycling, and carbon sequestration potential (Chapin et al., 2011; Malhi et al., 2011).

Stem growth is a key indicator that reflects the ecosystem's capacity to accumulate biomass and sequester carbon, which is essential for ecosystem health and stability (Malhi et al., 2011). Leaf litter production provides a measure of organic matter input into the soil, influencing nutrient cycling and soil fertility, critical for sustaining ecosystem processes (Vitousek & Sanford, 1986). Net primary productivity, representing the rate of carbon fixation by the ecosystem, is fundamental for

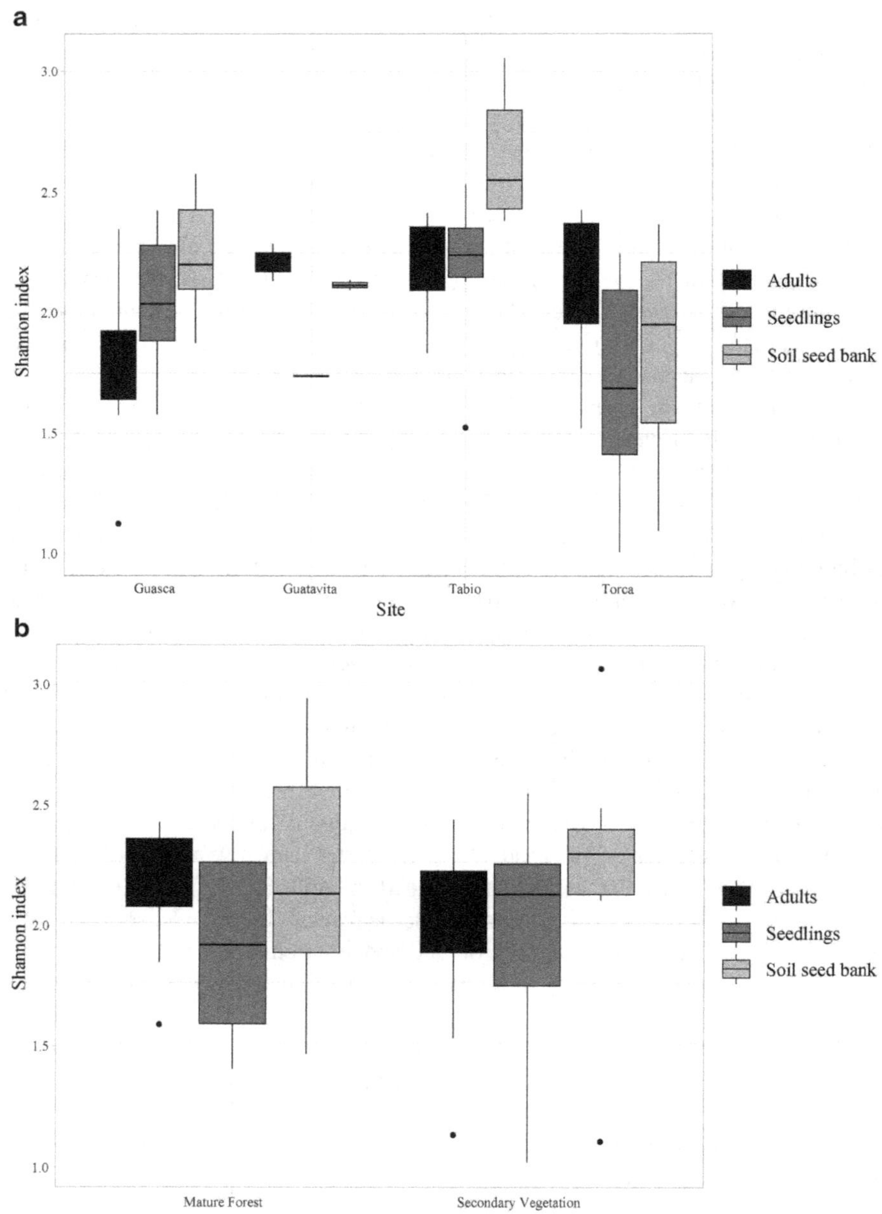

Fig. 5 Diversity comparison with the Shannon index for Adults, Seedlings, and soil seed banks along (**a**) study sites (Guasca; Guatavita; Tabio; Torca). (**b**) among mature forests and successional vegetation

understanding the overall functioning and resilience of the ecosystem, supporting diverse biotic communities (Chapin et al., 2011).

We used aboveground biomass as a proxy of stand age because it has been widely shown to increase asymptotically over time in chronosequence studies, making it a suitable structural descriptor of succession (Breugel et al., 2007).

Our findings demonstrate a relationship between aboveground biomass (AGB) and key indicators of ecosystem functioning in the peri-urban forests. We found that stem growth was related to AGB ($r2 = 0.45$, $p < 0.05$; Fig. 7a), indicating that mature forests with the highest AGB had greater stem growth than secondary forests with lower AGB.

Similarly, annual litter production and aboveground net primary productivity (ANPP) also increased with AGB ($r2 = 0.42$, $p < 0.05$; $r2 = 0.52$, $p < 0.05$, respectively; Fig. 7b, c). This suggests that mature forests, characterized by higher AGB, exhibited greater litter production and ANPP than the secondary forests. These findings highlight the importance of the forest succession stage in determining the rate of biomass accumulation and overall ecosystem functioning rather than site-specific factors alone (Chang & Turner, 2019).

The strong correlations between AGB and these key ecosystem attributes underscore the value of using AGB as a proxy for assessing the recovery of ecosystem functioning along the successional gradient. This information can inform management strategies to promote the restoration and resilience of these peri-urban forest ecosystems.

4 Conclusions

Our results showed that successional forest cover in the Colombian Andes has significantly increased during the last two decades. Successional forests have often been undervalued (e.g., Brook et al., 2006), yet their expanding extent presents an opportunity that aligns with global biodiversity conservation targets. Indeed, the Convention on Biological Diversity (UNCBD), the Convention to Combat Desertification (UNCCD), and the Convention on Climate Change (UNFCCC) are three United Nations frameworks on environmental policy that recognize successional forests and restoration as essential tools for addressing forest loss, biodiversity decline, desertification, and climate change (UNCBD c, 2012; UNCCD c, 2015; UNFCCC c, 2015). Additionally, more and more studies have shown that natural regeneration stands out as a highly cost-effective strategy for fulfilling restoration commitments at local and regional levels (Chazdon, 2017; Crouzeilles et al., 2017). In this sense, we need to give more visibility to regrowing forests in transformed landscapes as they are valuable allies in achieving the country's long-term restoration objectives (Echeverry-Galvis et al., 2019).

In the Andean region, where the intricate interplay between humans and ecosystems has deep historical roots, the ubiquity of successional forests is a legacy of these landscapes' dynamic nature. Our findings show that even peri-urban forests

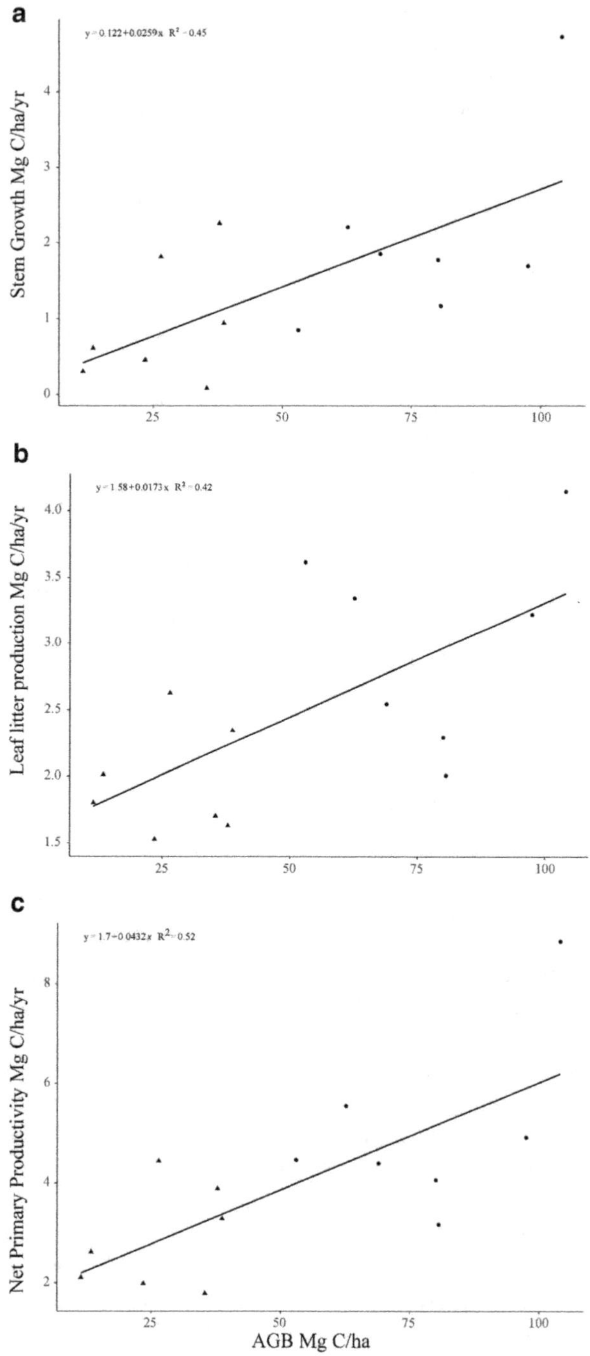

Fig. 7 Linear regressions of the productivity of the High Andean forests and the aboveground biomass as a measure of successional stage. (**a**) Stem growth for 14 plots. (**b**) Leaf litter for 14 plots. (**c**) Net primary productivity in 14 plots. Circles are early, and rhombuses are late successional stand

embedded in highly perturbed landscapes are valuable. They can be biodiversity reservoirs as they harbor threatened and endemic species and contribute to climate change mitigation by storing carbon. Therefore, municipality-level Stakeholders should integrate these forest fragments as crucial ecological elements in environmental policies and promote their study and conservation, regardless of their secondary status. Understanding the conservation status of regrowing forests, even in anthropogenic territories, and evaluating successional trajectories based on the manifold biological attributes that describe forest functioning (Poorter et al., 2021) should be a priority to develop successful conservation and management strategies that improve landscape connectivity and integrity in degraded ecosystems. As such, it is imperative to incorporate these forests into research agendas (Prach et al., 2016). This knowledge forms the basis for understanding and guiding restoration management practices (Suding & Hobbs, 2009; Walker & del Moral, 2009).

Articulating research in successional forests to policymaking is a considerable challenge and a significant opportunity for conserving and restoring key biodiversity areas, enhancing landscape connectivity, and increasing carbon storage. A three-step approach is recommended to harness these opportunities effectively. First, acknowledge the ecological value of successional forests. Early stages of natural regeneration often suffer from a negative reputation among landowners, managers, and decision-makers in tropical countries, frequently being labeled as 'degraded land' despite their regeneration potential (Chazdon & Guariguata, 2016; Hurtado-M et al., 2022). Contrary to this perception, successional forests provide crucial ecological functions in landscapes where mature forests are scarce, as shown here. Second, we require more systematic evaluations of the factors influencing succession under different environmental and social contexts. This is vital to prevent the disruption of recovery in successional forests and could bring essential elements to determine where restoration efforts have a high probability of success, ensuring that restoration actions are appropriately targeted and effective (Christmann et al., 2023). Finally, we urgently need to integrate social components into successional studies. Historically, research on successional forests has focused on the biological aspects of plant community reassembly (Poorter et al., 2023). However, they are permanently embedded in human-influenced landscapes, and their future depends on the local human populations. We, therefore, need to integrate the effect of management practices and socioeconomic variables as factors influencing successional trajectories (Vieira et al., 2014; Jakovac et al., 2021; Balvanera et al., 2021).

Addressing this challenge requires a comprehensive approach that considers each region's unique social and economic contexts, examines the ecological determinants of forest recovery across diverse ecosystems, and understands the factors contributing to deforestation in Colombian regions. By doing so, we will be able to balance the needs of human populations, who often rely on these landscapes for their livelihoods, with the imperative to preserve biodiversity, fostering local communities' involvement in conservation and sustainable land-use practices.

References

Aide, T. M., Clark, M. L., Grau, H. R., Lopez-Carr, D., Levy, M. A., Redo, D., Bonilla-Moheno, M., Riner, G., Andrade-Nuñez, M. J., & Muñiz, M. (2013). Deforestation and reforestation of Latin America and the Caribbean (2001–2010). *Biotropica, 45*(2), 262–271. https://doi.org/10.1111/j.1744-7429.2012.00908.x

Aide, T. M., Grau, H. R., Graesser, J., Andrade-Nuñez, M. J., Aráoz, E., Barros, A. P., Campos-Cerqueira, M., Chacon-Moreno, E., Cuesta, F., Espinoza, R., Peralvo, M., Polk, M. H., Rueda, X., Sánchez, A., Young, K. R., Zarbá, L., & Zimmerer, K. S. (2019). Woody vegetation dynamics in the tropical and subtropical Andes from 2001 to 2014: Satellite image interpretation and expert validation. *Global Change Biology, 25*(6), 2112–2126. https://doi.org/10.1111/gcb.14618

Armenteras, D., Rodríguez, N., Retana, J., & Morales, M. (2011). Understanding deforestation in montane and lowland forests of the Colombian Andes. *Regional Environmental Change, 11*, 693–705. https://doi.org/10.1007/s10113-010-0200-y

Asner, G. P., et al. (2013). Landscape-scale changes in forest structure and functional traits along an Andes-to-Amazon elevation gradient. *Biogeosciences Discussions, 10*, 15415–15454.

Balvanera, P., Paz, H., Arreola-Villa, F., Bhaskar, R., Bongers, F., Cortes, S., del Val, E., Garcıa-Frapolli, E., Gavito, M. E., & Gonzalez-Esquivel, C. E. (2021). Social-ecological dynamics of tropical secondary forests. *Forest Ecology and Management, 496*, 119369.

Bernal, R., Gradstein, S. R., & Celis, M. (Eds.). (2019). *Catálogo de plantas y líquenes de Colombia*. Instituto de Ciencias Naturales, Universidad Nacional de Colombia, Bogotá. http://catalogoplantasdecolombia.unal.edu.co

Breugel, M., Bongers, F., & Martínez-Ramos. (2007). Species dynamics during early secondary forest succession: Recruitment, mortality and species turnover. *Biotropica, 39*(5), 610–619.

Brook, B. W., Bradshaw, C. J. A., Pin Koh, L., & Sodhi, N. S. (2006). Momentum drives the crash: Mass extinction in the tropics. *Biotropica, 38*, 302–305.

Calbi, M., Clerici, N., Borsch, T., & Brokamp, G. (2020). Reconstructing long term high Andean forest dynamics using historical aerial imagery: A case study in Colombia. *Forests, 11*(8), 1–20. https://doi.org/10.3390/F11080788

Castellanos, H., Gómez, W. F. y Mayorga, N. Mapa nacional de coberturas de la tierra, escala 1:100.000, periodo 2018. 2021. Metodología Corine Land Cover adaptada para Colombia. Memoria técnica y resultados. Instituto de Hidrología, Meteorología y Estudios Ambientales (Ideam). Bogotá, D. C., Colombia.

Camelo, O. J., Urrego, L. E., & Orrego, S. A. (2017). Environmental and socioeconomic drivers of woody vegetation recovery in a human-modified landscape in the Rio Grande basin (Colombian Andes). *Restoration Ecology, 25*(6), 912–921. https://doi.org/10.1111/rec.12514

Chang, C. C., & Turner, B. L. (2019). Ecological succession in a changing world. *Journal of Ecology, 107*, 503–509. https://doi.org/10.1111/1365-2745.13132

Chapin, F. S., Matson, P. A., & Vitousek, P. M. (2011). *Principles of terrestrial ecosystem ecology*. Springer Science & Business Media.

Chazdon, R. (2017). Landscape restoration, natural regeneration, and the forests of the future. *Annals of the Missouri Botanical Garden, 102*, 251–257. https://doi.org/10.3417/2016035

Chazdon, R., & Guariguata, M. (2016). Natural regeneration as a tool for large-scale forest restoration in the tropics: Prospects and challenges. *Biotropica, 48*(6), 716–730. https://doi.org/10.1111/btp.12381

Chazdon, R., Broadbent, E. N., Rozendaal, D. M. A., Bongers, F., Al, E., Zambrano, A. M. A., Aide, T. M., et al. (2016). Carbon sequestration potential of second-growth forest regeneration in the LatinAmerican tropics. *Science Advances, 2*, e1501639–e1501639. https://doi.org/10.1126/sciadv.1501639

Christmann, T., Palomeque, X., Armenteras, D., Wilson, S. J., Malhi, Y., & Oliveras Menor, I. (2023). Disrupted montane forest recovery hinders biodiversity conservation in the tropical Andes. *Global Ecology and Biogeography, 32*(5), 793–808. https://doi.org/10.1111/geb.13666

Crouzeilles, R., Ferreira, M. S., Chazdon, R., Lindenmayer, D. B., Sansevero, J. B. B., Monteiro, L., Iribarrem, A., Latawiec, A. E., & Strassburg, B. B. N. (2017). Ecological restoration success is higher for natural regeneration than for active restoration in tropical forests. *Science Advances, 3*, e1701345. https://doi.org/10.1126/sciadv.1701345

Cuesta, F., Peralvo, M., & Valarezo, N. (2009). Los Bosques Montanos de los Andes Tropicales. Quito, Ecuador; Lima, Peru; La Paz, Bolivia: Programa Regional Ecobona-Intercooperation, Agencia Suiza para la Cooperación y el Desarrollo.

DANE--Departamento Administrativo Nacional de Estadistica. (2019). Censo Nacional de Población y Vivienda 2018. https://www.dane.gov.co/index.php/estadisticas-por-tema/demografia-y-poblacion/censo-nacional-de-poblacion-y-vivenda-2018

Davies, R. W., Edwards, D. P., & Edwards, F. A. (2020). Secondary tropical forests recover dung beetle functional diversity and trait composition. *Animal Conservation, 23*(5), 617–627.

Duque et al. (2021). Author correction: Mature Andean forests as globally important carbon sinks and future carbon refuges. *Nature Communications, 12*, 3617. https://doi.org/10.1038/s41467-021-23955-7

Echeverry-Galvis, M. Á., Unda, M., Bravo, M. P., García, N., Rubiano- Pinzón, G. A., & Palomino, J. V. (2019). Plan de Acción en Biodiversidad de la PNGIBSE, Metas Aichi y los Objetivos de Desarrollo Sostenible en Colombia: análisis y concordancia en el discurso entre metas propuestas para 2020. *Gestión y Ambiente, 22*(2), 207–234. https://doi.org/10.15446/ga.v22n2.82227

Etter, A., & van Wyngaarden, W. (2000). Patterns of landscape transformation in Colombia, with emphasis in the Andean region. *AMBIO: A Journal of the Human Environment, 29*(7), 432–439. https://doi.org/10.1579/0044-7447-29.7.432

Etter, A., McAlpine, C., & Possingham, H. (2008). Historical patterns and drivers of landscape change in Colombia since 1500: A regionalized spatial approach. *Annals of the Association of American Geographers, 98*, 2–23. https://doi.org/10.1080/00045600701733911

Fagua, J. C., & Ramsey, R. D. (2019). Geospatial modeling of land cover change in the Chocó-Darien global ecoregion of South America, one of the most biodiverse and rainy areas in the world. *PLoS One, 14*(2), e0211324. https://doi.org/10.1371/journal.pone.0211324

Hughes, C. E. (2016). The tropical Andean plant diversity powerhouse. *The New Phytologist, 210*(4), 1152–1154. https://www.jstor.org/stable/newphytologist.210.4.1152

Hurtado-M, A. B., Echeverry-Galvis, M. Á., Salgado-Negret, B., Muñoz, J. C., Posada, J. M., & Norden, N. (2021). Little trace of floristic homogenization in peri-urban Andean secondary forests despite high anthropogenic transformation. *Journal of Ecology, 109*(3), 1468–1478. https://doi.org/10.1111/1365-2745.13570

Hurtado-M, A. B., Muñoz, J. C., Echeverry-Galvis, M. Á., & Norden, N. (2022). Successional forests in Colombia: An opportunity for recovery of transformed landscapes. *Caldasia, 44*(2), 332–344. https://doi.org/10.15446/caldasia.v44n2.82255

IDEAM–Instituto de Hidrología Meteorología y Estudios Ambientales. (2010). Leyenda Nacional de Coberturas de la Tierra. Metodología CORINE Land Cover Adaptada para Colombia. Escala 1:100.000., Bogotá, D.C.

Jakovac, C. C., Junqueira, A. B., Crouzeilles, R., Peña-Claros, M., Mesquita, R. C. G., & Bongers, F. (2021). The role of land-use history in driving successional pathways and its implications for the restoration of tropical forests. *Biological Reviews*. https://doi.org/10.1111/brv.12694

Josse, C., Cuesta, F., Navarro, G., Barrena, V., Cabrera, E., Chacon-Moreno, E., Ferreira, W., Peralvo, M., Saito, J., & Tovar, A. (2009). *Ecosistemas de los Andes del Norte y Centro: Bolivia, Colombia, Ecuador, Peru y Venezuela*. Consortium for Sustainable Development of the Andean Ecoregion (CONDESAN).

Lennox, G. D., Gardner, T. A., Thomson, J. R., et al. (2018). Second rate or a second chance? Assessing biomass and biodiversity recovery in regenerating Amazonian forests. *Global Change Biology, 24*, 5680–5694. https://doi.org/10.1111/gcb.14443

Malhi, Y., Doughty, C., & Galbraith, D. (2011). The allocation of ecosystem net primary productivity in tropical forests. *Philosophical Transactions of the Royal Society B: Biological Sciences, 366*(1582), 3225–3245.

Malizia, A., Blundo, C., Carilla, J., Acosta, O. O., Cuesta, F., Duque, A., Aguirre, N., Aguirre, Z., Ataroff, M., Baez, S., Calderón-Loor, M., Cayola, L., Cayuela, L., Ceballos, S., Cedillo, H., Ríos, W. F., Feeley, K. J., Fuentes, A. F., Gámez Álvarez, L. E., et al. (2020). Elevation and latitude drives structure and tree species composition in Andean forests: Results from a large-scale plot network. *PLoS One, 15*(4), 1–18. https://doi.org/10.1371/journal.pone.0231553

Manrique, P., Alonso, J., Villacorte, T., & Alberto, W. (2008). Micropropagación de Ilex kunthiana Triana & Planchon (Aquifoliaceae), una especie de gran importancia en programas de revegetalización. *Revista Colombiana de Biotecnología, 10*(2), 72–84. Retrieved May 17, 2024, from http://www.scielo.org.co/scielo.php?script=sci_arttext&pid=S0123-34752008000200009&lng=en&tlng=

Mathez-Stiefel, S. L., Peralvo, M., Báez, S., Rist, S., Buytaert, W., Cuesta, F., et al. (2017). Research priorities for the conservation and sustainable governance of Andean forest landscapes. *Mountain Research and Development, 37*, 323–339.

Myers, N., Mittermeier, R. A., Mittermeier, C., Da Fonseca, G. A. B., & Kent, J. (2000). Biodiversity horspots for conservation priorities. *Nature, 403*, 853–858. https://doi.org/10.1038/468895a

Poorter, L., Amissah, L., Bongers, F., Hordijk, I., Kok, J., Laurance, S. G., et al. (2023). Successional theories. *Biological Reviews, 98*(6), 2049–2077.

Prach, K., Tichý, L., Lencová, K., Adámek, M., Koutecký, T., Sádlo, J., Bartošová, A., Novák, J., Kovář, P., Jírová, A., Šmilauer, P., & Řehounková, K. (2016). Does succession run towards potential natural vegetation? An analysis across seres. *Journal of Vegetation Science, 27*(3), 515–523. https://doi.org/10.1111/jvs.12383

Prach, K., & Walker, L. R. (2019). Differences between primary and secondary plant succession among biomes of the world. *Journal of Ecology, 107*, 510–516.

Poorter, L., Rozendaal, D. M. A., Bongers, F., de Jarcilene, S. A., Àlvarez, F. S., Luìs Andrade, J., Arreola Villa, L. F., Becknell, J. M., Bhaskar, R., Boukili, V., Brancalion, P. H. S., Cèsar, R. G., Chave, J., Chazdon, R. L., Colletta, G. D., Craven, D., de Jong, B. H. J., Denslow, J. S., Dent, D. H., … Westoby, M. (2021). Functional recovery of secondary tropical forests. *Proceedings of the National Academy of Sciences of the United States of America, 118*(49). https://doi.org/10.1073/pnas.2003405118

Rodríguez Eraso, N., Armenteras, D., & Retana Alumbreros, J. (2012). Land use and land cover change in the Colombian Andes: Dynamics and future scenarios. *Journal of Land Use Science*. https://doi.org/10.1080/1747423X.2011.650228

Rodríguez, N., Armenteras, D., Morales, M., & Romero, M. (2006). *Ecosistemas de los Andes colombianos* (p. 154). Segunda edición. Instituto de Investigación de Recursos Biológicos Alexander von Humboldt.

Rozendaal, D. M. A., Bongers, F., Aide, T. M., Alvarez-Dávila, E., Ascarrunz, N., Balvanera, P., et al. (2019). Biodiversity recovery of Neotropical secondary forests. *Science Advances, 5*. https://doi.org/10.1126/sciadv.aau3114

Rubiano, K., Clerici, N., Norden, N., & Etter, A. (2017). Secondary forest and shrubland dynamics in a highly transformed landscape in the Northern Andes of Colombia (1985-2015). *Forests, 8*(6), 216. https://doi.org/10.3390/f8060216

Sanchez-Cuervo, A. M., & Aide, T. M. (2013). Identifying hotspots of deforestation and reforestation in Colombia (2001-2010): Implications for protected areas. *Ecosphere, 4*(11). https://doi.org/10.1890/ES13-00207.1

Sánchez-Cuervo, A. M., Aide, T. M., Clark, M. L., & Etter, A. (2012). Land cover change in Colombia: Surprising forest recovery trends between 2001 and 2010. *PLoS One, 7*, e43943. https://doi.org/10.1371/journal.pone.0043943

Suding, K., & Hobbs, R. (2009). Threshold models in restoration and conservation: A developing framework. *Trends in Ecology & Evolution, 24*, 271–279. https://doi.org/10.1016/j.tree.2008.11.012

Tejedor-Garavito, N., Álvarez, E., Caro, S. A., Murakami, A. A., Blundo, C., Espinoza, T. E. B., & La Torre, M. A. (2012). Evaluación del estado de conservación de los bosques montanos en los Andes tropicales. *Ecosistemas: Revista Cietifica y Tecnica de Ecologia y Medio Ambiente, 21*, 148–166.

UNCBD–United Nations Convention on Biological Diversity. (2012). COP 11 decision XI/16: Ecosystem restoration. Recuperado de: https://www.cbd.int/decision/cop/default.shtml?id=13177.

UNCCD–United Nations Convention to Combat Desertification. (2015). *Land matters for climate: Reducing the gap and approaching the target*. Recuperado de https://www.unccd.int/Lists/SiteDocumentLibrary/Publications/2015Nov_Land_matters_For_ Climate_ENG.pdf

UNFCCC–United Nations Framework Convention on Climate Change. (2015). *Paris Agreement*. Recuperado de https://unfccc.int/files/essential_background/convention/application/pdf/english_paris_agreement.pdf

Van der Hammen, T. (1986). La Sabana de Bogotá y su lago en el Pleniglacial Medio. *Caldasia, 15*(71–75), 249–262.

Vieira, I. C. G., Gardner, T., Ferreira, J., Lees, A. C., & Barlow, J. (2014). Challen- ges of governing second-growth forests: A case study from the Brazilian Amazonian State of Pará. *Forests, 5*(7), 173–71752. https://doi.org/10.3390/f5071737

Vitousek, P. M., & Sanford, R. L. (1986). Nutrient cycling in moist tropical forest. *Annual Review of Ecology and Systematics, 17*(1), 137–167.

Walker, L., & del Moral, R. (2009). Transition dynamics in succession: Implications for rates, trajectories and restoration. In *New models for ecosystem dynamics and restoration*. Island Press.

Effects of Urban Growth and Local Disturbances on Extent and Composition of High Andean Forests

Mariasole Calbi, Francisco Fajardo Gutiérrez, Robert Lücking, Grischa Brokamp, and Thomas Borsch

1 Introduction

1.1 Review of High Andean Forests Studies

High Andean forests are one of the global conservation priorities, home to an extraordinary biodiversity that includes several endemic and threatened species (Mittermeier et al., 1999; Myers et al., 2000). These forests host more than 10,000 species of plants, 1200 species of birds, 400 species of frogs, 270 species of rodents and bats, and over 400 species of lichens (Rangel Churio et al., 1997; Soto Medina et al., 2021). However, conservation efforts in high Andean forests often do not adequately protect local biodiversity (Bax & Francesconi, 2019) and these forests' extent is extremely fragmented (Armenteras et al., 2003; Etter et al., 2008), occupying less than 50% of the original forest cover (Rubiano et al., 2017; Llambí et al., 2019). Moreover, the tropical Andes are considered a climate change hotspot, where strong ecosystem perturbations are expected to negatively impact provisioning of ecosystem services (Clerici et al., 2019a, b). Additionally, in this area the effects of climate change are likely to interact cumulatively with the effects of human disturbance and land transformation on one hand, and with the instable political environment on the other hand, which brings about armed conflicts (Clerici et al., 2019b), and poorly planned environmental strategies with inadequate goals (Eufemia et al., 2019).

Particularly in the surroundings of populous Andean cities, high Andean forests have suffered a wide array of disturbances and transformations, including timber extraction and fast urbanization since the beginning of the eighteenth century (Etter

M. Calbi (✉) · F. F. Gutiérrez · R. Lücking · G. Brokamp · T. Borsch
Botanischer Garten und Botanisches Museum Berlin, Freie Universität Berlin, Berlin, Germany
e-mail: mariasolecalbi@edu.unige.it

N. Clerici (ed.), *Conservation of Andean Forests*,
https://doi.org/10.1007/978-3-031-80805-0_5

& van Wyngaarden, 2000). These perturbations and alterations typically resulted in a complex matrix of urban areas, cultivated fields, secondary forests, and few small and isolated well preserved forest fragments (Aubad et al., 2008; Weigend et al., 2005). In this context, the capital city of Colombia, Bogotá, serves as an exemplar case to assess how land cover change interplays with other political and ecological factors in a fragile ecosystem setting (Etter & Villa, 2000), and to study the impact of human-induced alterations of high Andean forests. During the late twentieth century Colombia experienced a growing urbanization and industrialization process, being the Distrito Capital of Bogotá one of the fastest growing regions. Additionally, more than half a century of political instability and armed conflicts have had a severe impact on the country, adding up to increased pressure on Colombian ecosystems and their biodiversity (Fjeldså et al., 2005). The city of Bogotá heavily relies on the ecosystem services provided by the surrounding natural areas, (i.e., *páramos* and high Andean forests) for example for water purification and erosion control (Buytaert et al., 2006; Escobedo et al., 2015). High-Andean forest fragments near Bogotá are subjected to a wide range of local disturbances, including firewood and timber extraction, mining, intentional fires, and cattle husbandry (Etter & Villa, 2000; Armenteras et al., 2003; Etter et al., 2006), together with the introduction of an increasing number of invasive species. These perturbations also impact soil structure and thus hydrological regulation (Etter & van Wyngaarden 2000), causing negative consequences on water storage patterns and water quality (Borrelli et al., 2015).

The resulting composition of the vegetation nearby Bogotá is well studied as the characterization of high Andean plant communities has been carried out since the first half of the nineteenth century with the contributions of naturalists Francisco J. Caldas and Alexander von Humboldt (Van der Hammen & Rangel, 1997). Moreover, several phytosociological studies based on the Zürich-Montpellier methodology (Braun-Blanquet, 1932) investigated the Capital District vegetation (Cuatrecasas, 1934, 1958; Lozano-Contreras & Schnetter, 1976; Cleef, 1981; Sturm & Rangel Churio, 1985; Sánchez & Rangel, 1990; Van der Hammen, 1998; Franco & Betancur, 1999; Cortés-Sánchez, 2011; Van der Hammen, 2008; Higuera & Cuéllar, 2013; Avella et al., 2014). Lately, plot-based characterizations of the local flora and its relevant features, not based on the phytosociological approach, were carried out especially by local institutions (i.e., the Jardín Botánico de Bogotá and Universidad Distrital).

More recently, studies focusing on ecological features of communities were carried out to evaluate the stand structure of the woody vegetation (Vargas & Zuluaga, 1986), associated aboveground biomass (Cardozo & Schnetter, 1976) and its relationship to functional diversity and forest fragmentation (Rodríguez-Alarcón et al., 2018), and to characterize the floristic, ecological, and physiognomic changes in species assemblages (Cortés Sánchez, 2008). There is also a considerable number of plot-based studies focusing on habitat quality, vegetation dynamics and their evolution through time, including the characterization of regeneration stadia of *bosques altoandinos* (Rojas, 2017) or the role of plant functional groups during secondary succession (Castellanos-Castro & Bonilla, 2011). Hurtado-M et al. (2021)

performed a thorough assessment of the floristic composition and regeneration potential of the vegetation in permanent plots in various forest fragments in the north of Bogotá, as part of the Rastrojos project. Moreover, the work by Acuña Cabanzo (2013), Arias-Escobar and Barrera-Cataño (2007) and Valencia (2019), targeted high Andean ecosystem condition and regeneration potential after anthropogenic disturbance. The ecological connectivity, regeneration, and conservation status of *bosques altoandinos* near Bogotá were also assessed (Isaacs-Cubides et al., 2017). Lastly, several studies investigated habitat quality indicators such as the presence of invasive species, especially targeting the European species *Ulex europaeus* L. (Beltrán & Barrera-Cataño, 2014; Ocampo-Zuleta & Solorza-Bejarano, 2017) or species of the Poaceae family (Giraldo-Cañas, 2010). Nevertheless, high Andean forests are considered one of the least known ecosystems in the tropics (Garavito et al., 2012). However, the quantity of multi-scalar and multi-temporal studies focused on high Andean forests is still scarce (Young & Leon, 2007; Kappelle & Brown, 2003; Jørgensen et al., 2011; Restrepo Abadia, 2016). Detailed information on how past land cover changes impact the composition and structure of plant communities is currently not readily available at the regional scale.

Colombian high Andean forests extent and their changes through time have been investigated thoroughly (Etter & Villa, 2000; Armenteras et al., 2003; Sánchez-Cuervo et al., 2012; Rodríguez Eraso et al., 2013), specifically in the surroundings of Bogotá (Etter, 2002; Rubiano et al., 2017; Anselm et al., 2018; Calbi et al., 2020). However, the spatial extent of forest cover is not *per se* an indicator of the quality of local habitats in terms of condition and composition of the ecosystem. Insights into the degree of habitat degradation are rather provided by analyses of forest cover quality (e.g., plantations vs. natural secondary growth), local land-use history and present management, which may convey soil degradation and erosion, as well as invasive species penetration (Aide & Grau, 2004; Vanwalleghem et al., 2017). Accordingly, the focus changed to integrating studies on forest cover change with environmental and anthropogenic constraints, or to relating forest cover attributes with functional traits and associated ecosystem services (Borrelli et al., 2015; Clerici et al., 2016, 2019a, b; Salamanca-Fonseca et al., 2023).

Here, we jointly present and discuss the results from two analyses performed in the same study area to better understand the trajectories of forest fragment recovery by bridging forest cover and connectivity assessment through time with floristic surveys. We focused on six administrative localities situated in the high plains of Bogotá, connecting the observed land cover change with each locality's peculiar features of plant communities and history.

1.2 Study Area

The city of Bogotá is situated at 2650 m a.s.l. on a plateau within the Eastern Andean Cordillera. The capital region is the most densely populated area of the country, with nearly 9 million inhabitants and approximately 4500 people per km^2

(DANE, 2019). The city is surrounded by hills on its eastern and southern sides and enclosed to the west and north by the Bogotá river (Andrade et al., 2013). In the surroundings of Bogotá, high Andean forests can be found in the humid and relatively cold areas between ca. 2600 and 3600 m a.s.l. (Cuatrecasas, 1958), although the local altitudinal limits may vary according to their microclimate (Cortés-Duque et al., 2013). These forests share their lower limit with Andean Forests (*bosque andino*; ca. 2200–2600 m a.s.l.) and their upper limit often with a shrubby intermediating zone of dryer enclaves (*matorrales*; ca. 3200–3500 m a.s.l.) which lead to the emblematic *páramo*, or tundra-like tropical mountain grasslands (ca. 3200–4700 m a.s.l.). The *páramo* of Sumapaz, part of the Capital District of Bogotá, is the world's largest ecosystem of this kind.

In the vicinity of Bogotá, semi-natural areas are highly influenced by the city and its contiguous suburbs. Thus, sparse remnants of original vegetation are intermingled with areas of recent secondary growth (Rubiano et al., 2017). As a result, high Andean forest fragments near Bogotá display a heterogeneous stand structure, which ranges from young even-aged shrub-like patches, where early successional species are abundant, to more developed fragments with a more balanced and diverse species composition (Rodríguez-Alarcón et al., 2018). Usually, tree height is below 15–25 m and, in terms of species composition, high Andean forest remnants are often dominated by *Weinmannia* spp. (Cunoniaceae), *Myrsine* spp. (Myrsinaceae), *Myrcianthes* spp. (Myrtaceae), *Morella* spp. (Myricaceae), *Ageratina* spp., *Diplostephium* spp. (Asteraceae), *Hesperomeles* spp. (Rosaceae), *Oreopanax* spp. (Araliaceae), *Gaiadendron punctatum* (Ruiz & Pav.) G. Don. (Loranthaceae), *Daphnopsis caracasana* Meisn. (Thymelaeaceae), *Drimys granadensis* L.f. (Winteraceae), *Andesanthus lepidotus* (Humb. & Bonpl.) P.J.F. Guim. & Michelang., and *Miconia* spp. (Melastomataceae) (Restrepo Abadia, 2016; Cortés-Sánchez, 2011; Van der Hammen, 1998, Hurtado-M et al., 2021). Little disturbed and old forests are commonly characterized by the presence of tree species of the families Lauraceae (*e.g., Andea* spp., *Persea* spp.), Melastomataceae (*Meriania* spp., *Axinaea* spp.), Podocarpaceae (*Prumnopitys montana;* Humb. & Bonpl. ex Willd.) de Laub., *Retrophyllum rospigliosii* (Pilg.) C.N. Page) and Rosaceae (*Prunus* spp.), *Clusia* spp. (Clusiaceae), *Ceroxylon* spp. (Arecaceae), *Cedrela montana* Moritz (Meliaceae), *Meliosma* spp. (Sabiaceae), and some species of *Weinmannia* (Cunoniaceae) (Hurtado-M et al., 2021); Calbi and Fajardo-Gutiérrez, personal observation). On the other hand, secondary forests in the area are characterized by the dominance of Ericaceae (*e.g., Cavendishia* spp., *Macleania* spp.), *Rubus* spp. (Rosaceae), *Chaetogastra grossa* (L.f.) P.J.F. Guim. & Michelang., *Miconia* spp. (Melastomataceae), *Vallea stipularis* L.f. (Elaeocarpaceae), *Escallonia* spp. (Escalloniaceae), *Clethra* spp. (Clethraceae), and *Chusquea* spp. (Poaceae) (Cortés-Sánchez 2011; Hurtado-M et al., 2021).

High Andean forests typically host a dense understory stratum with abundant epiphytic plants, lichens and mosses that cover trees as well as the forest floor, forming thick layers (Bader, 1999). The understory is characterized by a high diversity of orchids, bromeliads, aroids, ericads, peperomiads, ferns, bryophytes, lichens, and other epiphytic groups (Bader, 1999; Gradstein, 2008). These epiphytic taxa play a fundamental role in regulating the terrestrial hydrological and nutrient cycles in cloud forests (Lai et al., 2020), and in Colombian cloud forests their biomass can be as high as 44 tons per hectare (Hofstede et al., 1993). Epiphyte diversity, abundance, and composition is also one of the features that can help in discriminating between undisturbed and disturbed fragments (Gradstein, 2008; Rios et al., 2015; Ramírez-Morán et al., 2016).

The study area investigated here encompasses six localities (Sumapaz, Pasquilla, Torca, Tabio, Guasca, Guatavita) across ca. 4600 km^2, spanning peri-urban and rural areas in the department of Cundinamarca and the administrative region of the city of Bogotá D.C. (Distrito Capital) (Fig. 1). These localities often have different

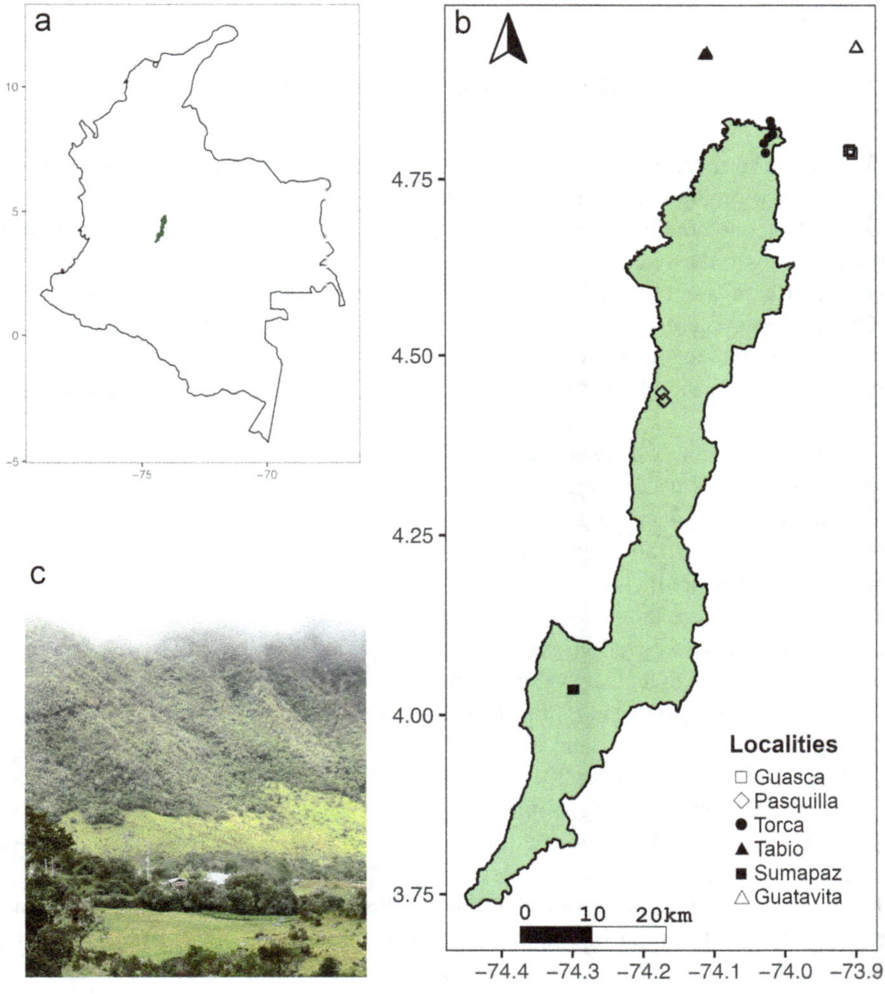

Fig. 1 Study area; (**a**) Colombia: the capital district of Bogotá (Bogotá D.C.), located in the geographical center of Colombia, in the Eastern Cordillera; (**b**) The Capital Distric of Bogotà: geographic position of the localities within and outside the administrative limits of Bogotá D.C.; (**c**) fragments of high Andean Forest vegetation in the locality of Sumapaz

environmental and topographic features and a different socio-economic history. The climate displays a marked isothermality with an annual mean temperature of around 14 °C and a mean annual precipitation between 600 and 1300 mm. There are two rainy seasons: from April to June and from September to November, with a drier and warmer season from January to March (Anselm et al., 2020). The topography is marked by a vast high plain, at around 2600 masl, where most of the urban and agricultural area is located, and by surrounding steep elevation gradients including mountains of up to 4100 m altitude. Andisols are the dominant soils in the study area (Etter, 2002).

2 Methods

2.1 *Land Cover and Landscape Connectivity Mapping*

Forty-three aerial greyscale photos of the six selected study localities were retrieved from the Colombian Geographical Institute Agustín Codazzi (IGAC). The photos were selected to obtain roughly one photo per decade and locality and spanned from the 1940s up to the year 2007. For the time series of each locality, a consistent geographical area was clipped and georeferenced using QGIS v 2.18.14-Las Palmas (QGIS Development Team, 2017), with thin plate spline as a transformation method due to its high precision (Boer et al., 2001). Then, the photos were resampled at 1 m pixel size. From the initial 43 scenes, 28 were retained as suitable to perform the analysis.

First, texture analysis was performed to enhance the pixel classification of the greyscale pictures using the glcm package (Zvoleff, 2020) in RStudio 1.0.153 (R Core team, 2018). A moving window size of 7 × 7 pixels was used as to avoid the drawbacks of either too large or too small window sizes (Huang et al., 2008; Wen et al., 2009). Texture metrics were calculated for the four directions and the obtained raster layers were examined. Based on the most informative metrics to visually distinguish our land cover classes, we retained three metrics for classification purposes: entropy, mean and second moment (Hall-Beyer, 2000).

Secondly, we performed object-based classification to efficiently delineate homogeneous polygons of land-cover/land-use based on panchromatic aerial imagery (Harb Rabia & Terribile, 2013). Object-based techniques consist of segmentation or the grouping of neighbouring pixels of similar properties to form objects (Blaschke et al., 2004). Segmentation for the object-based image analysis (OBIA) was carried out using the 'Seeded Region Growing Algorithm' (Adams & Bischof, 1994) on a raster stack of the selected three texture metrics and the original greyscale aerial picture with the OBIA module in SAGA GIS 2.1.4 (Conrad et al., 2015) to obtain better results than the image classification at the pixel level by reducing the within-class variability (Blaschke, 2010; Radoux & Bogaert, 2014). Minimum object size was five pixels according to the smallest recognizable entity, i.e., the average width of a tree crown in the study system.

To carry out the supervised classification, we visually selected our training polygons from the ones retrieved by the segmentation. We defined three land cover classes: B (bare ground/built-up), G (grass, crops, and low shrubs), and T (trees and higher shrubs). We could not identify and separate forest plantations since the quality of the aerial imagery did not allow for it. Supervised classification was performed using the maximum likelihood criterion in SAGA GIS 2.1.4. (Conrad et al., 2015) due to its classification efficacy when there are well separated land cover classes, as was the case in our study (Franklin et al., 2003).

The accuracy assessment included the calculation of overall accuracy (OA), class-specific user accuracy (UA) and producer accuracy (PA). To this aim, we employed a balanced, random sampling response design to obtain a more accurate estimate of the user's accuracy for the proportionally smallest class B and allocated an adequate sample size to it (Stehman, 2009). One-hundred (100) random points were generated for each of the three classes. Therefore, our response design consisted of 300 blocks (Stehman & Wickham, 2011) of 3 × 3 pixels around each random point. Inside each block, we adopted a majority rule. For a sound accuracy assessment, the visual validation of random blocks was performed on the original high-resolution greyscale aerial pictures. Stratified area estimators as well as the overall and class-specific accuracy were obtained following the equations proposed by Olofsson et al. (2013).

As a further step to obtain better insights into the quality of land cover change and to derive landscape features in relation to ecological structural connectivity, we paired the land cover classification with the morphological spatial pattern analysis (MSPA; Vogt et al., 2007). We focused on structural connectivity as its increase relates to physical connections between habitat patches and may in turn ensure higher chances of dispersal for both short-distance dispersed species and long-distance dispersed species (Saura et al., 2011).

We derived MSPA forest structural connectivity classes from the geometric pattern analysis within the T class to infer structural connectivity features of the forest cover at any of the studied time points. We opted for MSPA to derive a limited set of selected indicators in order to aid the interpretation of landscape features that can relate both to the needs of the local fauna and plant communities.

The MSPA was carried out using the MSPA QGIS plugin (Soille & Vogt, 2009). The MSPA uses a series of mathematical morphological operators to describe the geometrical properties of binary raster maps, e.g., forest (or foreground) = 1, non-forest (or background) = 0. As output, this analysis provides a raster layer with a minimum of seven morphological classes (core, islet, loop, bridge, perforation, edge, and branch) (Soille & Vogt, 2009) that trace the geometry and connectivity of the binary layer components' spatial arrangement (Vogt & Riitters, 2017). We performed the MSPA exclusively on the T class considering the other two classes (B and G) as background.

According to their geometrical attributes MSPA categories can be grouped into four main clusters: (i) core, (ii) islet, (iii) connectors (bridges and loops), and branches, and (iv) edge (McGarigal, 1995).

2.2 Community Sampling and Assessment of Forest Composition Drivers

The here employed sampling design was influenced by both the small size and the location of available and accessible forest fragments, making a widely scattered position of plots necessary to best represent the landscape (e.g., including differently inclined slopes). Plot locations belonged to privately owned protected areas and farms, for which we obtained the required permits of entry from the corresponding owners. We aimed at collecting data of the different vegetation strata. To this aim we complemented the data from the tree layer census of 20 plots obtained from the Rastrojos project with data from an additional set of 12 plots, autonomously established. We maintained a plot size of 20 × 20 m (0.04 ha) as used in the Rastrojos project (Acuña Cabanzo, 2013; Hurtado-M et al., 2021). In the totaling 32 plots located in six administrative regions of Bogotá D.C. and Cundinamarca we assessed the tree layer, the understory layer, and the epiphyte cover on selected representative trees individuals.

Following the protocol of Hurtado-M et al. (2021), we measured the diameter at breast height (DBH) and visually estimated tree height for every woody plant with basal diameter >5 cm (taken at 5 cm from the ground). This method is quite precise for lower canopies such as the ones of high Andean forests. For the understory assessment, in each plot, we sampled eight quadrants of 1 × 1 m. In each of these quadrants, all vascular plants, including tree seedlings, were surveyed, and mean height and total per species cover were recorded. In Addition, bare soil, leaf litter, bryophytes, lichens, and coarse woody debris covers where visually estimated in each quadrant. To correctly identify the species, plant material was collected and identified with the available literature (Gentry & Vasquez, 1993; Trelease & Yuncker, 1950; or webpages: https://plantasdecolombia.com), by comparison with herbarium specimens, digitized specimens available online (JBB: http://herbario.jbb.gov.co/especimen/avanzada; COL: http://www.biovirtual.unal.edu.co/en/collections/search/plants), or with additional help from local experts. Specimens were deposited in the herbarium of the Jardín Botánico de Bogotá José Celestino Mutis (JBB). To estimate the epiphyte cover we sampled 40 randomly selected trees in each plot. We assigned a categorical cover class (from 0 to 3) to each of the five major epiphyte groups (bryophytes, lichens, ferns, bromeliads, and orchids), separately for trunk and canopy branches for each of the selected phorophytes. Presence and abundance of all woody species (trees, shrubs, and lianas) were compiled for each plot. Relative abundance was calculated for the mean cover of the tree and understory layer.

For each plot, environmental variables were compiled from different sources in QGIS. Elevation, slope, and aspect (northness and eastness) were derived from an Aster Digital elevation model of the study area; for this, ASTGTM2_N04W075, ASTGTM2_N05W075, ASTGTM2_N05W074, and ASTGTM2_N04W074 data products were retrieved from the NASA Land Processes Distributed Active Archive Center (LP DAAC; https://lpdaac.usgs.gov/tools/data-pool, NASA/METI/AIST/Japan Spacesystems and U.S./Japan ASTER Science Team, 2009). Mean annual precipitation and mean and maximum temperature data for the years 1981–2010 were downloaded from each meteorological station that was the closest to each plot (http://www.pronosticosyalertas.gov.co/mapas-graficos-tiempo-clima/indicadores-climatologicos). Mean population density was calculated in two buffers (radius 1 km and 5 km) around the plots' centroids from the worldpop database for South America at 1 ha resolution (https://www.worldpop.org; Sorichetta et al., 2015).

We calculated functional diversity indices to provide further information on the position and relative abundances of species in a multidimensional functional space, allowing for detailed insights into the mechanisms bridging biodiversity and ecosystem function (Villéger et al., 2008). To obtain plot-level functional diversity indices, we measured three leaf functional traits (specific leaf area: SLA, leaf thickness: LT, and leaf dry matter content: LDMC) for each tree species, according to widely accepted protocols (Perez-Harguindeguy et al., 2016). Wood density for our species pool (WD) was obtained from Rodríguez-Alarcón et al. (2018) when available and complemented with the global wood density database (Chave et al., 2009) using the R package *biomass* (Réjou-Méchain et al., 2017). The final trait database was completed with data from the Rastrojos project and J. M. Posada (unpublished results). To calculate functional diversity we used SLA, LDMC, LT, WD, maximum recorded height in the study plots, and life form (tree or shrub). The traits were log10-transformed to reduce skewness, before calculating functional divergence, functional dispersion, functional richness, functional evenness, and Rao's quadratic entropy (FDiv, FDis, FRic, FEve and Rao's Q) as recommended by Villéger et al. (2008), with the R package *FD* (Laliberté et al., 2014). We also specified "corr + lingoes, m = 3" to reduce dimensionality. Using the 'PD' function in the R package *picante* (Kembel et al., 2010) we computed the functional diversity (FD) index with the R function 'hclust' as the total branch length of a functional dendrogram generated on a distance matrix of traits (Petchey & Gaston, 2002).

To obtain land cover maps for each plot locality, a Landsat 8 scene was downloaded from the US Geological Survey and analyzed in QGIS with the *SCP* plugin (Congedo, 2016). Landscape metrics were obtained for the forest class within a 1000 m diameter buffer zone around the plots with the *LecoS* plugin for QGIS (Jung, 2013) related to the shape, size, number, and position of land-use patches within a landscape. Distance to nearest roads (main, secondary, or track) was calculated with the *NNJoin* plugin on a road shapefile (DANE, 2018). Distances to closest houses or tracks were manually measured on the map in QGIS. Presence/absence of cattle and cultivated fields was surveyed in the field within concentrical buffers (0 m, 50 m, 100 m, or 500 m). Minimum age of the forest cover of each plot was estimated through the visual analysis of the 43 aerial pictures previously acquired from the IGAC. The minimum age was assigned based on the oldest documented continuous occurrence of closed forest at the plot centroid in available images.

Moreover, based on published information (Cuatrecasas, 1958; Cleef, 1981; Van der Hammen, 1998) and field observation, tree layer species were grouped into 'late successional', 'fast-growing' or 'exotic'. Exotic species' understory cover was also recorded. Species number and relative proportion of specimens (for trees) and the percentage of cover (for the understory) of exotic species were used as indicators of disturbance versus conservation. DBH and height variance was then computed in all plots, together with the total number of tree individuals and stems, mean stems per tree value and percentage of large trees (DBH > 30 cm) out of the total number of trees. Average understory height and cover was calculated, as well as average epiphyte cover.

Plot-based taxonomical alpha-diversity (Shannon's diversity, Simpson's and Pielou's evenness) was calculated with the R package *vegan* (Oksanen et al., 2007). To investigate structure of the phylogenetic community we computed phylogenetic diversity metrics based on a published angiosperm super tree (Phylomatic tree R20120829, available at https://github.com/camwebb/tree-of-trees/blob/master/megatrees/R20120829.new). First, we generated a regional pool tree with the Phylomatic webtool (Webb & Donoghue, 2005), to which we assigned branch lengths with the *bladj* algorithm (Wikström et al., 2001) in the software Phylocom 4.2 (Webb et al., 2008). Then, we computed phylogenetic diversity (PD), mean pairwise distance (MPD), mean nearest taxon distance (MNTD), and their standardized counterparts (sesPD, sesMPD, and sesMNTD) for both trees and the understory layer with the *picante* R package (Kembel et al., 2010). We calculated both abundance-weighted and standardized PD metrics, the latter expressing the difference between observed and average value in units of standard deviation (SD). Positive values indicate phylogenetic overdispersion (co-occurring species are more distantly related than expected by chance) and negative values phylogenetic clustering (co-occurring species are more closely related than expected by chance).

Standing tree layer biomass was calculated with the R package *biomass* (Réjou-Méchain et al., 2017). To deal with missing data, we performed imputation of missing DBH values (since missing data was below 5%) with the R package *mice* (Van Buuren & Groothuis-Oudshoorn, 2011). For height measurements, a regional diameter-height model was built also using the *biomass* R package. Error propagation was carried out using the AGBmonteCarlo function. Wood density error (errWD) was obtained with the 'getWoodDensity' function as prior values on the uncertainty on wood density values, obtained using the mean SD at the species, genus, and family levels of taxa having at least 10 wood density values in the Global Wood Density database (Chave et al., 2009). Height error (errH) was calculated as the RSE resulting from the local height-diameter models, as in Réjou-Méchain et al. (2017), and diameter measurements propagation error (Dpropag) was set to "chave2004" (Réjou-Méchain et al., 2017). To calculate mean aboveground biomass (AGB) and the 95%-credibility interval following the error propagation we used the allometric equation from Chave et al. (2014). Lastly, mean AGB per tree was calculated by dividing the total AGB value of each plot by the number of tree individuals.

2.3 Statistical Data Aanalysis

To filter the variables with the aim to retain only dominant predictor variables, we first performed a nonmetric multidimensional scaling (NMDS) on plot-based species data in *vegan* using the 'mds' function with Bray-Curtis distances, specifying three as maximum number of axes. Secondly, we fitted all the variables with the 'envfit' function and then examined their NMDS scores, to retain only the variables that were strongly correlated with species composition. The NMDS filtering was carried out separately for the tree and the understory layer. After this preliminary step, we obtained a subset of variables that correlated with the main axes above a given threshold ($Rsq > 0.35$). Then, we checked for correlation among selected variables ($r > |0.7|$), with a Variance Inflation Factor (VIF) analysis (setting the threshold to 10).

A partial redundancy analysis (pRDA) was fitted using the *vegan* R package, both for the tree and the understory layer. We also used the 'ordistep' function in the *vegan* package for an automatic stepwise model building for constrained ordination methods (Oksanen et al., 2007). Complementary, to investigate spatial autocorrelation and its impact on explained variance, we established "locality" as condition. We performed the pRDA on Hellinger-transformed community data, as recommended by Legendre and Gallagher (2001). To fit GLMs we put together a set of predictors inspecting the NMDS and pRDA graphs and retaining uncorrelated response variables based on the direction of the arrows in the graphs. We further removed highly correlated and spatially autocorrelated variables ($r > |0.7|$) and performed a VIF analysis (setting the threshold to 10), to gather a small set of predictor variables. In addition, we selected meaningful response variables, including diversity metrics and indicators (Table 1).

For each of these response variables, we tested the best conditional distribution and then we used the 'regsubsets' function in the *leaps* R package (Lumley & Lumley, 2013) with four as maximum number of predictors for automated selection of the best fitting GLM. A "log" link was specified in the family argument and predictors were scaled. Since canopy characteristics may be influencing understory response variables, we first investigated species diversity indices for trees and the understory (species richness, Shannon and Pielou's), phylogenetic diversity and structure (trees sesPD, sesMPDABU, sesMNTDABU; understory sesPD, sesMPDABU, sesMNTDABU), functional diversity (FDiv, FRic, FEve, FDis), and aboveground biomass per plot (AGBplot) as response variables. Secondly, we fitted another round of GLMs using AGB and understory diversity metrics as responses, with tree diversity indices and direct disturbance indicators as secondary predictors (Table 1).

Lastly, we analyzed conjunctly plot-based species data and all retained variables for each locality to link forest cover and connectivity trends to assessed community composition in each locality. We performed an indicator species (IS) analysis for each locality's tree and understory layer, with the function 'multipatt' of the *indicspecies* R package. We then plotted relevant cover and structural connectivity variables by locality as a basis to discuss the underlying differences in relation to each locality's forest cover trend and history.

Table 1 Summary of the variables selected through NMDS and pRDA after removal of highly correlated, spatially autocorrelated variables and VIF analysis

Predictors		Responses	
Abbreviation	Name of the variable	Abbreviation	Name of the variable
north	Northness	***Tree layer diversity***	
slope	Slope	TSR	Trees species richness
mean_prec	Mean annual precipitation	TPielou	Trees Pielou's evenness
mean_T	Mean annual temperature	Tshann	Trees Shannon's diversity[a]
cult_50m	Cultivated fields in 50 m buffer	TsesPD	Trees standardized phylogenetic diversity[a]
cattle	Presence of cattle		
logg	Logging signs	TsesMPDABU	Abundance-weighted trees standardized mean pairwise distance[a]
protected	Protection status of the area		
path_dist	Distance to the closest path	TsesMNTDABU	Abundance-weighted trees standardized mean nearest taxon distance[a]
house_dist	Distance to the closest house		
track_dist	Distance to the closest track		
edge	Located at the edge of the forest fragment	Fdis	Functional Dispersion[a]
		Feve	Functional Evenness[a]
age	Minimum age of the plot	Fdiv	Functional Divergence[a]
cattle_100	Presence of cattle in 100 m buffer	Fric	Functional Richness[a]
median_patch	Median forest patch size in 1 km buffer	AGBplot	plot above-ground biomass[a]
n_inv_sp_T	Number of invasive species of trees	***Understory diversity***	
n_FST_sp_T	Number of fast-growing species of trees		
n_FST_ind_T	Number of individuals of fast-growing species of trees	HSR	Understory species richness
		Hpielou	Understory Pielou's evenness
%_n_CON_spT	Percentage of species of trees associated with conserved forests	Hshann	Understory Shannon's diversity index
n_large_trees	Number of trees with DBH > 30 cm	HsesPD	Understory standardized phylogenetic diversity
n_stems	Number of stems	HsesMPDABU	Abundance-weighted understory standardized mean pairwise distances
n_trees	Number of trees		
n_ sp > 10DBH	Number of species with DBH > 10 cm	HsesMNTDABU	Abundance-weighted understory standardized mean nearest taxon distances

[a]Variables used first as response and later as secondary predictors in the GLMs for understory diversity metrics responses

3 Results and Discussion

3.1 Land Cover and Connectivity Temporal Analysis

The accuracy results of the supervised land cover classifications were overall satisfactory, retrieving relatively high OA values (i.e. > 0.9). Class-specific accuracy values (UA and PA) were satisfactory for T and G classes. Class B had satisfactorily UA values but had low PA values in some cases due to underlying omission errors. That means that the classification may show a tendency to underestimate the cover bare ground/buildings, roads, etc., in some cases; but it showed high accuracy for the dominant classes of trees (T) and grasslands (G).

For the period between 1940 and 2007, we detected a general increase of forested surface in five of the six study localities (Fig. 2). This overall forest recovery in the Bogotá high plain, first described by Etter (2002) for the years 1940–1996, was also recently corroborated by studies in areas north of Bogotá for the period 1985–2015 (Rubiano et al., 2017), and in general for the Colombian Andes for the period 2001–2010 (Sánchez-Cuervo et al., 2012). Here, we added two localities in the southern area of the Bogotá high plain, further supporting those findings. Pasquilla was the only locality displaying a net loss of forested surface (−6.77 ha). Oppositely, Torca displayed the highest increase in forest (+289.40 ha). A notable forest increase was detected in Guasca, Guatavita and Tabio (+141.19 ha, +115.61 ha and +99.64 ha, respectively).

Enhanced landscape connectivity together with an increase in surface of forested fragments is pivotal to restore freedom of movement and availability of resources to the local flora and fauna (Bierregaard et al., 1992; Dunning et al., 1992; Ferraz et al., 2007; Aubad et al., 2010; Tinoco et al., 2013), and can aid in limiting the ecological damage by past deforestation and land degradation (Guariguata & Ostertag, 2001; Chazdon, 2008; Sánchez-Cuervo et al., 2012). Accordingly, in the MSPA analysis, the *core* category was found to be the most abundant after the *background* category, increasing its net value through time in all localities (Fig. 3). The most striking example for an increase of core area was found in Torca between 1940 and 1962 (+122%). However, a decreasing core was highlighted for particular time frames in Pasquilla (1941–1961), Guatavita (1958–1962), and Torca (1978–1993). The *islets* proportion diminished through time in every locality, except for Sumapaz and Tabio, where it increased slightly after the 1990s (+23%) and markedly (+85%) in the 2000s.

Connectors (loops, bridges, and branches), a category including potential connecting elements (bridges) bridging core areas (Vogt et al., 2009), diminished only in Guasca and Pasquilla. The *edge* group fraction (*perforations* and *edge* classes), an indicator of increasing edge effect (Vogt et al., 2009; De Paula et al., 2016), steadily increased in Guatavita, Tabio, and Sumapaz.

Dramatic deforestation takes place mostly in highly economically exploited zones (Hecht, 2010), e.g. in some regions of the Amazon basin, in the Andean lowlands (Armenteras et al., 2006, 2011; Etter et al., 2006) or in southern Ecuador

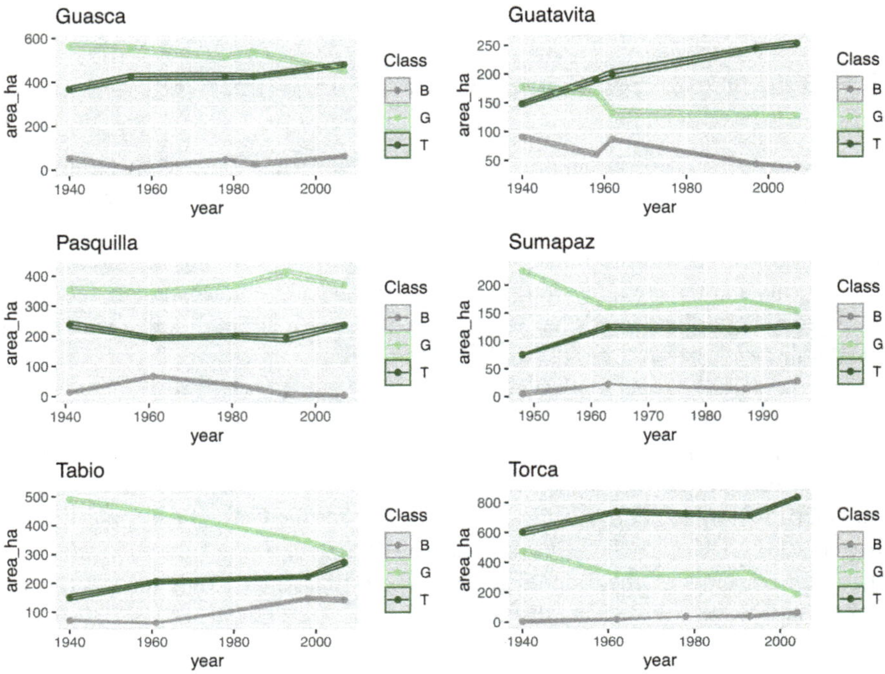

Fig. 2 Land cover changes for the six localities under study; forest cover (T), grasses, pastures, and agriculture cover (G), and bare ground, buildings, and local roads (B)

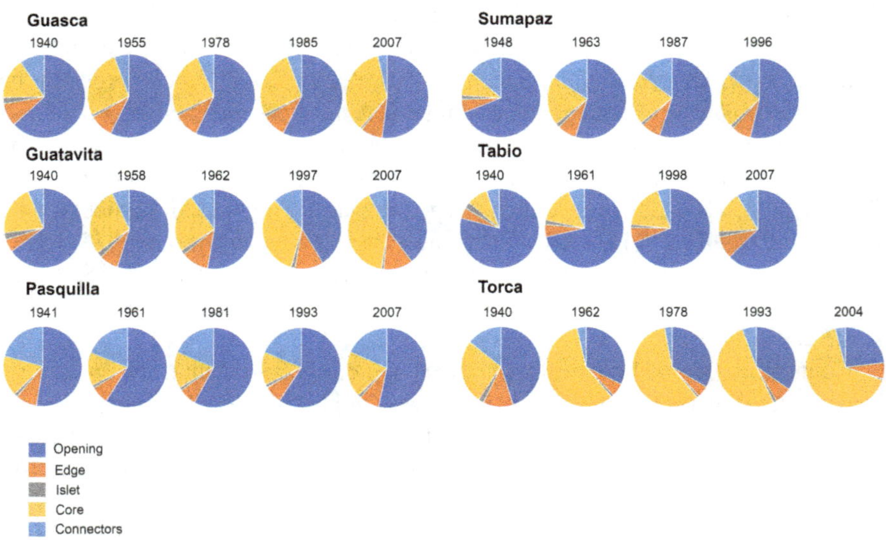

Fig. 3 Proportions of the MSPA classes through time, for each locality

(Curatola Fernández et al., 2015; Tapia-Armijos et al., 2015), where population increase, linear infrastructure growth, and inadequate protection regimes favor timber and firewood exploitation (Tapia-Armijos et al., 2015). In the surroundings of heavily populated Andean cities, several factors play a part in forest regrowth including the inaccessibility of some mountainous forest fragments (Etter, 2002; Rubiano et al., 2017). In particular, in our study system, the underlined forest recovery is mostly caused by rural depopulation leading to the abandonment of rural practices (Aide & Grau, 2004; Rubiano et al., 2017; Etter, 2002; Sánchez-Cuervo et al., 2012).

3.2 Community Structure and Composition in Relation to Anthropic Factors

In the plot-based community assessment we recorded 9841 tree individuals corresponding to 98 taxa from 64 genera and 41 families. For the understory we recorded 326 taxa from 174 genera and 82 families. Tree species richness was comparable to the 90 taxa sampled by Rodríguez-Alarcón et al. (2018) in a similar study area, also close to Bogotá.

Species richness numbers for the understory layer are not available for a comparison with our results, since this component is often overlooked, although our findings are similar to what Van der Hammen (1998) reported for similar sized plots in high Andean forest in the watershed of the Rio Bogotá (50–60 species, 20–30 of which were trees and shrubs). Moreover, our results further corroborate that high Andean forests nearby Bogotá are far away from being floristically and structurally homogeneous, displaying a high species diversity, and also a high beta diversity, mainly in the herbaceous layer.

As to the drivers of community composition, from the 25 and 38 variables with $R^2 > 0.35$ retained after the NMDS for the tree layer and the understory respectively, we obtained a subset of 10 variables for the tree layer (elevation, presence of logging, Shannon's landscape diversity, mean temperature, presence of cattle, presence of cultivated fields in a 100 m buffer, minimum fragment age, distance to roads, and presence of cattle in a 50 m buffer) and 10 variables for the understory (elevation, number of trees, edge density, Shannon's landscape diversity, mean tree aboveground biomass, presence of cultivated fields in a 500 m buffer, minimum fragment age, distance to roads, people density in a 5 km buffer, and fragment size).

Tree layer and Understory pRDA

We retained seven variables for the tree layer (elevation, presence of logging, Shannon's landscape diversity, mean temperature, presence of cattle, minimum fragment age, and distance to roads) and seven variables for the understory (elevation, edge density, Shannon's landscape diversity, mAGBT, presence of cultivated fields in a 500 m buffer, distance to roads, and people density in a 5 km buffer) (Fig. 4).

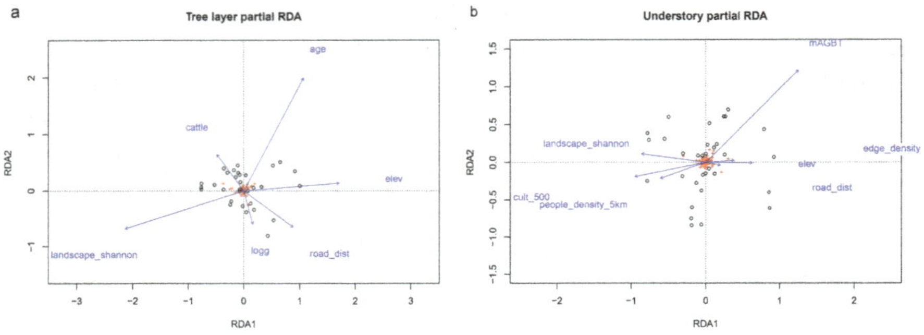

Fig. 4 **Partial RDA** (pRDA) of tree and understory layer. The tree layer pRDA had an R^2 of 0.23 and adjusted R^2 of 0.17. The proportional conditional variance explained was 0.45, while the constrained explained variance was 0.24. The unconstrained explained variance was 0.31. For the understory, the pRDA had an R^2 of 0.26 and adjusted R^2 of 0.11. The proportional conditional explained variance was 0.29, while the constrained explained variance was 0.26. The unconstrained explained variance was 0.45

In both the tree layer and understory ordination the high proportion of conditional explained variance (linked to the "locality") underlining the differences in species composition between the study sites probably is due to the topographic variation across small scale gradients in high Andean ecosystems, which furthers species turnover and structural differences (Homeier et al., 2010; Lopez & Duque, 2010).

In general, our findings exhibit the impact of both environmental parameters and anthropic activities-related variables on forest fragments' structure and diversity. The thorough variable selection process applied in this study underpins a limited suite of factors with elevation, mean temperature and relative humidity as environmental variables, and presence of cultivated fields and cattle in the immediate surroundings of the plots, population density, Shannon's landscape diversity, and forest edge density as disturbance-related variables.

To build the GLMs 15 primary predictors, 17 secondary predictors, and 17 responses were used (Tables 1 and 2).

Concerning disturbance-related predictors, several important predictors already reported in the literature were also retained in our multi-step variable selection process, i.e., minimum age of the fragment, distance to houses, edge effect, and within-plot presence and nearby presence of cattle and cultivated fields. Regarding forest fragmentation metrics, the only one which was retained as relevant for community diversity, was median forest patch size, as also reported by Fahrig (2003).

Among selected responses, tree layer diversity metrics were not particularly sensitive, with only one GLM with a good fit, linking lower tree layer diversity (Shannon's) and AGB to an increase in slope. Slope-related factors such as soil erosion, water drainage and thus drought may act as environmental filters limiting the number and aboveground productivity of taxa (Stegen et al., 2011; Parton et al., 2012). However, the link between higher slope and lower AGB could also be due to the different success of agricultural practices or timber extraction between more or

Table 2 Best fitted GLMs with p values <0.0001, for variable abbreviations see Table 1

			Coefficients				Pseudo R^2	
Response variable	Best model	Predictor variables	Estimate	*SE*	t value	p	Mc Fadden	Nagelkerke
Tshann	Tshann ~ slope+ mean_T + house_dist+ cult_50m	(Intercept)	0.704	0.025	27.699	**<0.001**	**0.753**	**0.815**
		slope	−0.087	0.027	−3.211	**.003**		
		mean_T	−0.027	0.029	−0.935	.358		
		house_dist	−0.047	0.026	−1.822	.080		
		cult_50m	−0.040	0.029	−1.395	.174		
Hshann	Hshann ~ track_dist+ mean_prec + age+ house_dist	(Intercept)	0.896	0.022	40.441	**<0.001**	**0.670**	**0.760**
		track_dist	0.069	0.029	2.379	**.025**		
		mean_prec	0.088	0.029	3.061	**.005**		
		age	−0.105	0.024	−4.389	**.0001**		
		house_dist	−0.035	0.023	−1.523	.139		
AGB	AGBplot ~ slope+age + catle+ cult_50m	(Intercept)	1.715	0.072	23.734	**<0.001**	0.150	**0.526**
		slope	−0.173	0.078	−2.219	.036		
		age	0.317	0.078	4.058	**.0003**		
		cattle	0.232	0.098	2.370	**.025**		
		cult_50m	−0.231	0.096	−2.395	**.024**		

(continued)

Table 2 (continued)

			Coefficients				Pseudo R^2	
Response variable	Best model	Predictor variables	Estimate	*SE*	*t* value	*p*	Mc Fadden	Nagelkerke
Hshann	Hshann ~ FDis+FRic + n_stems+ n_inv_sp_T	(Intercept)	0.898	0.025	35.825	**<0.001**	**0.371**	0.474
		FDis	−0.076	0.035	−2.180	**.0382**		
		FRic	0.078	0.031	2.540	**.0172**		
		n_stems	0.031	0.025	1.262	.218		
		n_inv_sp_T	−0.047	0.029	−1.626	.116		
HsesMPDABU	HsesMPDABU ~ FDiv+FRic + n_sp.10DBH+ TsesMPDABU	(Intercept)	0.827	0.071	11.605	**<0.001**	**0.261**	**0.582**
		FDiv	0.245	0.082	2.974	**.006**		
		FRic	−0.262	0.085	−3.088	**.005**		
		n_sp.10DBH	0.129	0.079	1.636	.114		
		TsesMPDABU	0.157	0.060	2.634	**.014**		
AGB	AGBplot ~ TsesMPD+n_large_trees + n_trees+%n_CON_sp_T	(Intercept)	1.714	0.075	22.955	**<0.001**	0.152	**0.529**
		TsesMPD	−0.069	0.118	−0.582	.565		
		n_large_trees	0.364	0.112	3.246	**.003**		
		n_trees	0.225	0.104	2.157	**.040**		
		%n_CON_ sp_T	0.299	0.114	2.621	**.014**		

The signs of coefficients and *t* values are key to understand the effect of the predictor variable in each model. GLMs with either a good fit (McFadden $R^2 > 0.2$) or a high Nagelkerke value (variance explained >0.50) and corresponding significant variables are shown in bold less easily accessible areas in the study region (Etter et al., 2008; Etter & van Wyngaarden, 2000).

Understory Shannon's diversity was sensitive to human disturbance and its increase was linked to an increase in mean precipitation and to distance to tracks. Positive impact of higher precipitation on understory diversity is probably linked to the lack of an environmental filter related to water availability. Distance to tracks can be related to inaccessibility of the plots and may reduce the extraction of orchids and bromeliads that are common in some high Andean forests. Moreover, AGB values were higher in relation to the presence of cattle within the plot but decreased with vicinity to cultivated fields. Notably, the link between AGB and vicinity to cultivated fields suggest a connection between factors promoting fragmentation and lower biomass accumulation. The fact that the presence of cattle was linked to higher AGB values may be peculiar for our study system, where often small and depauperated 'old' forest fragments with some very large trees are present close to farms. Cattle tends to seek shelter from heavy rain or solar exposure in forest fragments with bigger trees, while the agricultural fields are commonly places of work with high human presence, which may facilitate firewood and timber extraction for the fabrication of tools and fences.

This situation is still alarming since grazing and trampling of cattle can disrupt species recruitment and bring about higher soil erosion, leading to reduced productivity over time even in this 'old' last standing fragments (Nepstad et al., 2002). Also, AGB showed a positive relationship to increasing minimum age of the plot and an increasing number of trees and large trees and proportion of late successional tree species was linked to higher AGB values. Nevertheless, the positive relationship between AGB and fragment age, the number of large trees, and proportion of late successional tree species points out that AGB is positively driven by the abundance of slow-growing species that stock large amounts of carbon (Aldana et al., 2017; Álvarez-Dávila et al., 2017). Oppositely, fragmentation and environmental stressors may increase the mortality of large trees (D'Angelo et al., 2004; Laurance et al., 2000), promote the fast growth of gap pioneer species, which have lower wood density, in turn leading to reduced AGB (Laurance & Bierregaard, 1997; De Paula et al., 2016; Berenguer et al., 2014).

Phylogenetic clustering within the understory was also sensitive to human disturbance. Higher tree layer functional richness was linked to higher understory diversity (Shannon's) while on the other hand an increase in plot minimum age and functional dispersion corresponded to lower understory diversity (Shannon's) values. That means that the understory of older forests has less species and tends to be dominated by only few species of herbs and juvenile woody plants. Increasing tree layer functional divergence and tree layer abundance-weighted phylogenetic mean pairwise distances corresponded to higher understory abundance-weighted phylogenetic mean pairwise distances. On the contrary, increasing functional richness in the canopy was linked to lower understory abundance-weighted phylogenetic mean pairwise distances. The fact that the increase of tree functional divergence and tree phylogenetic dispersion was connected to a more phylogenetically dispersed understory, could be due to higher intra-stand heterogeneity, affecting the presence of more diverse microhabitat understory species (Vockenhuber et al., 2011). This is similar to the results of Ampoorter et al. (2014) and Evy et al. (2016), who linked multi-tree species mixtures to a higher understory diversity. The fact that the higher functional richness values are linked to a phylogenetically less dispersed understory might signify that the number of functional groups in the canopy is not the most important and explicative parameter regarding tree functionality, instead the functional divergence better represents the difference between the life history traits of the canopy affecting the understory PD.

3.3 Local Communities' Features in Regard to Local History

Our results from the plot-based community structure and composition assessment are similar to a study that found profound differences in community composition between different municipalities in the region (Hurtado-M et al., 2021), pinpointing the importance of underlying landscape heterogeneity to address the effects of human activities. On the other hand, from the results of our land cover change assessment it is evident that identified locality-specific differences in forest cover trends resulting from local patterns and drivers of deforestation are likely to vary according to the individual history and geographical features of each locality (Hecht, 2010).

3.4 Localities' History and Land Cover Change Impact on Communities' Composition and Structure

In our study, we encompassed places within a short distance but also with different backgrounds. More in detail, for each locality we observed substantial differences in both the analysis of the vegetation and the land cover changes, highlighting the importance of assessing these two aspects at once. In the case of Guasca, the minimum stand age was of 73 ± 7.7 years, the forest displayed intermediate/high DBH and AGB values, as well as one invasive species in the tree layer (*Ulex europaeus*). In addition, *Huperzia hippuridea* was the indicator species for the understory. Moreover, in the sampled forest stands there was the conjoint occurrence of both a high number of fast-growing species and many individuals of few late successional species. Overall tree and understory species belonged to phylogenetically closely related families. Regarding forest cover trends, we underlined a steady increase of forest cover and core areas through the assessed timeframe, but the effect of the severe disturbance brought about by the local limestone caves in the now protected area are evident through the presence of *U. europaeus* and several fast-growing species that had the chance to penetrate into the forest.

In Tabio, the minimum stand age was 56 ± 27.3 years. These stands displayed average values of forest structure metrics, and relatively high numbers of invasive and fast-growing species. Moreover, the stands had a very high tree and understory phylogenetic diversity and functional evenness (Fig. 5). In this locality we highlighted a steady increase of forested areas but a decrease of core areas and an increase of edges and islets since the 1990s. The sampled fragments were heterogeneous in structure and composition: some of them were relatively old and with a complex species assemblage but displaying the signs of constant human intervention. The fact that the medicinal species *Cardamine ovata* was the understory indicator species also points towards a constant presence of human settlements and usage of the forest. In Tabio, it is also quite common that the newly arrived inhabitants, who have recently migrated from the city to the countryside, carry out forest restauration practices, reforestation or silvo-pastoral activities that can aid to explain the recent change of land cover and the increase of islets in the last decades.

In Guatavita, the minimum stand age was 22 years. This locality had the lowest values in structural, compositional and diversity metrics but invasive species were not present. The indicator species for the tree layer was a rare high-Andean late successional species (i.e., *Maytenus laxiflora*). This locality saw an increase of tree cover and core areas during the past 30 years together with steady edges and a decrease of connectors. The sampled forest fragments display many features of relatively undisturbed early stage recovering forested areas.

Moreover, localities such as Guasca, Tabio, and Guatavita, did not experience profound negative effects from the armed conflict in Colombia and overall display a relatively plain topography. Here, the identified forest cover recovery is likely due to a mix of an abandonment of rural activities (Aide & Grau, 2004), a joint increase in low-density settlements of second houses, not adding much stress on the nearby forest fragments (Rubiano et al., 2017), and the ongoing efforts in natural area protection that started in the 1970s (Armenteras et al., 2011).

In Torca, the minimum stand age was 65 ± 19.6 yrs. The fragments were heterogeneous, displaying low to high values of structural metrics and some scattered invasive species. The stands had low to high numbers of late successional species and late successional species' individuals and low phylogenetic diversity and functional richness. *Clusia multiflora*, a typical high Andean species that is capable of enduring perturbation, was the indicator species for the tree layer. In this locality we highlighted an increase of forested areas and core areas starting in the 1990s, after a prior period of decrease. The forest fragments were located in a heterogeneous landscape, very close to the city and the highway but with inaccessible areas of preserved forest. Within a general forest regrowth context, Torca stands out due to its contiguity to the city of Bogotá but still hosts many forested areas with a diverse species composition, and partially a good conservation status, especially in its steeper areas (Hurtado-M et al., 2021; Calbi et al., 2021). In the last century, agriculture and animal husbandry decreased dramatically in Torca, leaving place for the settlement of middle- and upper-class residents that do not engage in rural practices (Benavides Escobar, 2017). In the decade of 1950s, the city suffered important transformations like the official annexation of Usaquén, the administrative locality

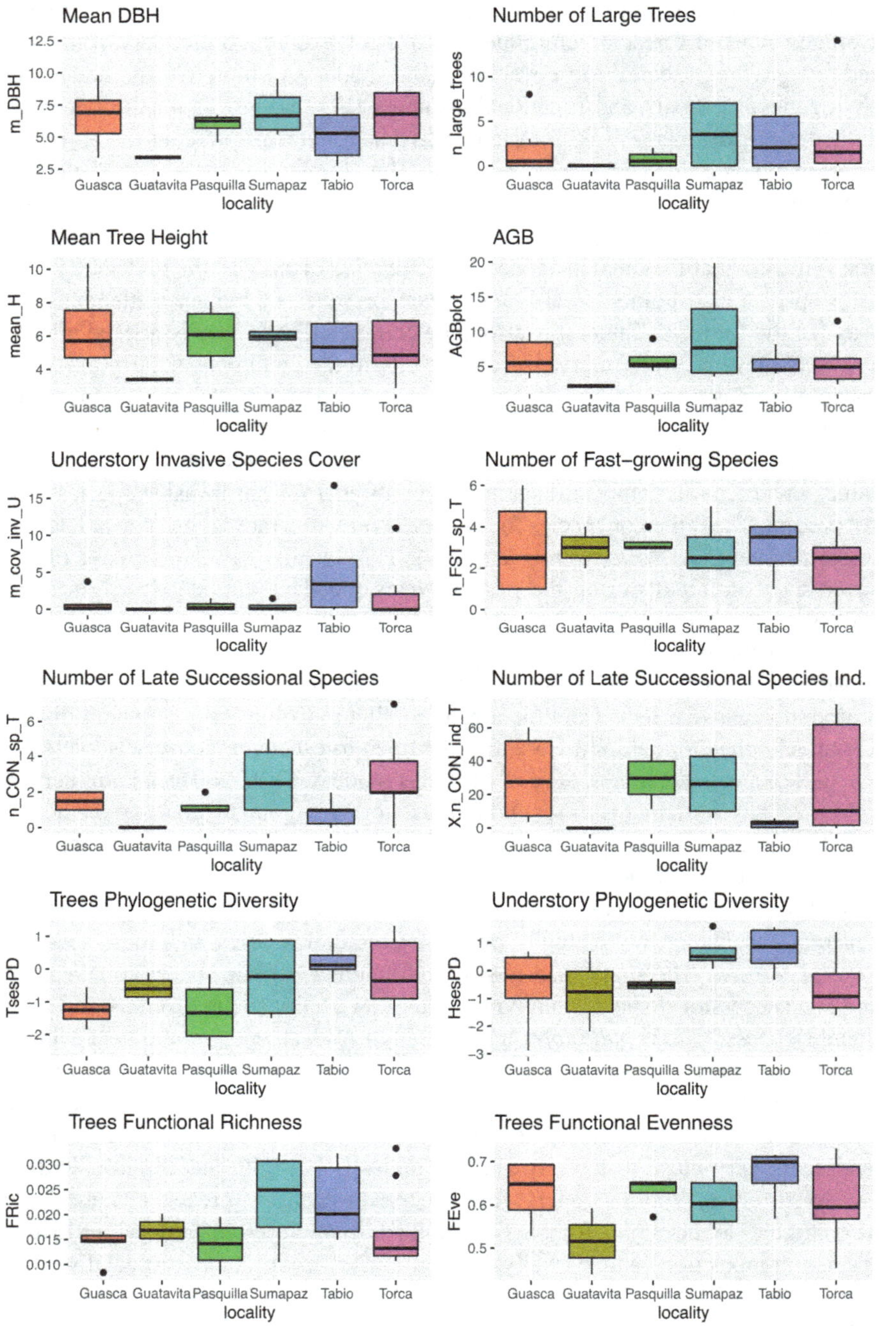

Fig. 5 Boxplots of selected response variables by locality. Acronyms on y-axes are explicitated in the title of each plot

which includes Torca, as part of Bogotá's 'Special District'; Usaquén was previously a separated municipality. The construction of the highway 'Autopista Norte' fragmented the Torca wetland in the high plain and the building of urban households close by promoted the movement of inhabitants from rural to urban spaces in Usaquén. This, jointly with both the implementation of the protected area "Bosque Oriental de Bogotá" in the 1970s and the large portion of very steep areas, lead to forest recovery. Indeed, most of the areas showing forest recovery are found in lower elevations, next to the main motorways, which often were cleared in the past.

For Pasquilla, the minimum stand age was 77 ± 0. Here we find low values of forest structural metrics except tree height and height variance that suggest a high degree of stand heterogeneity. Here, we also found a good amount of late successional individuals belonging to few species and an overall low phylogenetic diversity. The indicator species for the understory is the invasive European *Digitalis purpurea*. These areas display an increase of forested areas and connectors starting in the 90s and the forest fragments are now mostly protected but stand by a continuous agricultural matrix and are still subjected to forest product usage/management by the locals. It still is a well populated rural landscape nearby Bogotá. Pasquilla is the only locality in our study with an overall net forest loss. In the 1940s, the increasing violence in the country and the migration towards the cities brought many residents to Pasquilla which contributed to agricultural exploitation of its surroundings (Julio, 2003). Accordingly, we highlighted an initial decrease in forest cover in Pasquilla, starting from the 1940s, followed by a slow recovery phase starting around the 1980s. Moreover, the shift that took place between 1995 and 2009 from agriculture to animal husbandry is visible in our results with the increased area of size class G. This land use shifts brought about an increase in azonal *páramos* (which are especially affecting the forest cover and its ecological integrity (Hernández-Gómez et al., 2013)) and thus more openings within the forest areas, reachable by cattle, in turn exacerbating forest perforations and reducing the size of fragments, compromising the regenerative potential of the forest (Etter & Villa, 2000; Forman, 2014). Lastly, the observed recovery of forest core areas in 2006 may result from implemented protection efforts, together with a growing interest for environmental commitment by the residents, which led to the establishment of the protected area "Los Encenillales" in 2004 (Hernández-Gómez et al., 2013).

Lastly, in Sumapaz the minimum stand age was 52 ± 18.4 years. Here we find high structural metric values, the highest AGB values, fewer fast-growing species, and many late successional species. Overall, the fragments are characterized by high values of tree and understory diversity. In Sumapaz, the forest cover has been increasing since the 80s and core areas since the 60s, following a period of heavy decline. The stands here display compositional and structural heterogeneity and some of them are very well-preserved ones. Sumapaz also suffered the impact of the 1940s migrations on forest cover changes. In these mountainous areas, the armed conflict begun already in 1920 and caused the dramatical displacement of the local population (Cadavid, 2020). In the results for the locality Sumapaz, accordingly we underline an initial increase in forested areas taking place in the 1950s and 1960s, a possible outcome of rural depopulation in Sumapaz, with the farmers migrating towards the city of Bogotá or other less affected areas as Pasquilla. The increase of islets can be related to the prevalence of cattle raising over agricultural production, since it is most common to leave some trees in the cattle pastures (a type of of silvo-pastoral system), leading to the formation of islets of vegetation, meanwhile it is rather difficult to avoid felling of trees during the mechanical preparation of the land for cultivation.

4 Conclusions and Future Directions

Composition and structure of high Andean forests has rarely been studied in regard to urbanization impacts, and the here presented investigation is one of the first to assess the influence of multiple anthropogenic factors on observed plant communities' dissimilarities in the areas surrounding the metropolis of Bogotá at different geographical scales. Our multiscale investigation on forest cover and structural connectivity change, species composition and structure of high Andean forest fragments in six different administrative localities of the Bogotà high plain underscored the heterogeneity of plant communities and the lack of floristic homogenization processes in the study area, but also the multiple facets of the interactions between disturbance drivers and plant communities. While an overall forest cover recovery was apparent in all localities for the selected period, connectivity pattern and species composition and structure varied largely as the history of each locality varies.

Our study area has been undergoing continuous anthropic disturbance and land cover change for few centuries now. In this context the here underlined forest regrowth and connectivity restoration, can be taken as a positive sign of the region's resilience potential.

Nevertheless, the increase of disturbance related predictors resulted in general in a negative impact on aboveground biomass, while in the understory, higher disturbance was associated with more diversity. Thus, aboveground biomass seems to be a sensitive indicator of human disturbance intensity and was also linked to the community proportion of late successional tree species. These two factors may be successful proxies of anthropic impact instead of diversity metrics which showed ambiguous reactions to disturbance. Loss of aboveground biomass and the disappearance of late successional tree species should be interpreted as a warning, and protection efforts and restoration measures are needed to counteract this fact.

Good evidence is that a high plant diversity can still be hosted in relatively small and isolated high Andean forest fragments, which can also be used as temporary refugial areas for the local fauna within the intensely modified agriculture-urbanization matrix. More efforts to aid the restoration of forest connectivity should be actively promoted by local decision-makers. Future studies targeting high Andean forests should use permanent plot networks to monitor productivity and species turnover through time in relation to ongoing disturbance but also as to local land use history, addressing natural and human-induced causes of plant community changes.

References

Acuña Cabanzo, A. (2013). *Potencial de regeneración de rastrojos y bosques secundarios en la sabana de Bogotá*. [Bachelor's thesis, Pontificia Universidad Javeriana].

Adams, R., & Bischof, L. (1994). Seeded region growing. *IEEE Transactions on Pattern Analysis, 16*, 641–647.

Aide, T. M., & Grau, H. R. (2004). Globalization, migration, and Latin American ecosystems. *Science, 305*(5692), 1915–1916.

Aldana, A. M., Villanueva, B., Cano, Á., Correa, D. F., Umaña, M. N., Casas, L. F., Cárdenas, S., Henao-Diaz, L. F., & Stevenson, P. R. (2017). Drivers of biomass stocks in Northwestern South American forests: Contributing new information on the Neotropics. *Forest Ecology and Management, 389*, 86–95.

Álvarez-Dávila, E., Cayuela, L., González-Caro, S., Aldana, A. M., Stevenson, P. R., Phillips, O., Cogollo, Á., Peñuela, M. C., von Hildebrand, P., Jiménez, E., Melo, O., Londoño-Vega, A. C., Mendoza, I., Velásquez, O., Fernández, F., Serna, M., Velázquez-Rua, C., Benítez, D., & Rey-Benayas, J. M. (2017). Forest biomass density across large climate gradients in northern South America is related to water avail-ability but not with temperature. *PLoS One, 12*(3), e0171072.

Ampoorter, E., Baeten, L., Koricheva, J., Vanhellemont, M., & Verheyen, K. (2014). Do diverse overstoreys induce diverse understoreys? Lessons learnt from an experimental–observational platform in Finland. *Forest Ecology and Management, 318*, 206–215.

Andrade, G. I., Remolina, F., & Wiesner, D. (2013). Assembling the pieces: A framework for the integration of multi-functional ecological main structure in the emerging urban region of Bogotá, Colombia. *Urban Ecosystems, 16*, 723–739.

Anselm, N., Brokamp, G., & Schütt, B. (2018). Assessment of land cover change in peri-urban high andean environments South of Bogotá, Colombia. *Land, 7*(2), 75.

Anselm, N., Rojas, O., Brokamp, G., & Schütt, B. (2020). Spatiotemporal variability of precipitation and its statistical relations to ENSO in the 2128 lCALBI etAL.high Andean Rio Bogotá watershed, Colombia. *Earth Interactions, 24*(3), 1–1.

Arias-Escobar, M. A., & Barrera-Cataño, J. I. (2007). Caracterización florística y estructural de la vegetación vascular en áreas con diferente condición de abandono en la cantera soratama, localidad de Usaquén, Bogotá. *Universitas Scientiarum, 12*(ll), 25–45.

Armenteras, D., Gast, F., & Villareal, H. (2003). Andean forest fragmentation and the representativeness of protected natural areas in the eastern Andes, Colombia. *Biological Conservation, 113*(2), 245–256.

Armenteras, D., Rudas, G., Rodriguez, N., Sua, S., & Romero, M. (2006). Patterns and causes of deforestation in the Colombian Amazon. *Ecological Indicators, 6*, 353–368.

Armenteras, D., Rodríguez, N., Retana, J., & Morales, M. (2011). Understanding deforestation in montane and lowland forests of the Colombian Andes. *Regional Environmental Change, 11*, 693–705.

Aubad, J., Aragón, P., Olalla-Tárraga, M. Á., & Rodríguez, M. Á. (2008). Illegal logging, landscape structure and the variation of tree species richness across North Andean forest remnants. *Forest Ecology and Management, 255*(5–6), 1892–1899.

Aubad, J., Aragón, P., & Rodríguez, M. A. (2010). Human access and landscape structure effects on Andean forest bird richness. *Acta Oecologica, 36*, 396–402.

Avella, A., Torres, S., Gómez, W., & Pardo, M. (2014). Los páramos y bosques altoandinos del pantano de Monquentiva o pantano de Martos (Guatavita, Cundinamarca, Colombia): caracterización ecológica y estado de conservación. *Biota Colombiana, 15*(1), 3–39.

Bader, M. (1999). *A study on the distribution of vascular epiphytes in a secondary cloud forest, Central Cordillera, Colombia*. [Doctoral thesis/ Wageningen University and Research Center].

Bax, V., & Francesconi, W. (2019). Conservation gaps and priorities in the Tropical Andes biodiversity hotspot: Implications for the expansion of protected areas. *Journal of Environmental Management, 232*, 387–396.

Beltrán-G, H. E., & Barrera-Cataño, J. I. (2014). Characterization of Ulex europaeus L. invasions of different ages, as a tool for ecological restoration of Andean forests, Colombia. *Biota Colombiana, 15*(2), 3–26.

Benavides Escobar, M. C. (2017). *Una mirada a la gentrificación. El caso Bogotá* (Doctoral dissertation).

Berenguer, E., Ferreira, J., Gardner, T. A., Aragão, L. E. O. C., De Camargo, P. B., Cerri, C. E., Durigan, M., Oliveira, R. C. D., Vieira, I. C. G., & Barlow, J. (2014). A large-scale field assessment of carbon stocks in human-modified tropical forests. *Global Change Biology, 20*(12), 3713–3726.

Bierregaard, R. O., Lovejoy, T. E., Kapos, V., dos Santos, A. A., & Huntchings, R. W. (1992). The biological dynamics of tropic rainforest fragments: A prospective comparison of fragments and continuous forest. *Bioscience, 42*, 859–866.

Blaschke, T. (2010). Object based image analysis for remote sensing. *ISPRS Journal of Photogrammetry, 65*, 2–16.

Blaschke, T., Burnett, C., & Pekkarinen, A. (2004). Image segmentation methods for object-based analysis and classification. In *Remote sensing image analysis: Including the spatial domain* (pp. 211–236). Springer.

Boer, E. P., de Beurs, K. M., & Hartkamp, A. D. (2001). Kriging and thin plate splines for mapping climate variables. *International Journal of Applied Earth Observation and Geoinformation, 3*(2), 146–154.

Borrelli, P., Armenteras, D., Panagos, P., Modugno, S., & Schütt, B. (2015). The implications of fire management in the Andean páramo: A preliminary assessment using satellite remote sensing. *Remote Sensing, 7*(9), 11061–11082.

Braun-Blanquet, J. (1932). *Plant sociology. The study of plant communities. Plant sociology*. McGraw-Hill.

Buytaert, W., Célleri, R., De Bièvre, B., Cisneros, F., Wyseure, G., Deckers, J., & Hofstede, R. (2006). Human impact on the hydrology of the Andean páramos. *Earth-Science Reviews, 79*(1–2), 53–72.

Cadavid, E. Historia de la Guerrilla en Colombia. Universidad Federal de Juiz de Fora. Available online: www.ecsbdefesa.com.br/defesa/fts/HGC.pdf. Accessed on 20 January 2020.

Calbi, M., Clerici, N., Borsch, T., & Brokamp, G. (2020). Reconstructing long term high Andean forest dynamics using historical aerial imagery: A case study in Colombia. *Forests, 11*(8), 788.

Calbi, M., Fajardo-Gutiérrez, F., Posada, J. M., Lücking, R., Brokamp, G., & Borsch, T. (2021). Seeing the wood despite the trees: Exploring human disturbance impact on plant diversity, community structure, and standing biomass in fragmented high Andean forests. *Ecology and Evolution, 11*(5), 2110–2172.

Cardozo, C. H., & Schnetter, M. L. (1976). Estudiosecologicos en el Paramo de Cruz Verde, Colombia. *Caldasia, II*(54), 85–91.

Castellanos-Castro, C., & Bonilla, M. A. (2011). Grupos funcionales de plantas con potencial uso para la restauración en bordes de avance de un bosque altoandino. *Acta Biológica Colombiana, 16*(1), 175–184.

Chave, J., Coomes, D., Jansen, S., Lewis, S. L., Swenson, N. G., & Zanne, A. E. (2009). Towards a worldwide wood economics spectrum. *Ecology Letters, 12*(4), 351–366.

Chave, J., Réjou-Méchain, M., Búrquez, A., Chidumayo, E., Colgan, M. S., Delitti, W. B., et al. (2014). Improved allometric models to estimate the aboveground biomass of tropical trees. *Global Change Biology, 20*(10), 3177–3190.

Chazdon, R. L. (2008). Beyond deforestation: Restoring forests and ecosystem services on degraded lands. *Science, 320*, 1458–1460.

Cleef, A. M. (1981). The vegetation of the páramos of the Colombian Cordillera Oriental. *Mededelingen Van Het Botanisch Museum En Herbarium Van De Rijksuniversiteit Te Utrecht, 481*(1), 1–320.

Clerici, N., Rubiano, K., Abd-Elrahman, A., Posada Hoestettler, J. M., & Escobedo, F. J. (2016). Estimating aboveground biomass and carbon stocks in periurban Andean secondary forests using very high resolution imagery. *Forests, 7*(7), 138.

Clerici, N., Salazar, C., Pardo-Díaz, C., Jiggins, C. D., Richardson, J. E., & Linares, M. (2019a). Peace in Colombia is a critical moment for Neotropical connectivity and conservation: Save the northern Andes–Amazon biodiversity bridge. *Conservation Letters, 12*(1), e12594.

Clerici, N., Cote-Navarro, F., Escobedo, F. J., Rubiano, K., & Villegas, J. C. (2019b). Spatio-temporal and cumulative effects of land use-land cover and climate change on two ecosystem services in the Colombian Andes. *Science of the Total Environment, 685*, 1181–1192.

Congedo, L. (2016). Semi-automatic classification plugin documentation. *Release, 4*(0.1), 29.

Conrad, O., Bechtel, B., Bock, M., Dietrich, H., Fischer, E., Gerlitz, L. M., Wehberg, J., Wichmann, V., & Böhner, J. (2015). System for Automated Geoscientific Analyses (SAGA) v.2.1.4. *Geoscientific Model Development, 8*, 1991–2007.

Cortés Sánchez, S. P. (2008). Caracterización fisionómica, estructural y florística de algunas comunidades vegetales en la cuenca media del río Tunjuelo. *Pérez-Arbelaezia, 19*, 85–114.

Cortés Sánchez, S. P. (2011). *Caracterización fisionómica, estructural y florística de algunas comunidades vegetales en la cuenca media del río Tunjuelo.*

Cortés-Duque, J., Sarmiento Pinzón, C. E., & Suárez Mejía, A. P. (2013). *Aportes a la conservación estratégica de los páramos de Colombia: Actualización de la cartografía de los complejos de páramo a escala 1: 100.000.* Instituto de Investigación de Recursos Biológicos Alexander von Humboldt.

Cuatrecasas, J. (1934). *Observaciones geobotánicas en Colombia.* Junta para Ampliación de Estudios e Investigaciones Científicas.

Cuatrecasas, J. (1958). Aspectos de la vegetación natural de Colombia. *Revista De La Academia Colombiana De Ciencias Exactas, Físicas Y Naturales, 10*(40), 221–268.

Curatola Fernández, G. F., Obermeier, W. A., Gerique, A., López Sandoval, M. F., Lehnert, L. W., Thies, B., & Bendix, J. (2015). Land cover change in the Andes of Southern Ecuador—Patterns and drivers. *Remote Sensing, 7*, 2509–2542.

DANE Departamento Administrativo Nacional de Estadística (DANE). (2018). Retrieved from: http://www.dane.gov.co/?descarga-marco-geoestadistico-nacional.

DANE Departamento Administrativo Nacional de Estadística (DANE). (2019). Retrieved from https://geoportal.dane.gov.co/?estimaciones-proyecciones

D'Angelo, S. A., Andrade, A. C., Laurance, S. G., Laurance, W. F., & Mesquita, R. C. (2004). Inferred causes of tree mortality in frag-mented and intact Amazonian forests. *Journal of Tropical Ecology, 20*(2), 243–246.

De Paula, M. D., Groeneveld, J., & Huth, A. (2016). The extent of edge effects in fragmented landscapes: Insights from satellite measurements of tree cover. *Ecological Indicators, 69*, 196–204.

Dunning, J., Danielson, B., & Pulliam, H. (1992). Ecological processes that affect populations in complex landscapes. *Oikos, 65*, 169–175.

Escobedo, F. J., Clerici, N., Staudhammer, C. L., & Corzo, G. T. (2015). Socio-ecological dynamics and inequality in Bogotá, Colombia's public urban forests and their ecosystem services. *Urban Forestry & Urban Greening, 14*(4), 1040–1053.

Etter, A. (2002). Multitemporal analysis (1940–1996) of land cover changes in the southwestern Bogota highplain (Colombia). *Landscape and Urban Planning, 59*(3), 147–158.

Etter, A., & van Wyngaarden, W. (2000). Patterns of landscape transformation in Colombia, with emphasis in the Andean region. *Ambio: A Journal of the Human Environment, 29*(7), 432–439.

Etter, A., & Villa, L. A. (2000). Andean forests and farming systems in part of the Eastern Cordillera (Colombia). *Mountain Research and Development, 20*(3), 236–245.

Etter, A., McAlpine, C., Wilson, K., Phinn, S., & Possingham, H. (2006). Regional patterns of agricultural land use and deforestation in Colombia. *Agriculture, Ecosystems & Environment, 114*(2–4), 369–386.

Etter, A., McAlpine, C., & Possingham, H. (2008). Historical patterns and drivers of landscape change in Colombia since 1500: A regionalized spatial approach. *Annals of the Association of American Geographers, 98*(1), 2–2.

Eufemia, L., Bonatti, M., Castro-Nunez, A., Lana, M., Morales, H., & Sieber, S. (2019). Colombia's inadequate environmental goals. *Science, 364*(6439), 444–445.

Evy, A., Federico, S., Harald, A., Lander, B., Sigrid, B., Elisa, C., Andrea, C., Mariangela, F., Kalliopi, R., Nurlaila, S. N., Margot, V., & Kris, V. (2016). Driving mechanisms of overstorey–understorey diversity relation-ships in European forests. *Perspectives in Plant Ecology, Evolution and Systematics, 19*, 21–29.

Fahrig, L. (2003). Effects of habitat fragmentation on biodiversity. *Annual Review of Ecology, Evolution, and Systematics, 34*(1), 487–515.

Ferraz, G., Nichols, J. D., Hines, J. E., Stouffer, P. C., Bierregaard, R. O., & Lovejoy, T. E. (2007). A large scale deforestation experiment: Effects of patch area and isolation on Amazon birds. *Science, 315*, 238–240.

Fjeldså, J., Álvarez, M. D., Lazcano, J. M., & Leon, B. (2005). Illicit crops and armed conflict as constraints on biodiversity conservation in the Andes region. *Ambio: A Journal of the Human Environment, 34*(3), 205–211.

Forman, R. T. (2014). *Land Mosaics: The ecology of landscapes and regions 1995* (p. 217). Cambridge University Press.

Franco, P., & Betancur, J. (1999). La flora del alto Sumapaz (Cordillera oriental, Colombia). *Revista de la Academia colombiana de ciencias exactas, físicas y naturales, 23*, 53–78.

Franklin, J., Rogan, J., Phinn, S. R., & Woodcock, C. E. (2003). Rationale and conceptual framework for classification approaches to assess forest resources and properties. In *Remote sensing of forest environments* (pp. 279–300). Springer.

Garavito, N. T., Álvarez, E., Caro, S. A., Murakami, A. A., Blundo, C., Espinoza, T. B., et al. (2012). Evaluación del estado de conservación de los bosques montanos en los Andes tropicales. *Ecosistemas: Revista Cietifica y Tecnica de Ecologia y Medio Ambiente, 21*(1–2).

Gentry, A. H., & Vasquez, R. (1993). A field guide to families and genera of woody plants of Northwest South America. Conservation International, Washington, DC Givnish, TJ 1999. On the causes of gradients in tropical tree diversity. *Journal of Ecology, 87*, 193–210.

Giraldo-Cañas, D. (2010). Distribución e invasión de gramíneas C3 y C4 (Poaceae) en un gradiente altitudinal de los Andes de Colombia. *Caldasia, 32*(1), 65–86.

Gradstein, S. R. (2008). Epiphytes of tropical montane forests-impact of deforestation and climate change. *Göttingen Centre for Biodiversity and Ecology. Biodiversity and Ecology Series, 2*, 51–65.

Guariguata, M. R., & Ostertag, R. (2001). Neotropical secondary forest succession: Changes in structural and functional characteristics. *Forest Ecology and Management, 148*, 185–206.

Hall-Beyer, M. (2000). *GLCM texture: A tutorial. National Council on Geographic Information and Analysis Remote Sensing Core Curriculum v.3*. Available online: https://ojskrede.github.io/inf4300/additional_resources/uc_calgary_glcm_tutorial.pdf (accessed on 3 March 2019).

Harb Rabia, A., & Terribile, F. (2013). Semi-automated classification of gray scale aerial photographs using geographic object based image analysis (GEOBIA) technique. In *EGU general assembly conference abstracts* (Vol. 15). EGU General Assembly.

Hecht, S. (2010). The new rurality: Globalization, peasants and the paradoxes of landscapes. *Land Use Policy, 27*, 161–169.

Hernández-Gómez, A., Rojas-Robles, R., & Sánchez-Calderón, F. V. (2013). Cambios en el uso del suelo asociados a la expansión urbana y la planeación en el corregimiento de Pasquilla, zona rural de Bogotá (Colombia). *Cuadernos de Geografía: Revista Colombiana de Geografía, 22*, 257–271.

Higuera, E. E., & Cuéllar, M. (2013). Diversidad y caracterización florística de la vegetación natural en tres sitios de los cerros orientales de Bogotá D.C. *Colombia Forestal, 16*(2), 228–256.

Hofstede, R. G., Wolf, J. H., & Benzing, D. H. (1993). Epiphytic biomass and nutrient status of a Colombian upper montane rain forest. *Selbyana, 14*, 37–45.

Homeier, J., Breckle, S. W., Günter, S., Rollenbeck, R. T., & Leuschner, C. (2010). Tree diversity, forest structure and productivity along altitudinal and topographical gradients in a species-rich Ecuadorian montane rain forest. *Biotropica, 42*(2), 140–148.

Huang, Y., Yue, A., Wei, S., Li, D., Luo, M., Jiang, Y., & Zhang, C. (2008). Texture feature extraction for land-cover classification of remote sensing data in land consolidation district using semi-variogram analysis. *WSEAS Transactions on Computers, 7*, 857–866.

Hurtado-M, A. B., Echeverry-Galvis, M. Á., Salgado-Negret, B., Muñoz, J. C., Posada, J. M., & Norden, N. (2021). Little trace of floristic homogenization in peri-urban Andean secondary forests despite high anthropogenic transformation. *Journal of Ecology, 109*(3), 1468–1478.

Isaacs-Cubides, P. J., Trujillo-Ortiz, L. N., & Jaimes, V. (2017). Zonificación de alternativas de conectividad ecológica, restauración y conservación en las microcuencas Curubital, Mugroso, Chisacá y Regadera, cuenca del río Tunjuelo (Distrito Capital de Bogotá), Colombia. *Biota Colombiana, 18*, 70–88.

Jørgensen, P. M., Ulloa Ulloa, C., León, B., León-Yánez, S., Beck, S. G., Nee, M., et al. (2011). *Regional patterns of vascular plant diversity and endemism. Climate change and biodiversity in the tropical Andes* (pp. 192–203). Inter-American Institute for Global Change Research (IAI) and Scientific Committee on Problems of the Environment (SCOPE).

Julio, M. Á. (2003). *Ciudad Bolívar: Gestión ambiente y territorio*. [Bachelor's Thesis/ Universidad Nacional de Colombia].

Jung, M. (2013). LecoS-A QGIS plugin for automated landscape ecology analysis (No. e116v2). *PeerJ PrePrints*.

Kappelle, M., & Brown, A. D. (2003). *Mountain biodiversity in neotropical cloud forests: Distribution, status and trends*. CBD Technical Series No. 8, 54.

Kembel, S. W., Cowan, P. D., Helmus, M. R., Cornwell, W. K., Morlon, H., Ackerly, D. D., et al. (2010). Picante: R tools for integrating phylogenies and ecology. *Bioinformatics, 26*(11), 1463–1464.

Lai, G. Y., Liu, H. C., Kuo, A. J., & Huang, C. Y. (2020). Epiphytic bryophyte biomass estimation on tree trunks and upscaling in tropical montane cloud forests. *PeerJ, 8*, e9351.

Laliberté, E., Legendre, P., Shipley, B., & Laliberté, M. E. (2014). Package 'fd'. Measuring functional diversity from multiple traits, and other tools for functional ecology, *1*, 0–12.

Laurance, W. F., & Bierregaard, R. O. (1997). *Tropical forest remnants: Ecology, management, and conservation of fragmented communities*. University of Chicago Press.

Laurance, W. F., Delamônica, P., Laurance, S. G., Vasconcelos, H. L., & Lovejoy, T. E. (2000). Conservation: Rainforest fragmentation kills big trees. *Nature, 404*(6780), 836.

Legendre, P., & Gallagher, E. D. (2001). Ecologically meaningful transformations for ordination of species data. *Oecologia, 129*, 271–280.

Llambí, L. D., Becerra, M. T., Peralvo, M., Avella, A., Baruffol, M., & Díaz, L. J. (2019). Monitoring biodiversity and ecosystem services in Colombia's high Andean ecosystems: Toward an integrated strategy. *Mountain Research and Development, 39*(3), A8.

Lopez, W., & Duque, M. Á. J. (2010). Beta diversity in Neotropical Mountain forests. *Caldasia, 32*(1), 175–189.

Lozano-Contreras, G., & Schnetter, R. (1976). Estudios ecológicos en el páramo de Cruz Verde, Colombia II. Las comunidades vegetales. *Caldasia*, 53–68.

Lumley, T., & Lumley, M. T. (2013). Package 'leaps'. Regression subset selection. Thomas Lumley Based on Fortran Code by Alan Miller. Available online: http://CRAN.R-project.org/package=leaps. Accessed on 18 March 2018.

McGarigal, K. (1995). *FRAGSTATS: Spatial pattern analysis program for quantifying landscape structure*; US Department of Agriculture, Forest Service, Pacific Northwest Research Station.

Mittermeier, R. A., Myers, N., Mittermeier, C. G., & Robles Gil, P. (1999). *Hotspots: Earth's biologically richest and most endangered terrestrial ecoregions*. CEMEX, SA, Agrupación Sierra Madre, SC.

Myers, N., Mittermeier, R. A., Mittermeier, C. G., Da Fonseca, G. A., & Kent, J. (2000). Biodiversity hotspots for conservation priorities. *Nature, 403*(6772), 853–858.

Nepstad, D. C., Moutinho, P., Dias-Filho, M. B., Davidson, E., Cardinot, G., Markewitz, D., Figueiredo, R., Vianna, N., Chambers, J., Ray, D., & Guerreiros, J. B. (2002). The effects of partial throughfall exclusion on canopy processes, aboveground production, and biogeochemistry of an Amazon forest. *Journal of Geophysical Research: Atmospheres, 107*(D20), LBA-53.

Ocampo-Zuleta, K., & Solorza-Bejarano, J. (2017). Banco de semillas de retamo espinoso Ulex europaeus L. en bordes del matorral invasor en un ecosistema zonal de bosque altoandino, Colombia. *Biota Colombiana, 18*(1 Sup), 89–98.

Oksanen, J., Kindt, R., Legendre, P., O'Hara, B., Stevens, M. H. H., Oksanen, M. J., & Suggests, M. A. S. S. (2007). The vegan package. *Community Ecology Package, 10*(631–637), 719.

Olofsson, P., Foody, G. M., Stehman, S. V., & Woodcock, C. E. (2013). Making better use of accuracy data in land change studies: Estimating accuracy and area and quantifying uncertainty using stratified estimation. *Remote Sensing of Environment, 129*, 122–131.

Parton, W., Morgan, J., Smith, D., Del Grosso, S., Prihodko, L., LeCain, D., Kelly, R., & Lutz, S. (2012). Impact of precipitation dynamics on net ecosystem productivity. *Global Change Biology, 18*(3), 915–927.

Perez-Harguindeguy, N., Diaz, S., Garnier, E., Lavorel, S., Poorter, H., Jaureguiberry, P., et al. (2016). Corrigendum to: New handbook for standardised measurement of plant functional traits worldwide. *Australian Journal of Botany, 64*(8), 715–716.

Petchey, O. L., & Gaston, K. J. (2002). Functional diversity (FD), species richness and community composition. *Ecology Letters, 5*(3), 402–411.

QGIS Development Team. QGIS Geographic Information System, Version 2.18.14-Las Palmas. Open Source Geospatial Foundation. 2017. Available online: http://www.qgis.org/. Accessed on 20 January 2018.

R Core Team. (2018). *R: A language and environment for statistical comput-ing. R Foundation for Statistical Computing*. Retrieved from https://www.r-project.org/.

Radoux, J., & Bogaert, P. (2014). Accounting for the area of polygon sampling units for the prediction of primary accuracy assessment indices. *Remote Sensing of Environment, 142*, 9–19.

Ramírez-Morán, N. A., León-Gómez, M., & Lücking, R. (2016). Uso de biotipos de líquenes como bioindicadores de perturbación en fragmentos de Bosque Altoandino (Reserva Biológica" Encenillo", Colombia). *Caldasia, 38*(1), 31–52.

Rangel Churio, J. O., Van der Hammen, T., Velázquez, A., Lowy Cerón, P. D., Aguilar Puentea, M., & Garzón, A. (1997). *Colombia diversidad biótica II: Tipos de vegetación en Colombia*. Academia Colombiana de Ciencias Exactas, Físicas y Naturales.

Restrepo Abadia, J. F. (2016). *Caracterización vegetal del bosque altoandino en diferentes estados sucesionales de la reserva biológica Encenillo, Guasca-Cundinamarca*. [Master's thesis/ Pontificia Universidad Javeriana].

Réjou-Méchain, M., Tanguy, A., Piponiot, C., Chave, J., & Hérault, B. (2017). Biomass: An r package for estimating above-ground biomass and its uncertainty in tropical forests. *Methods in Ecology and Evolution, 8*(9), 1163–1167.

Rios, A. I. A., Moncada, B., & Lücking, R. (2015). Epiphyte homogenization and de-diversification on alien Eucalyptus versus native Quercus forest in the Colombian Andes: A case study using lirellate Graphidaceae lichens. *Biodiversity and Conservation, 24*, 1239–1252.

Rodríguez Eraso, N., Armenteras-Pascual, D., & Alumbreros, J. R. (2013). Land use and land cover change in the Colombian Andes: Dynamics and future scenarios. *Journal of Land Use Science, 8*(2), 154–174.

Rodríguez-Alarcón, S., Rodríguez-Eraso, N., Pineda-Rincón, I., & López-Camacho, R. (2018). Effects of fragmentation on functional diver-sity associated with aboveground biomass in a high Andean forest in Colombia. *Landscape Ecology, 33*(11), 1851–1864.

Rojas, S. L. (2017). Estructura y composición florística de la vegetación en proceso de restauración en los Cerros Orientales de Bogotá (Colombia). *Caldasia, 39*(1), 124–139.

Rubiano, K., Clerici, N., Norden, N., & Etter, A. (2017). Secondary for-est and shrubland dynamics in a highly transformed landscape in the Northern Andes of Colombia (1985–2015). *Forests, 8*(6), 216.

Salamanca-Fonseca, M., Aldana, A. M., Vargas-Martinez, V., Acero-Gomez, S., Fonseca-Tellez, J., Gutierrez, S., et al. (2023). Effects of urban, peri-urban and rural land covers on plant functional traits around Bogotá, Colombia. *Urban Ecosystems*, 1–10.

Sánchez-Cuervo, A. M., Aide, T. M., Clark, M. L., & Etter, A. (2012). Land cover change in Colombia: Surprising forest recovery trends between 2001 and 2010. *PLoS One, 7*, e43943.

Sánchez-M, R., & Rangel, J. O. (1990). Estudios ecologicos en la Cordillera Oriental colombiana v. Analisis fitosociologico de la vegetacion de los depositos turbosos paramunos de los alrededores de Bogota. *Caldasia*, 155–191.

Saura, S., Vogt, P., Velázquez, J., Hernando, A., & Tejera, R. (2011). Key structural forest connectors can be identified by combining landscape spatial pattern and network analyses. *Forest Ecology and Management, 262*(2), 150–160.

Soille, P., & Vogt, P. (2009). Morphological segmentation of binary patterns. *Pattern Recognition Letters, 30*, 456–459.

Sorichetta, A., Hornby, G. M., Stevens, F. R., Gaughan, A. E., Linard, C., & Tatem, A. J. (2015). High-resolution gridded population datasets for Latin America and the Caribbean in 2010, 2015, and 2020. *Scientific Data, 2*(1), 1–12.

Soto Medina, E., Diaz, D., & Montaño, J. (2021). Biogeography and richness of lichens in Colombia. *Revista de la Academia Colombiana de Ciencias Exactas, Físicas y Naturales, 45*(174), 122–135.

Stegen, J. C., Swenson, N. G., Enquist, B. J., White, E. P., Phillips, O. L., Jørgensen, P. M., Weiser, M. D., Monteagudo Mendoza, A., & Núñez Vargas, P. (2011). Variation in above-ground forest biomass across broad climatic gradients. *Global Ecology and Biogeography, 20*, 744–754.

Stehman, S. V. (2009). Sampling designs for accuracy assessment of land cover. *International Journal of Remote Sensing, 30*, 5243–5272.

Stehman, S. V., & Wickham, J. D. (2011). Pixels, blocks of pixels, and polygons: Choosing a spatial unit for thematic accuracy assessment. *Remote Sensing of Environment, 115*, 3044–3055.

Sturm, H., & Rangel Churio, J. O. (1985). *Ecología de los páramos andinos: una visión preliminar integrada*. Universidad nacional de Colombia.

Tapia-Armijos, M. F., Homeier, J., Espinosa, C. I., Leuschner, C., & de la Cruz, M. (2015). Deforestation and forest fragmentation in South Ecuador since the 1970s–losing a hotspot of biodiversity. *PLoS One, 10*, e0133701.

Tinoco, B. A., Astudillo, P. X., Latta, S. C., Strubbe, D., & Graham, C. H. (2013). Influence of patch factors and connectivity on the avifauna of fragmented Polylepis forest in the Ecuadorian Andes. *Biotropica, 45*, 602–611.

Trelease, W., & Yuncker, T. G. (1950). *The Piperaceae of Northern South America* (Vol. 1). University of Illinois Press.

Valencia, O. M. (2019). *Efectos de la minería en el ensamblaje de anuros de un bosque Alto Andino*. [Master's thesis/Universidad del Valle].

Van Buuren, S., & Groothuis-Oudshoorn, K. (2011). mice: Multivariate imputation by chained equations in R. *Journal of Statistical Software, 45*, 1–67.

Van der Hammen, T. (1998). *Plan ambiental de la cuenca alta del río Bogotá: Análisis y orientaciones para el ordenamiento territorial*. Corporación Autónoma Regional de Cundinamarca Bogotá.

Van der Hammen, T. (2008). *La Cordillera Oriental Colombiana: Transecto Sumapaz*. J. Cramer.

Van der Hammen, T., & Rangel, J. O. (1997). *El estudio de la vegetación en Colombia (Recuento histórico-tareas futuras)*. Colombia diversidad biótica II, 17–57. Universidad .

Vanwalleghem, T., Gómez, J. A., Amate, J. I., De Molina, M. G., Vanderlinden, K., Guzmán, G., et al. (2017). Impact of historical land use and soil management change on soil erosion and agricultural sustainability during the Anthropocene. *Anthropocene, 17*, 13–29.

Vargas, O., & Zuluaga, S. (1986). Clasificación y ordenación de comunidades vegetales de páramo. *Perez-Arbelaezia, 1*(2), 125–143.

Villéger, S., Mason, N. W., & Mouillot, D. (2008). New multidimensional functional diversity indices for a multifaceted framework in functional ecology. *Ecology, 89*(8), 2290–2301.

Vockenhuber, E. A., Scherber, C., Langenbruch, C., Meißner, M., Seidel, D., & Tscharntke, T. (2011). Tree diversity and environmental con-text predict herb species richness and cover in Germany's larg-est connected deciduous forest. *Perspectives in Plant Ecology, Evolution and Systematics, 13*(2), 111–111.

Vogt, P., & Riitters, K. (2017). GuidosToolbox: Universal digital image object analysis. *European Journal of Remote Sensing, 50*, 352–361.

Vogt, P., Riitters, K. H., Estreguil, C., Kozak, J., Wade, T. G., & Wickham, J. D. (2007). Mapping spatial patterns with morphological image processing. *Landscape Ecology, 22*, 171–177.

Vogt, P., Ferrari, J. R., Lookingbill, T. R., Gardner, R. H., Riitters, K. H., & Ostapowicz, K. (2009). Mapping functional connectivity. *Ecological Indicators, 9*, 64–71.

Webb, C. O., & Donoghue, M. J. (2005). Phylomatic: Tree assembly for applied phylogenetics. *Molecular Ecology Notes, 5*(1), 181–183.

Webb, C. O., Ackerly, D. D., & Kembel, S. W. (2008). Phylocom: Software for the analysis of phylogenetic community structure and trait evolution. *Bioinformatics, 24*(18), 2098–2100.

Weigend, M., Rodríguez, E. F., & Arana, C. (2005). The relict forests of northwest Peru and southwest Ecuador. *Revista Peruana de Biología, 12*(2), 185–194.

Wen, C., Zhang, Y., & Deng, K. (2009). Urban area classification in high resolution SAR based on texture features. In J. Zhang, Z. Li, P. Cheng, & Q. Yan (Eds.), *Proceedings of the ISPRS 2009 Congress, Beijing, China, 14–16 September 2009* (pp. 281–285). ISPRS.

Wikström, N., Savolainen, V., & Chase, M. W. (2001). Evolution of the angiosperms: Calibrating the family tree. *Proceedings of the Royal Society of London Series B: Biological Sciences, 268*(1482), 2211–2220.

Young, K. R., & Leon, B. (2007). Tree-line changes along the Andes: Implications of spatial patterns and dynamics. *Philosophical Transactions of the Royal Society B: Biological Sciences, 362*(1478), 263–272.

Zvoleff, A. (2020). Package 'glcm'. Calculate Textures from Grey-Level Co-Occurence Matrices (GLCMs). Available online: https://cran.r-project.org/web/packages/glcm/index.html. Accessed on 25 October 2022.

A Multi-Scale Ecological Approach for the Conservation and Restoration of Venezuelan Andean Cloud Forests

Carlos García-Núñez, Fermin Rada, Ana Quevedo-Rojas, Mauricio Jerez-Rico, Luis D. Llambí, Carlos E. Pacheco-Angulo, Luis E. Gámez, and Emilio Vilanova

1 Introduction

The Andean eco-region of South America is one of the most prominent biodiversity hotspots in the world (Myers et al., 2000). This can be attributed to the existence of contrasting environments that are largely linked to complex latitudinal and altitudinal gradients from tropical to sub-tropical and temperate regions and from the foothills to the snow-capped peaks. This together with the presence on both flanks of the mountain range of extensive depositional plains, make the Andean landscape a mosaic of strongly interlocked ecosystems. The different precipitation patterns, both in amount and temporal distribution (i.e., seasonality) also contribute to generating favorable conditions for high biodiversity in which we can often find at the same elevation ecosystems as contrasting as montane forests, dry thorny shrublands or paramos depending on whether it is a humid slope or an arid pocket (Ataroff & Sarmiento, 2003).

C. García-Núñez (✉)
Instituto de Ciencias Ambientales y Ecológicas (ICAE), Facultad de Ciencias, Universidad de Los Andes, Mérida, Venezuela

Center for Urban and Global Studies (CUGS), Trinity College, Hartford, CT, USA

F. Rada
Instituto de Ciencias Ambientales y Ecológicas (ICAE), Facultad de Ciencias, Universidad de Los Andes, Mérida, Venezuela

Laboratorio de Ecología y Fisiología Vegetal, Universidad de Los Andes (UniAndes), Bogotá, Colombia

A. Quevedo-Rojas
Facultad de Ciencias Forestales y Ambientales, Escuela Técnica Superior Forestal (ETSUFOR), Universidad de Los Andes, Mérida, Venezuela

M. Jerez-Rico · C. E. Pacheco-Angulo
Facultad de Ciencias Forestales y Ambientales, Centro de Estudios Forestales y Ambientales de Postgrado (CEFAP), Universidad de Los Andes, Mérida, Venezuela

N. Clerici (ed.), *Conservation of Andean Forests*,
https://doi.org/10.1007/978-3-031-80805-0_6

An important feature of the Andean Mountain landscape is the presence of diverse and structurally complex forests, which are fundamental in the provision of ecosystem services mainly linked to water provision and regulation and protection against erosion, ability to absorb and store a substantial amount of carbon in the form of soil organic matter (SOM) and plant biomass, and high levels of biodiversity and endemism (Hostettler, 2002; Raich et al., 2006; Bruijnzeel et al., 2010). Among these, tropical mountain cloud forests (TMCFs) occur in the cloud immersion zone which encompasses a lower limit between 1500 and 2750 m asl and an upper limit from 2400 to 3300 m asl depending on latitude (Stadtmüller, 1987). The presence of frequent and persistent cloud immersion and/or fog, results in a distinct environment which influences the structure and function of the vegetation (Fahey et al., 2016). The global extension of TMCFs is about 215,000 km^2 which represents 1.4% of the total area of global tropical forests (Oliveira et al., 2014).

In Venezuela, the Andean ecoregion is geographically located in the western-central zone, covering approximately 10% of the land area of the country. The section of the Andean Cordillera in Venezuela is located from 7° 30′ to 10° 10′ N and 69° 20′ to 72° 50′ W. It extends for about 450 km in a northeast direction, from the Colombian border to the central-western region of Venezuela. As a part of the Andes, the "Cordillera de Mérida" constitutes the longest continuous mountain range in the country and includes the highest elevations (i.e., maximum summit is the "Pico Bolivar" at 4978 m asl) (Ataroff, 2001). The main ecological units that have been defined are submontane forests, semi-deciduous forests, cloud forests, dry evergreen forests, thorny shrublands and páramos (Ataroff & Sarmiento, 2003).

In the "Cordillera de Mérida", cloud forests represent about 20–30% of the vegetation formations and have been classified by its elevation range into lower montane (between 1700 and 2200 m asl) and upper montane (between 2200 and 3000 m asl). Overall, species richness is lower in the upper montane forest (ca. 1.7 times) and there is an increase in epiphyte cover and decrease in tree height, canopy stratification and leaf area index with increasing elevation (Ataroff, 2001) (Fig. 1).

In Venezuela, the studies of mountain forests date back to the beginning of the nineteenth century following the work of naturalists such as Alexander Von Humboldt and Aimé Bonpland, who explored the different altitudinal levels of vegetation on the mountains of the American tropics. Since then, there has been an

L. D. Llambí
Instituto de Ciencias Ambientales y Ecológicas (ICAE), Facultad de Ciencias, Universidad de Los Andes, Mérida, Venezuela

Consorcio para el Desarrollo Sostenible de la Ecoregión Andina-CONDESAN, Quito, Ecuador

L. E. Gámez
Laboratorio de Dendrología, Facultad de Ciencias Forestales y Ambientales, Universidad de Los Andes, Mérida, Venezuela

E. Vilanova
Wildlife Conservation Society (WCS), Bronx, NY, USA

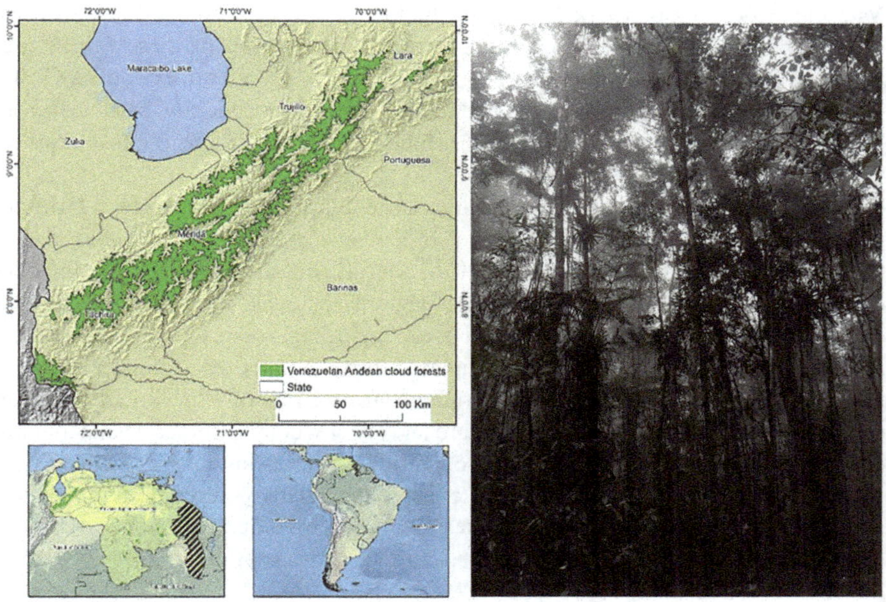

Fig. 1 Left panel: Approximate distribution of cloud forests in the Venezuelan Andes for the year 2020. Map elaborated using the Hansen et al. (2013) product of global forest cover including the gains and losses of cover since the year 2000; Right panel: a typical image of Venezuelan Andean cloud forests, taken at La Carbonera. (Photo: Emilio Vilanova)

increasing number of studies that have contributed to the understanding of cloud forests focusing on aspects related to species richness (i.e., floristic inventories), and different environmental aspects (geology, geomorphology, soils and climate). Systematic studies in the cloud forests of Venezuela, mainly in the Coastal Cordillera (e.g., ecological studies at the Rancho Grande biological station, Henri Pittier National Park, Aragua State) started in the 1950s, with the arrival of the Austrian botanist Volkmar Vareschi, who contributed to the creation of the School of Biology of the Central University of Venezuela (UCV), and its program of Plant Ecology largely with a floristic and functional focus (Lindorf, 2013).

At the same time, the first phytosociological studies of the Andean cloud forests began, mainly in La Mucuy (LM) and La Carbonera (LC), in the Cordillera de Mérida, and was led by two Swiss botanists, Hans Lamprecht and Jean Pierre Veillon, hired by the School of Forestry Sciences of the Universidad de Los Andes (ULA). Since the 1980s, with the consolidation of the different research groups and graduate programs in Venezuela, the scientific knowledge of our mountain forests covered different aspects, from its structure and floristic composition to its functional ecology (i.e., eco-hydrology, productivity and ecophysiological aspects related to carbon, water and nutrient relations at plant scale level). For example, detailed information is available on their distribution, structure and floristic composition (Lamprecht, 1954; Veillon, 1965, 1985, 1989; Huber, 1986a, b; Medina, 1986;

Vareschi, 1992; Bono, 1996; Michelangeli, 1998; Ataroff, 2001, 2003; Ataroff & García-Núñez, 2013). Ecological studies on the functioning of these forests are more limited and concentrated in specific localities of the Cordillera de La Costa (Altos de Pipe and Rancho Grande) and in the Cordillera de Mérida, particularly in La Mucuy (LM) and La Carbonera (LC) (Huber, 1986a, b; Medina, 1986; Ataroff & García-Núñez, 2013).

Studies on ecosystem productivity and nutrient cycling (Fassbender & Grimm, 1981a, b; Grimm & Fassbender, 1981a, b; Delaney et al., 1997; Vilanova et al., 2018) and ecohydrology (Steinhardt, 1979; Ataroff & Rada, 2000) have also been carried out in LM and LC. These studies show a relatively high accumulation of carbon in the aboveground biomass and in the superficial soil, as well as its regulating role of the water balance when compared to pastures as the main replacement system. Likewise, for these localities (mainly for LC) there is information at the community level related to forest structure and species composition associated with environmental factors (e.g., soil water conditions; contrasting environments along gradients determined by tree fall gaps considering relative humidity, temperature and light availability) (Hetsch & Hoheisel, 1976; García Núñez et al., 1995; Acevedo et al., 2001, 2003; Dávila, 2009; Quiroz, 2010; Schwarzkopf et al., 2011; Naranjo, 2013; Quevedo et al., 2015, 2016, 2018).

For many tropical mountain forests there is limited evaluation of the impacts of habitat fragmentation at all ecological levels (Soh et al., 2019), and this is particularly evident in the case of the Andean region, including Venezuela. There is a great value from many of the aforementioned studies in that they could serve as the basis for planning conservation and restoration programs, particularly important for Andean cloud forests (ACFs), which have been subjected to severe deforestation and fragmentation processes in the last 70 years (Rodríguez-Morales et al., 2009). Yet, both theoretical and applied studies in ecological restoration are incipient. For example, ecological restoration studies based on ecophysiological characteristics of shade tolerance of cloud forest native species have been conducted in LC, illustrating the usefulness of this functional approach for restoration planning (Quevedo-Rojas & Jerez-Rico, 2020). Furthermore, this approach is useful for the development and implementation of mechanistic and process-based models to improve our understanding and prediction of ecosystem function under future scenarios of land use and climate change (see Nagy et al., 2023).

In this chapter, we present a review of ecological studies conducted at different spatial scales, from the landscape to plant levels, to describe the main structural and functional characteristics of the Andean cloud forests in Venezuela. In doing so, we place special emphasis on the research carried out at *La Carbonera* (LC) and *La Mucuy* (LM) in the Mérida Cordillera, two of the most intensively studied sites in the Venezuelan Andes, which also include long-term monitoring plots part of the Andean Forest Network (https://redbosques.condesan.org). While reviewing an important research base about how these ecosystems function and their current state, we also explore how this cumulative body of knowledge may provide guidelines for conservation, restoration and adaptation planning. We conclude with a discussion of the main challenges for future research towards an understanding and better prediction of the dynamics of Venezuelan Andean forests under scenarios of global change. We specifically address the following questions:

1. What are the main environmental characteristics of the Andean cloud forests in Venezuela?
2. How forest cover has changed across Venezuelan Andean cloud forests in the last two decades and what are the main drivers of change?
3. How the ecohydrology of Andean cloud forests in Venezuela respond to changes from intact forest stands to pastures?
4. How forest structure in Andean cloud forests respond to light and soil-water gradients, and changes due to fragmentation processes?
5. What are the recent trends in the long-term dynamics of tree turnover and productivity in Venezuelan cloud forests?
6. How an ecophysiological approach could be used as a tool for ecological restoration of Andean cloud forests?

2 Environmental Characteristics of the Andean Forests of Venezuela

Most of the cloud forests in the Venezuelan Andes are composed by mosaics of evergreen forests, with their structure and composition varying depending on abiotic (e.g., elevation, temperature, soil, etc.) and biotic elements (e.g., disturbance regime), and the history of land use and anthropogenic disturbances. Josee et al. (2009) indicate that there are four main types of forests ecosystems in this region, where the Humid Montane Forest and the Seasonal Montane Forest of the Northern Andes are the most prominent and are characterized by a great natural variability and biological diversity (Rodríguez & Rojas-Sánchez, 1999; Hokche et al., 2008; Huérfano et al., 2020). In the case of vascular plants, from the approximately 7500 species that have been identified broadly for the Venezuelan Andes (Huérfano et al., 2020), there are more than 4100 species specifically associated to forests, which in many cases lack in-depth ecological studies. In fact, in the latest version of the Red Book of the Flora of Venezuela (Huérfano et al., 2020), only 196 Andean Forest species are reported with information about their Conservation Status, with the rest categorized as Data Deficient.

2.1 *Site Description: Location, Climate, and Soils*

Because of their ecological representativeness of the wider montane cloud forest ecosystem of the Venezuelan Andes, we use La Carbonera (LC) and La Mucuy (LM) sites as models to describe the environmental characteristics of these ecosystems (Fig. 2). Both sites represent the most intensive studied locations of the cloud

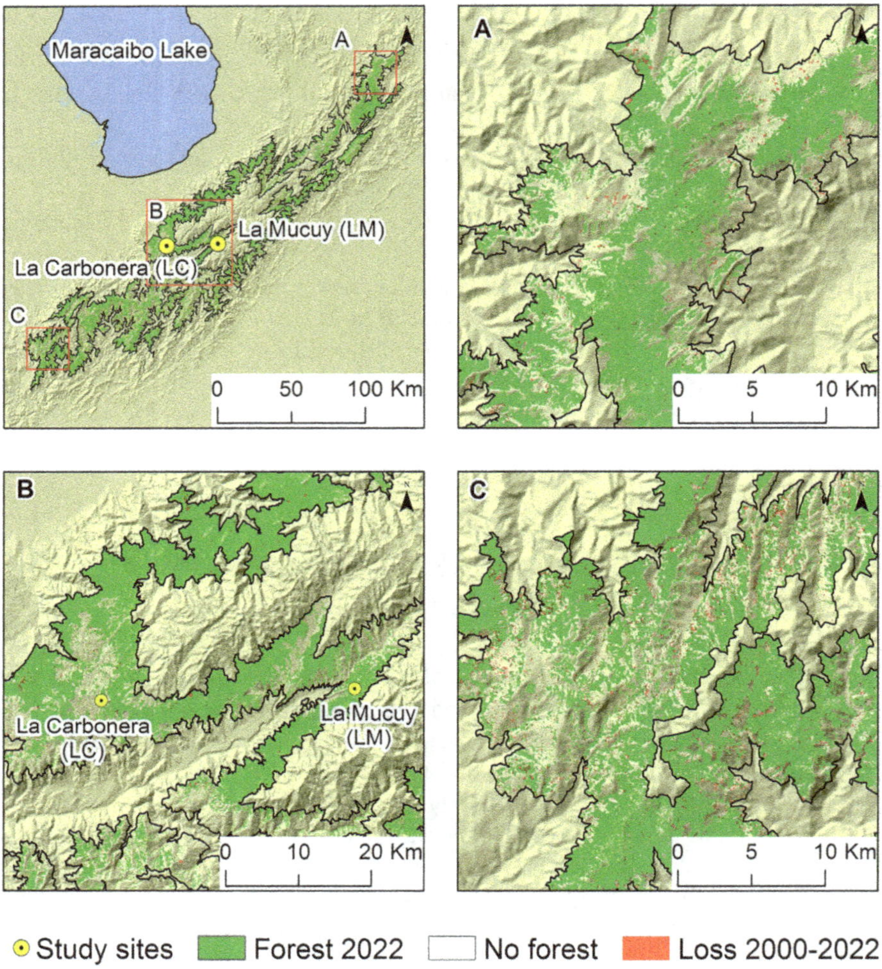

Fig. 2 Dynamics of forest cover in the Venezuelan cloud forests between 2000 and 2022. The different map insets in A, B and C show how cover has changed or remained stable in three locations. In B, the location of LC and LM is included for reference

forests in the Andes of Venezuela and encapsulate the rich biodiversity, soils, and environmental dynamics of the wider region. First, LC (8° 37′ 00″ N, 71° 21′ 00″ W) includes the San Eusebio University Forest (SEUF), the experimental station of the University of Los Andes established in 1950 for silvicultural and ecological research. The elevation ranges from 2200 to 2500 m, and the site receives an annual precipitation between 1400 and 1560 mm with a short dry season (December–February). The annual average temperature is 14.9 °C with little variation throughout the year. The soils are derived from the Colón Cretacean formation, characterized by stratified, massive, black, non-calcareous (i.e., argillaceous) shales with conchoidal fractures. The landscape consists of rounded hills, with mild to steep slopes

(Marquez, 1988). Overall, soils are mainly silty-clay, greatly variable in depth, structure and drainage condition, highly acidic (pH 4.2–5) and with high levels of exchangeable aluminum, low in base saturation, and cation exchange capacity (12–18 meq/100 g soil), high organic matter in the A1 layer (5–8% C content) of a depth of 26–33 cm; presenting an udic moisture regime (Grimm & Fassbender, 1981a; Marquez, 1988).

La Mucuy (LM) (08° 38′ N, 71° 02′ W) is located in the Sierra Nevada National Park. The vegetation extends from the lower limit of the upper montane cloud forest (2200 m asl) to the limit of the forest-paramo line (3065 m asl). Mean annual temperature at the lower limit of the cloud forest is 14 °C, with little variation throughout the year, while at the upper limit is 11 °C. Mean annual precipitation at 2300 m ranges between 2800 and 3400 mm. The precipitation pattern shows two annual peaks, one between April and May, and the other between October and November. Low clouds and fog are the main ecological factor acting on these forests most of the year, producing a limited radiation (4–7 hours of insolation) and very humid climate, with relative humidity between 80% and 100% throughout the day (Ataroff 2003; Linares, 2008). This forest presents a very rugged relief with slopes from medium to very strong (40–80%) and a NW orientation. The geology is complex and diverse, dominated by the "Iglesias Complex", characterized by pre-Cambrian igneous and metamorphic rocks. Soils are entisols (Psamments udipsamments) with loamy, sandy loam and/or clay loam textures, acid (pH 3.5–4.5), with good drainage and high stone content; surface soil has dark colors (brown to black) and high organic matter content 6.6–35.5%; average depth of the soil organic layer is 34 cm (Lamprecht, 1954) (Table 1).

2.2 General Characteristics of the Vegetation

La Carbonera and La Mucuy are both broadly classified as upper montane cloud forest sites. Early studies on the structure and floristic composition of these sites started in the 1950s (Lamprecht, 1954; Veillon, 1965), including subsequent floristic inventories (total cormophytes species, including spermatophytes and pteridophytes) in LC (Vareschi, 1992; Engwald, 1999), and in LM (Linares, 2008). In LM, above 2900 m, canopy height declines (up to 10–15 m in height), with low stem density and with characteristics of a stumpy forest with abundant bamboo species (*Chusquea spencei* Ernst) (Lamprecht, 1954). In general, both study sites share structural features in relation to the number of tree species, tree density and basal area. However, LC displays a taller canopy and higher LAI values (Table 2). Overall, at elevations of 2200–2500 m, the forest communities are characterized by a high diversity of epiphytes, mosses and lichens, structurally composed by two to three layers with irregular open canopies at 20–30 m, a medium stratum (trees 6–15 m in height), and a low stratum of small trees and shrubs (1–6 m in height). The families with the highest number of tree species reported in the Andean forests of Venezuela are:

Table 1 Geographic location and environmental characteristics of two representative Andean cloud forests locations in Venezuela

Localities	Geographic location	Altitude (m asl)	Rainfall (mm)	Temperature (°C)	Radiation/ Insolation (cal cm^{-2}) [h]	Soil characteristics
La Carbonera	8° 37′ N,[b] 71° 21′ W	2200–2500[b]	1400–1560[b]	14.9[b]	296[a] (247–350) [7.2][a]	Mainly silty-clay, greatly variable in depth, structure and drainage condition; highly acidic, low in base saturation and cation exchange capacity, high organic matter.[b]
La Mucuy	8° 38′ N,[c] 71° 02′ W	2200–3000[c]	2800–3400[c]	14.0[c]	300[a] (200–400) [7.2][a]	Loamy, sandy loam and/or clay loam, acids, with good drainage and high stone content and high organic matter content.[c]

Altitudinal range, mean annual rainfall range, mean annual temperature, mean daily solar radiation (range in parenthesis), mean daily solar insolation [in brackets]
Source:
[a]Veillon (1989)
[b]Vareschi (1992)
[c]Ataroff (2001)

Lauraceae, Rubiaceae, Melastomataceae, Myrtaceae, Euphorbiaceae, Piperaceae and Solanaceae (Lamprecht, 1954; Gentry, 1993; Ataroff, 2001; Estrada, 2003; Schneider et al., 2003; Hokche et al., 2008; Linares, 2008; Gutiérrez et al., 2012). In relation to this, Gentry (1993) mentions that the composition of Andean forests is determined by a set of families and genera predominantly of Laurasic origin, which explains the existence of several of these families at elevations above 1500 meters. The most common genera in these cloud forests include: *Alchornea, Axinaea, Billia, Beilschmiedia, Centronia, Cestrum, Clusia, Cyathea, Eugenia, Ficus, Hedyosmum, Miconia, Myrcia, Nectandra, Ocotea, Oreopanax, Palicourea, Psychotria, Solanum, Tetrorchidium* and *Piper* (Estrada, 2003), including some species of palms (e.g., *Chamaedorea pinnatifrons* (Jacq.) Oerst.), shrubs and herbs (Linares, 2008; Quevedo et al., 2015).

Table 2 General characteristics of the vegetation for La Carbonera and La Mucuy in the Venezuelan Andes

Site	N° total spp/families of pteridophytes and spermatophyta	N° tree spp ha^{-1} (≥10 cm dbh)	N° ind. trees ha^{-1} (≥10 cm dbh)	Basal area (m^2 ha^{-1})	Canopy height (m)	Leaf Area Index (LAI)
La Carbonera	369/77[d]	40[a, b]	400–850[a, b]	30–69[h]	20–30[a, b]	4.4–5.4[j]
La Mucuy	208/71[g]	35–48[b, c]	272–723[b, c]	32–61[b, c]	20–25[i]	1.6–2.3[e, f]

Source:
[a]Lamprecht (1954)
[b]Veillon (1965)
[c]Lamprecht, 1980
[d]Engwald (1999)
[e]Ataroff and Rada (2000)
[f]Acevedo et al. (2001)
[g]Linares (2008)
[h]Quiroz (2010)
[i]Schwarzkopf et al. (2011)
[j]Quevedo et al. (2016)

2.3 *The Light Environment and Its Relationship with Forest Structure*

For both study sites, investigations focusing on the light conditions and the relation with forest structure include measurements of global radiation through standard meteorological stations (open field) (Veillon, 1989), and measurements of the light conditions of the forest floor under undisturbed forest, canopy gaps (gap area of 100 and 200 m^2 for LC and LM, respectively) and open field using data loggers with quantum sensors (photosynthetic photon flux density (PPFD), and pyranometers (global solar radiation) (García Núñez et al., 1995; Vilanova, 1996; Cabrera, 1999; Quevedo et al., 2018). Hemispherical photographs have also been used to derive different parameters of the forest floor light environment (e.g., the percentage of Canopy Openness (%CO), a Light Index (LI) calculated from the relative proportions of direct and diffuse light transmitted to the understory, and the Leaf Area Index - LAI) (Acevedo et al., 2001, 2003; Schwarzkopf et al., 2011; Quevedo et al., 2015, 2016, 2018).

In general, these sites are characterized by a daily insolation pattern in which the maximum radiation values are reached between mid-morning and noon, being the forest profusely covered by low clouds and fog from about 14:00 h in the afternoon onwards. Daily measurements showed that an open field site and the center of a canopy gap site (gap area = 200 m^2) received mean maximum values of PPFD between 1561 and 1800 μmol m^{-2} s^{-1} for LC and LM, respectively. In both sites, mean PPFD in the forest understory ranged between 15 and 26 μmol m^{-2} s^{-1} (≅

1–2% of the radiation measured in the open field) with mean maximum values between 214 and 240 μmol m^{-2} s^{-1} during brief sun flecks (García Núñez et al., 1995; Vilanova, 1996; Cabrera, 1999).

The studies on the quantity and quality of the understory light environment have shown that these forests have relatively low canopy openness (% CO ≤ 5%) (Acevedo et al., 2001, 2003; Quevedo et al., 2016). Higher LAI values were measured in LC (5.4 and 4.4 for undisturbed forest and gaps, respectively) compared to LM (1.6 and 2.3) (Table 2). For LC, Quevedo et al. (2016) showed that significant differences were found between seasons, as well as when comparing different gap sizes and under closed canopy conditions (Table 3) (Ataroff & Rada, 2000; Acevedo et al., 2001; Schwarzkopf et al., 2011; Quevedo et al., 2016). Studies carried out in Venezuelan Andean cloud forests show lower LAI values (1.6–5.9; Ataroff & Rada, 2000; Quevedo et al., 2016; Schwarzkopf et al., 2011) when compared to lowland humid forests (6.9–12.0; Grubb, 1977). Low LAI has been attributed to low light input, and as a response to increasing slope angle and elevation (Kitayam & Aiba, 2002; Leuschner et al., 2007; Moser et al., 2007). For example, structural comparisons between cloud forest sites in the "Cordillera de Merida" indicate that LAI is correlated with slope orientation (i.e., LAI showed higher values at south facing plots) (Schwarzkopf et al., 2011).

Table 3 Estimates of the light and canopy structure variables for undisturbed forest (UF) and gaps in San Eusebio University Forest (n = 288, UF = 120, gaps = 168) in each season

Variables	Season	Undisturbed forest	Gaps	P-value
LAI	Dry	5.2 (0.1)[b]	4.0 (0.1)[b]	<0.01
	Wet	5.6 (0.1)[a]	4.4 (0.1)[a]	0.01
%CO	Dry	2.3 (0.1)[a]	4.9 (0.2)[a]	<0.01
	Wet	1.5 (0.1)[b]	4.1 (0.2)[b]	<0.01
%TDIR	Dry	4.0 (0.2)[a]	9.1 (0.5)[a]	0.04
	Wet	3.1 (0.2)[a]	6.5 (0.4)[b]	<0.01
%TDIF	Dry	3.2 (0.1)[a]	8.0 (0.4)[a]	<0.01
	Wet	2.5 (0.2)[a]	6.6 (3.3)[b]	<0.01
SFD	Dry	15.7 (1.0)[a]	34.6 (2.9)[a]	<0.01
	Wet	13.9 (0.9)[a]	34.1 (2.6)[a]	<0.01
SFF	Dry	12.3 (0.7)[a]	21.5 (1.2)[a]	<0.01
	Wet	4.9 (0.3)[b]	7.9 (0.5)[b]	0.12

Table from Quevedo et al. (2016) (CC By Open Access license)
Standard errors in parenthesis. Small letters indicate statistically significant differences
LAI: Leaf area index; %CO: percentage of canopy openness; %TDIR: percentage of transmitted direct light; %TDIF: percentage of transmitted diffuse light; SFD: average duration in minutes of sun-flecks; SFF: sun-fleck frequencies. Different letters in the cells indicate statistically significant differences between seasons ($P < 0.05$). The P-value corresponds to the signification level for the comparisons between UF and gaps

3 Dynamics of Forest Cover in the Cloud Forests of the Venezuelan Andes

As defined earlier, the Andean cloud forests located in western Venezuela are inserted in a high and narrow mountain range (1800–3000 m asl) that extends from northeast to southwest over about 450 km, and from northwest to southeast over a distance of 66 km, covering an area of ~800,000 ha within two long mountain ranges and two long intermontane valleys (Chacón-Moreno et al., 2021) (Fig. 1).

Using information for the year 2000 from the Global Forest Change (GFC) product in its version 1.10 (Hansen et al., 2013), the cloud forest ecosystem covered about 53.6% (429,777 ha) of their potential range in the Venezuelan Andes. Between 2000 and 2022, about 1.1%, equivalent to 8112 ha of cover were lost over this 22-year period, with an annual average of −369 ha (−0.1% per year). As of 2022, the cloud forests covered ~421,666 ha. Annual losses between 2000 and 2022 ranged from as low as 80 ha (2014–2015) to as high as 973 ha (2011–2012). 71.4% (5792 ha) of these losses were concentrated between 1800 and 2200 meters, 20.6% (1674 ha) between 2200 and 2600 meters above sea level, and 8.0% (646 ha) between 2600 and 3000 meters above sea level (Fig. 2).

To understand the main drivers of deforestation in this ecosystem, the criteria proposed by Geist and Lambin (2002) for tropical countries and Pacheco et al. (2011) for Venezuela were considered. With these criteria, a causal model of deforestation was generated to characterize and group deforestation drivers into five major classes: agropastoral expansion, infrastructure extension, logging, mining exploitation, and the presence of social conflicts. These drivers were identified in the literature review and subsequently analyzed and quantified based on the frequency of occurrence in the various sources analyzed and within each study period. This allowed for an analysis of each group of main drivers to determine, based on their percentage, their specific influence on deforestation. This analysis revealed a complex interaction between the primary and underlying causes acting independently or via different combinations varying in frequency and magnitude depending on historical and spatial contexts to drive the observed decline of forest cover in the Venezuelan Andes. Nevertheless, the primary cause with the greatest influence on deforestation has been the expansion of agricultural and cattle raising activities (37.3%,) followed by infrastructure development (29.4%), and logging (23.5%). The most common interaction pattern between primary and underlying causes in the last 20 years stemmed from the occupation of private and public lands for agriculture. This occupation process has been further incentivized by political and institutional factors through agricultural instruments, benefits, and inputs provided by national agricultural institutions, thereby increasing demographic pressure to achieve better economic benefits.

An additional but often overlooked process of change is related to forest degradation, which refers to the gradual deterioration of forest's health, structure, or functioning, often resulting from human activities such as logging, unsustainable land use, or the impacts of climate change, but that does not involve a total loss of cover.

These activities are responsible for the reduction of ecological integrity (Grantham et al., 2020), which ultimately results in a decline of important ecosystem services. Anthropogenic pressures that have contributed to forest degradation, many of which are also responsible for the loss of forest cover, include forest encroachment and fragmentation for agricultural and cattle expansion. In this regard, using the definition of forest degradation proposed by Vancutsem et al. (2021), degraded cloud forests in Venezuelan Andes covered close to 58,000 ha or 13% of the total area up to 2022, an area significantly higher than the reported for deforestation, which highlights the importance of considering degradation as a major process of change in the Venezuelan Andes.

4 Ecohydrological Effects of the Loss of Intact Cloud Forests in the Venezuelan Andes

Cloud forests are of special importance as water collectors and reservoirs, and for the maintenance of stable hydrological systems (Stadtmüller, 1987; Bruijnzeel et al., 2011). An important aspect which differentiates cloud forests from other lower elevation montane forests is the persistent presence of fog and low clouds which supply an additional input of water to the vegetation. This "extra" water varies considerably between different tropical mountain regions (0.2–4.0 mm day^{-1}) (Bruijnzeel et al., 2011).

In the Cordillera de Mérida, research has focused on the main water fluxes in the cloud forests and the impact that deforestation can have when these intact ecosystems are replaced by pastures, mostly of *Pennisetum clandestinum* (Hochst. *ex* Chiov.) Morrone, and *Melinis minutiflora* P. Beauv. (Ataroff & Rada, 2000). In LM for example, measured cloud water interception (309 mm) under intact conditions is equivalent to an extra month of rain (mean annual precipitation of 3124 mm) (Ataroff & Rada, 2000). In relation to rainfall interception, LM presents higher values (49–51% of total input) (Ataroff & Rada, 2000), compared to LC (2300 m asl), with a lower rainfall input (1400–1600 mm), where interception ranges between 20 and 35% (Steinhardt, 1979; Montilla, 2009). These values of water interception represent a very wide range reflecting the differences found in these forests in both rainfall patterns and the structure of the vegetation.

Calculated evapotranspiration (Steinhardt, 1979) accounts for almost 62% of rainfall in LC, while in LM, Ataroff and Rada (2000) estimated this to be ~73% of rainfall partitioned as: 16% transpiration, 51% foliage evaporation and 6% evaporation from the soil. Overall, of the infiltrated fraction, 26% in LM and 37% in LC becomes subsurface runoff, eventually contributing to the maintenance of streams and rivers (Ataroff and Rada 2000). Additionally, the lower proportion in the case of surface runoff (1%) throughout the year reduces the risks of flooding in streams of forested areas. Water budget studies of *P. clandestinum* pastures in LM indicate that as much as 91% of incoming water reaches the soil's upper surface (about

2843 mm yr^{-1}). Compared to the forest, surface runoff in the pasture was slightly higher, showing an average value of 2% while transpiration represented 66% of the incoming water (2070 mm yr^{-1}), a high value compared to the intact forest. The different components of the water dynamics of both cloud forests are shown in Fig. 3 (Grimm & Fassbender, 1981b; Ataroff & Rada, 2000).

Given these marked differences between intact forests and pastures, it is important to highlight that the increased pressure to convert forests into pasture for cattle grazing in the high Andean basins in Venezuela (Rodríguez-Morales et al., 2009) under drier/warmer climate scenarios, would potentially lead to lower soil storage capacity and rainfall interception, and higher transpiration rates in pastures compared to intact forests (Ataroff & Rada, 2000). These transformations may increase soil erosion, lower streamflow, and increase fires due to added flammable material (i.e., dry pastures) with the consequent losses of SOM, increased CO_2 emissions, and higher risks of tree mortality at the edges of forest patches (Bonnesoeur et al., 2019; Breshears et al., 2013; Brodribb et al., 2020; Choat et al., 2018; Dávila, 2009; Herold et al., 2011; Roman-Cuesta et al., 2014; Oliveras et al., 2017; Soh et al., 2019).

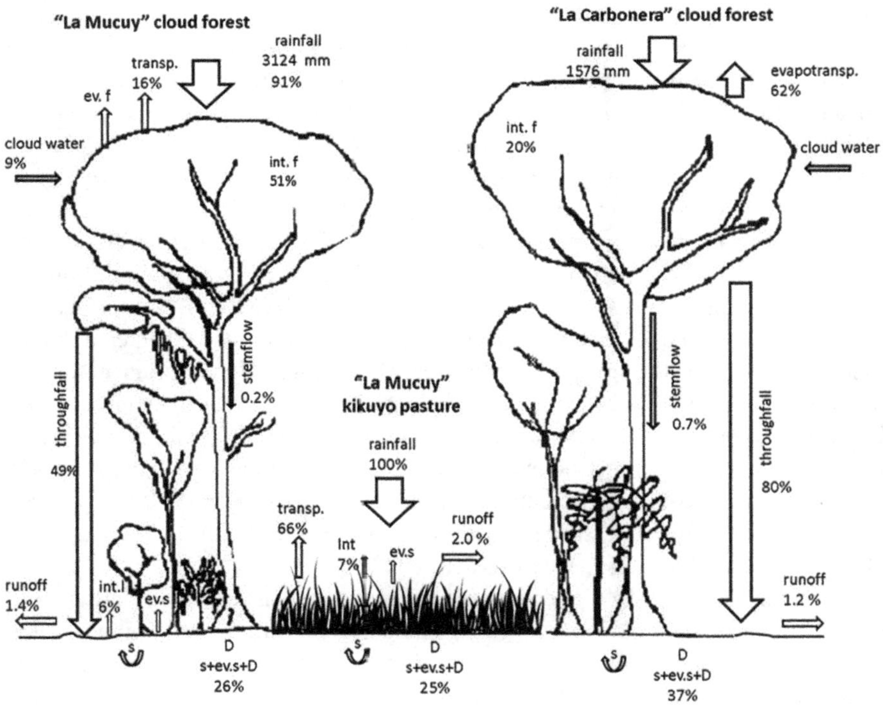

Fig. 3 Water dynamics for undisturbed upper montane cloud forests (La Mucuy and La Carbonera), and a kikuyu pasture at the Cordillera de Merida, Venezuela. Abbreviations: transp. = transpiration rate; ev.f = foliage evaporation; int.f = foliar interception; int.l = litter interception; ev.s = soil evaporation; s = soil storage; D = drainage by throughflow

5 The Effects of Light and Soil-Water Gradients on Forest Structure

5.1 Structure of Juvenile Tree Communities in Relation to the Understory Light Environment

The light-limited environment of the wet tropical forest understory is very heterogeneous at different spatial and temporal scales (Chazdon & Fetcher, 1984; Endler, 1993). In the case of tropical montane cloud forests, the frequent cloudiness together with high topographic variations accentuate this heterogeneity, which imposes energetic limitations and affects other factors such as evapotranspiration, temperature and humidity (Hogan & Machado, 2002; Schwarzkopf et al., 2011).

Understanding the factors that drive regeneration and species composition of these forests, and in particular how light availability affects the patterns of juvenile tree distribution in the understory, is critical for informing conservation and restoration programs. Since light is a limiting factor for plants in these ecosystems, regeneration and subsequent growth is largely limited in the understory, where most juveniles of tree species often must take advantage of small canopy openings (Chazdon et al., 1996). This light gradient partitioning therefore plays a major role in determining the tree community structure (Quevedo et al., 2015). In a study conducted in LC, the average range size of the cumulative distribution of the abundance of juveniles of the most important species was analyzed along the light gradient represented by the percentage of canopy openness (%CO), and the average pairwise overlap of these ranges was estimated (Quevedo et al., 2015). Figure 4 shows the range size distributions and the degree of overlap of 20 juvenile tree species along a gradient of canopy openness (%CO). More than 75% of the plots studied showed a %CO $\leq$ 4.5% and 50% had a %CO $\leq$ 2.2%; maximum %CO was 12.6% (Quevedo et al., 2015). In general, the abundance-based distribution of tree species was non-random in relation to the light environment, but with a high overlap within the interquartile range of %CO. In fact, depending on the tree species, juveniles growing in this usually low light environment show a differential distribution along the light gradient, which allows them to be categorized into functional groups according to their light requirements using evidence from physiological and morphological features measured in a shade house experimental study (Quevedo et al., 2015, 2016, 2018; Quevedo-Rojas & Jerez-Rico, 2020).

Although 15 of the 20 studied species showed a non-random distribution along the canopy openness gradient, suggesting some degree of light partitioning, the test based on the bounded null model indicates that the community fits within the confidence interval for the null model, meaning that species' range overlap did not deviate significantly from a random pattern (Fig. 5). This species distribution pattern indicate that low canopy openness is the rule in these forests, despite the scattered tree-fall gaps, and suggests that light partitioning does not determine the tree community structure at the juvenile stage, but rather there is a strong niche overlap in relation to light availability (Lieberman et al., 1995; Lusk et al., 2006; Quevedo et al., 2015).

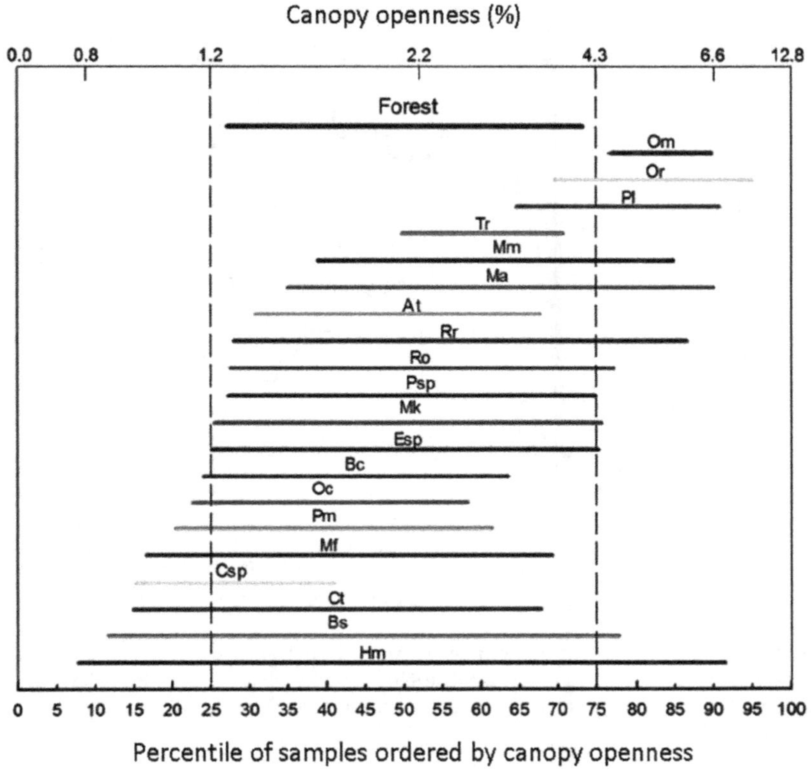

Fig. 4 Interquartile ranges of juvenile tree species' distribution on the rank transformed gradient of percentage of canopy openness (280 plots, 1975 individuals), in high dense forest plots of the San Eusebio University Forest in LC. Letters identifying species abbreviations and their position corresponds to median (midpoint) of the abundance distribution of juvenile trees. Figure from Quevedo et al. (2015) (CC by NC-4.0). Species are: Hm *Hyeronima moritziana* (Euphorbiaceae); Om *Ocotea macropoda* (Laureaceae); Bs *Beilshmiedia sulcata* (Laureaceae); Tr *Tetrorchidium rubrivenium* (Euphorbiaceae); Rr *Retrophyllum rospigliosii* (Podocarpaceae); Ma *Myrcia acuminata* (Myrtaceae); Oc *Oreopanax capitatus* (Araliaceae); Bc *Billia colombiana* (Hippocastanaceae); Csp *Cyathea spp* (Cyatheaceae); Or *Oreopanax reticulatus* (Araliaceae); Pl *Passiflora lindeniana* (Passifloraceae); Mm *Miconia meridensis* (Melastomataceae); Ct *Casearia tachirensis* (Flacourtiaceae); Pm *Prunus moritziana* (Rosaceae); Esp *Eugenia sp.* (Myrtaceae); Mf *Myrcia fallax* (Myrtaceae); Ro *Roupala obovata* (Proteaceae); Mk *Myrcianthes karsteniana* (Myrtaceae); Psp *Psicotria spp.* (Rubiaceae); At *Aegiphila terniflora* (Verbenaceae).

5.2 *Characteristics of Forest Structure in Small Fragments and at the Treeline Ecotone*

Forest fragmentation in the Andean basins of Venezuela have resulted in a mosaic of small fragments (often <10 ha) with strong edge effects from the surrounding grassland matrix (Dávila, 2009; Rodríguez-Morales et al., 2009). An evaluation of the Capaz River watershed (where LC site is located), showed that since the 1950s the

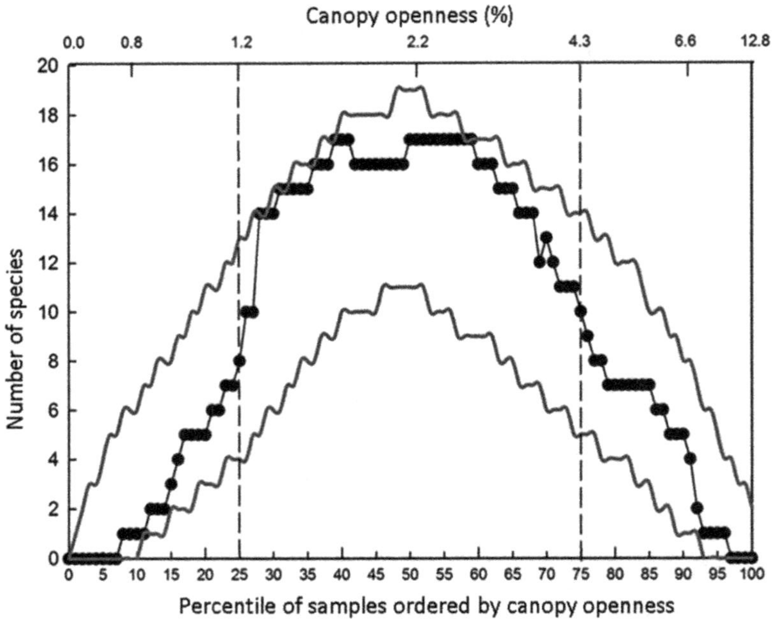

Fig. 5 Empirical distribution of species richness along the canopy openness gradient (dotted line) and 95% confidence limits for the average distribution predicted by the bounded null model (continuous lines) in La Carbonera cloud forests, Venezuelan Andes. Figure from Quevedo et al. (2015) (CC by NC-4.0))

landscape has been fragmented into small forest patches (fragments ranging from 356 ha in 1952 to 80 ha in 1997), with most of these patches surrounded by a matrix of grasslands and crops (Rodríguez-Morales et al., 2009). Edge effects produce distinct changes in biotic and microclimatic conditions between forest borders and the interior, with these changes being more severe as the forest fragments become smaller (Hansen et al., 2020).

The brighter and drier conditions of the forest edges in small fragments are a strong selective pressure for the dominance of gap-specialist tree species (Laurance, 2000). In general, the small forest patches studied in the Capaz River watershed (2228–2336 m asl), although floristically diverse and with a composition relatively similar to the undisturbed forest (Dávila, 2009; Naranjo, 2013), tend to be dominated by species described as early successional colonizers of gaps (e.g., *Myrcia acuminata* DC., *Alchornea triplinervia* (Spreng.) Müll.Arg., *Clusia multiflora* Kunth, *Tetrorchidium rubrivenium* Poepp. & Endl.) (García Núñez et al., 1995; Jaimes, 2015; Quevedo et al., 2018). The small fragments are characterized by a higher tree density (>2.5 cm diameter) and recruitment rates comparable to forests stands in undisturbed areas (Naranjo, 2013). These results suggest that protecting multiple cloud forest patches is essential to conserve landscape-level tree diversity, and even small forest reserves can contribute significantly to biodiversity conservation (Wilson & Rhemtulla, 2018). Ultimately, this means that actions are needed to enhance connectivity to favor dispersal and establishment, hence increasing the likelihood of forest recovery.

Similar edge effects have been observed in the upper limit of continuous forests in the treeline ecotone (3000–3400 m of elevation). In these areas, small forest fragments/islands occur naturally above the forest line, and they are dominated, specially at their edges, by small pioneer tress (e.g., *Diplostephium venezuelense*, *Vallea stipularis*), which sometimes are also able to colonize the open páramo grasslands of the lower alpine páramo belt (Ramirez et al., 2009; Llambí et al., 2013). In turn, when these treeline forests are cleared for agriculture or cattle grazing, trees face severe limitations for establishment, generating open patches dominated by páramo grassland species (González et al., 2011). Edge effects are also key in these successional areas, with most forest regeneration occurring near the borders of these open grassland patches within the forest matrix (Bueno & Llambí, 2015).

5.3 The Effects of Soil Water Content on Forest Structure

Hetsch and Hoheisel (1976) described vegetation communities at LC based on slope and soil characteristics, primarily physical features related to soil water content, proposing the hydro-sequence concept as the main ecological gradient which may explain the differences in forest structure, productivity and reserves of organic and inorganic soil matter. The soil - hydric sequences consisted of seven site-types according to soil water tension measurements along the year at 25 and 80 cm in soil depth (Table 4). These authors indicated that the accumulation of organic matter in

Table 4 Site characteristics of the San Eusebio University Forest in LC from the hydro-sequence proposed by Hetsch and Hoheisel (1976)

Site	Slope (%)	[a]Soil water tension (cm)	[b]Soil types	Limitations	Average Wood volume (m^3/ha)
1–2	>25	101–1000	troperthents, dystropepts, humitropepts	Frequently saturated from 90 cm in depth	490
3–4	10–25	11–100	humitropepts, troperthents, dystropepts	Saturation starting at 60 cm in depth	530
5–6	5–10	≅0	tropaquepts, haplaquepts, dystropets	Saturation starting at 30 cm in depth	170
7	0–5	≅0	tropaquepts, haplaquepts	Frequently saturated from the soil surface	–

[a]Tensions close to 0 cm indicate water saturation; between 100 and 330 cm, field capacity; 1000 cm, drying levels for sensitive species to water deficit
[b]Valcarcel (1982)

above-ground biomass (AGB) and mineral soil reserves (e.g., mineral nitrogen) are strikingly reduced in soils with slight slopes with marked hydromorphism. Sites at LC have been mapped according to topography and water saturation (Valcarcel, 1982), where less sloped sites (0–10%) were classified as: tropaquepts, haplaquepts, dystropets; the more inclined sites (>10%) as: troperthents, dystropepts, humitropepts. The highest basal area (m^2 ha^{-1}) and trunk biomass volume (m^3 ha^{-1}) were found on well drained sites (medium slopes) and the lowest in valley bottoms (Hetsch & Hoheisel, 1976; Quiroz, 2010). In this regard, Quiroz (2010) found a mean tree density (tree species ≥5 cm DBH) between 1050 ind ha^{-1} in valley bottom sites, and 2140 ind ha^{-1} for top of hills; mean basal area ranged between 29.5 and 68.5 m^2 ha^{-1} (medium slopes with 23–47% of inclination, respectively).

In spatially heterogeneous environments, like the mountain forests of the Andes, topography and the competitive environment can influence the severity of water stress experienced by individual trees, leading to individual-level variation in drought impacts (Schwartz et al., 2020).

These studies underline the variability of factors that control stand structure and species composition in montane forests (e.g., hydro-sequence and/or nutrient gradients) and, at the same time, the need to further understand the ecosystem functional and structural responses in the face of a changing climate (Bellingham & Tanner, 2000; Bruijnzeel et al., 2010; Homeier et al., 2010; Werner & Homeier, 2015).

6 The Long-Term Dynamics of Tree Turnover and Productivity in Venezuelan Cloud Forests

Several studies over the last three decades have shown that mature tropical forests have experienced changes in stand structure, species composition, and species interactions with disturbance and the environment. Some of these changes include tree growth exceeding losses from tree death, thus increasing productivity and the overall aboveground biomass both in the Amazon (Phillips et al., 2004) and African forests (Lewis et al., 2009). The interpretations of the potential causes behind these long-term demographic processes have been diverse. Increased resource availability and improved conditions for growth (i.e., increase in CO_2 levels and higher temperatures) have been argued as a possibility for increasing levels of dynamism (Lewis et al., 2009; Phillips & Lewis, 2014). Yet, some forests might be responding to past natural or anthropogenic disturbances (Wright, 2005; Muller-Landau, 2009; McMichael et al., 2017), thus stand development and maturation processes (i.e., competition followed by gap-replacement dynamics) could also be driving these changes. In a study based on data collected from permanent plots across multiple forest-types in Venezuela, including the cloud forests of La Carbonera, Vilanova et al. (2018) demonstrated how elevation and low temperatures are key drivers of a 'slow dynamics' in which the turnover rate (calculated as the mean between recruitment and mortality) of these Andean cloud forests (mean of 1.61 ± 0.09% y^{-1}) is significantly lower than the case of more dynamic and productive sites of the Western lowland plains (mean of 2.74 ± 0.17% y^{-1}). This difference has important implications on how turnover and stem dynamics can drive biomass accumulation, and how different demographic mechanisms, especially tree mortality may affect forest structure, which in turn may affect forest biomass, with cloud forests attaining higher biomass stocks comparable to those observed in other carbon-rich regions of the tropics such as the Guiana Shield (Baker et al., 2004).

On the temporal trends, Vilanova (2019) reported for multiple intact forest sites in Venezuela that between 1960 and 2012 overall stem recruitment (r) and mortality (m) rates both showed significantly increasing trends, simultaneously showing a positive net change in aboveground biomass of 0.71 ± 0.24 Mg C ha^{-1} y^{-1}, with 80% of the plots analyzed nationally gaining biomass through time (mean relative change in AGB = 0.39% $year^{-1}$), thus acting as a net carbon sink. Along these lines, in a recent analysis comparing the forest dynamics between 2016 and 2022 in permanent monitoring plots in San Eusebio and La Mucuy, both part of the Andean Forest Network, a general increase in tree density, basal area and AGB in all plots was found. However, it is possible that different patterns may emerge as we continue the monitoring of these sites in the future.

7 An Ecophysiological Approach as a Tool for Ecological Restoration of Andean Cloud Forests

For tropical cloud forests in the Venezuelan Andes, ecophysiological studies on photosynthetic response to light environment (i.e., light gradients determined by canopy structure and gap openings) or edge-to-interior conditions in small forest fragments (<1 ha), and studies on the ability to acclimate to sudden changes in light intensity (García Núñez et al., 1995; Vilanova, 1996; Cabrera, 1999; Dulhoste, 2002; Bracho, 2014; Jaimes, 2015; Quevedo et al., 2018), show that species that dominate in gaps and small forest patches have relatively high maximum photosynthetic rates, comparable to sun leaves and pioneer species in lowland tropical moist forests (Strauss-Debenedetti & Bazzaz, 1996). On the contrary, shade-tolerant species have low photosynthetic rates, with lower foliar nitrogen content. Species with an acquisitive strategy have high assimilation rates and fast growth, related to larger and thinner leaves with a high specific leaf area (SLA), as well as low wood density. Conversely, resource investments of conservative species are directed primarily toward building perennial tissues that allow for longevity and resistance to biological and abiotic stress (e.g., low SLA and high wood density) (Wright et al., 2004, 2010; Chave et al., 2009; Reich, 2014; Díaz et al., 2016).

The study of the community structure of juvenile trees conducted in LC helped identify representative species from different shade tolerance groups along wide light gradients in the understory (Quevedo et al., 2015). This approach shed light on the physiological tolerance limits of species, revealing where they lie along successional gradients determined by the dynamics of natural disturbances (e.g., falling trees creating gaps in the canopy). The functional characteristics of shade tolerance (i.e., plant functional traits) are linked to life history traits related to the regeneration process, including relationships with the germination processes over time and space (e.g., seed size, reserve capacity, dispersal type, light responses, dormancy, etc.), water relations (e.g., hydraulic conductivity, water potentials at turgor loss) and nutritional status (e.g., leaf nitrogen content which correlates with photosynthetic capacity and relative growth rate). In turn, these characteristics are linked with morpho-anatomical features of the species (e.g., shoot/root ratio, leaf area ratio, specific leaf area, stomatal density, wood density, etc.) that can explain biotic interactions (e.g., leaf nitrogen content correlates with specific leaf area and its palatability). In summary, these traits indicate how plants use and acquire resources, signaling a compromise between growth and mortality (Kitajima, 1994; Lohbeck et al., 2013), and involve a spectrum of life history strategies from acquisitive to conservative species (Reich, 2014).

This combined understanding of the community and the ecophysiological processes can inform the development of restoration strategies for degraded or deforested areas by directly answering a key question such as which and where to introduce a particular species during restoration. A pioneering study for the Venezuelan Andean cloud forest that focused on the application of knowledge derived from research at the community and ecophysiological levels is the work of Quevedo-Rojas and Jerez-Rico (2020) who studied the establishment of mixed plantations for ecological restoration purposes in LC. This is a clear example of an assisted restoration approach aiming at facilitating the process of community recovery (Chazdon, 2008; Chazdon et al., 2021; Christmann & Oliveras, 2021; Christmann et al., 2023; Chua & Potts, 2018; Werden et al., 2022). As Quevedo-Rojas and Jerez-Rico (2020) pointed out, mixed plantations could be successful as a restoration option for creating a forest cover under certain conditions, where forests cannot regenerate by natural successional mechanisms as is the case in many degraded pasturelands.

In mixed plantations, two or more species are established in the same stand, which may vary in their functional characteristics (i.e., conservative vs. acquisitive species) according to their growth habits (i.e., shade tolerance, succession stage), tree architecture and potential use. Species arrangements can vary from monospecific patches, rows or intimate assemblages with several species of the same age or at different times to create uneven-aged stands. Silviculture and management of mixed species with native tropical cloud forests species is complex because the lack of knowledge on growing these species (e.g., most tree species seedlings are characterized by slow growth and low tolerance to direct sunlight), plant availability, production in nurseries, plantation establishment, and complex silvicultural management, particularly in species mixtures.

In the study of Quevedo-Rojas and Jerez-Rico (2020) twenty-five native tree species with different shade tolerance conditions were planted. This was done in mixed assemblages (excluded from cattle) on degraded areas originally covered by exotic pastures, including planting some isolated trees (sites at 2250 to 2300 m asl., slope

30–80%, with soils poor in organic matter and nutrients). Species were selected initially based on the findings from Quevedo et al. (2015, 2016) (about the distribution of juveniles along light gradients see Sect. 5.1 and Figs. 4 and 5) and landowner's preferences. The species were categorized in functional groups based on shade tolerance and fitted into three categories: (a) shade intolerant (SI), (b) partially shade tolerant (PT), and (c) shade tolerant (ST) (Table 5). Planting areas were subdivided into plots (~200 m^2 each, ~100 trees per plot). Plots were assigned to levels of light intensity (LI) based on %CO (measured with hemispherical photographs) at planting time. Three levels of LI were differentiated: (a) High light (HL, above 50% CO); (b) Medium light-(ML, 40–50% CO); and (c) Low light intensity (LL <40% CO). Overall, survival was high for all categories of shade tolerance

Table 5 List of species used for restoration in LC cloud forest in the Venezuelan Andes according to their shade tolerance and potential uses

Scientific name	Family	Shade Tolerance[a]	Uses[b]
Tetrorchidium rubrivenium	Euphorbiaceae	SI	WDF, FD, FA
Alchornea grandiflora	Euphorbiaceae	SI	WDF, MTB, FA
Montanoa quadrangularis	Asteraceae	SI	WDF, FD; FA
Ruagea pubescens	Meliaceae	SI	MTB; OR, FA
Cedrela montana	Meliaceae	SI	MTB, OR, FA
Inga oerstediana	Leguminosae	SI	MTB; FA, NF
Ocotea macropoda	Laureaceae	SI	MTB; FA
Miconia meridensis	Melastomataceae	SI	MTB, FA
Cecropia telenitida	Urticaceae	SI	MTB; OR
Hieronyma moritziana	Euphorbiaceae	PT	MTB, FA
Billia columbiana	Hippocastanaceae	PT	MD, MTB, FA
Casearia tachirensis	Flacourtiaceae	PT	HWR, MTB, FA
Beilschmiedia sulcata	Laureaceae	PT	HWR, FA
Nectadra laurel	Laureaceae	PT	MTB, FA
Prunus moritziana	Rosaceae	PT	MTB, OR, FA
Retrophyllum rospigliosii	Podocarpaceae	PT	MTB, OR, FA
Myrcia acuminata	Myrtaceae	PT	TB, FW, FA
Vochysia meridensis	Vochysiaceae	PT	MTB, OR, FA
Eugenia tamaensis	Myrtaceae	ST	MWR; FA
Myrcianthes karsteniana	Myrtaceae	ST	HWR, OR, MD, FA
Miconia resimoides	Melastomataceae	ST	MTB; OR
Podocarpus oleifolius	Podocarpaceae	ST	MTB, OR
Myrcia fallax	Myrtaceae	ST	MTB; MD
Aegiphila terniflora	Verbenaceae	ST	MTB, HWR, FA
Eschweilera tenax	Lecytidaceae	ST	HWR, FA

From: Quevedo-Rojas and Jerez-Rico (2020) (CC By Open Access license)

[a]Species shade tolerance group: *SI* shade intolerant, *PT* partially tolerant, *ST* shade tolerant

[b]Species uses: *FD* fodder, *FA* fauna attraction, *MD* medicinal, *FW* firewood, *NF* nitrogen fixing, *OR* ornamental, *MTB* multipurpose timber, *HWR* hardwood for building roofs, *WDF* wooden fences

within different levels of light intensity, averaging over 70% for all plots after 4.5 years. The high survival rates for most species were attributed to a relatively intense monitoring and maintenance within the first two years after planting, with cattle exclusion, irrigation during the first dry season after planting and the control of aggressive weeds and grasses at least for two years, being highly significant factors for survival. As expected, shade-intolerant species displayed faster height and root-collar diameter growth.

Shade-tolerant species had larger crown ratios due to persistence of lower branches, whereas shade-intolerant showed signs of crown recession at age 7. The positive results of the restoration project carried out in LC by Quevedo-Rojas and Jerez-Rico (2020) have motivated landowners to establish plantations of native species in critical areas with their support. Tree planting with native species is an attractive approach to integrate the ecological knowledge generated as a product of many years of research of the Venezuelan Andean Cloud Forests for testing predictions of performance at the species level. Successful trials based on a scientific foundation with the participation of local communities during their establishment and management provide strong evidence of the feasibility of using such method at small-medium scale to solve tangible and current environmental problems, which is evidence that ecological restoration projects are relevant, useful and desired by local communities (Quevedo-Rojas & Jerez-Rico, 2020).

8 Conclusions

The conservation efforts of Venezuelan ecosystems have largely focused on the creation of a system of extensive protected natural areas (Areas Under Special Administration - ABRAEs), covering approximately 46% of the territory with administrative figures such as national parks (14.8% of the territory), natural monuments, forest reserves, biosphere reserves (although none is located in the Andes), and other categories. About 72% of these areas are covered by forests (Gfw-Ven, 2002; Bevilacqua et al., 2006; Torres-Lezama et al., 2008). Among several challenges in forest conservation is maintaining and, when possible, improving the integrity, connectivity and long-term monitoring of these areas. In the Venezuelan Andes, there are several protected areas (e.g., National Parks: Sierra Nevada, Sierra La Culata, Tapo-Caparo, Paramos Batallon y La Negra, Chorro del Indio) that include a considerable area of the Andean cloud forest ecosystem. On the other hand, unprotected forest areas that are under public or private management could be integrated into special plans based on a multiple use approach (e.g., agriculture-forestry-livestock, eco-tourism, etc.), including ecological restoration projects.

This review shows that, although deforestation rates in the Andean zone seem to have stabilized and have been relatively low in recent years, fragmentation and forest degradation continue to be important threats. Thus, the long-term monitoring of the impacts that these anthropogenic forces have on the structure, dynamics and function is of primary importance. It is also important to highlight that the research conducted in La Carbonera and La Mucuy over many years, emphasizes that deforestation and the replacement of intact forests by pastures limit the capacity of these landscapes to regulate water flows, fog catchment, soil water storage and deep drainage, putting at risk the water provision for many neighboring communities.

The use of an ecophysiological lens has also allowed us to improve our understanding of the functional characteristics of the most abundant species that compose these complex forests. This research has shown that at least in the seedling–sapling stage, many tree species comprise a broad functional group of shade-tolerant generalists and are characterized by a limited capacity of photosynthetic acclimation to high radiation conditions. This is important knowledge to better understand the dynamics of regeneration in undisturbed forests in favor of practices that can contribute to their recovery and restoration.

Long-term forest studies show that these low-temperature high elevation montane forests are characterized by a slow dynamics with low turnover rates and woody productivity having important implications on the carbon cycle of these forests. The studied plots in LC and LM are characterized by high biomass stocks comparable to those observed in other carbon-rich regions of the tropics. Moreover, recent results of forest dynamics in LC and LM have shown that Andean cloud forests are important hotspots for climate change mitigation acting as net carbon sinks. Nevertheless, projections of forest dynamics under warmer and drier climate scenarios forecast an increase in tree mortality rates and a potential shift in species range from lowland areas towards the Andes that could displace cloud forest species from their current altitudinal belt (i.e., "thermophilization") (Foster, 2001; Willis & MacDonald, 2011; Oliveira et al., 2014; Fadrique et al., 2018). This, in turn, could affect the current carbon sink provided by these forests. Therefore, to understand the dynamics of these ecosystems in climate change scenarios and derive adaptation strategies, it is necessary to continue monitoring the dynamics and ecology of these forests, including the collection of functional trait data that can serve to identify their potential for climate change mitigation. Finally, the positive results of the restoration project carried out in LC serves as a solid starting point that has motivated local landowners to establish plantations of native species in critical areas with their support. This bottom-up approach (from local ecological restoration to landscape restoration) constitutes a strong support to the incorporation of alternative restoration approaches such as assisted regeneration and passive restoration, which must be considered in forest landscape restoration plans.

As shown in this chapter, Andean cloud forests are ecosystems with a complex ecology and the evidence collected from two representative long-term research and monitoring sites, although limited in geographical scale, can help understanding how tropical cloud forests may respond to climate change and how management activities can contribute to increase their resilience and adaptive capacity. This review was an attempt to consolidate a body of knowledge while highlighting the potential gaps and needs for future research.

Acknowledgments The different research projects analyzed in this review have been supported by the Institute of Environmental and Ecological Sciences (ICAE), Faculty of Sciences, and the

Faculty of Forestry and Environmental Sciences, Universidad de Los Andes, Mérida, Venezuela. CG-N thanks the Institute of International Education's Scholar Rescue Fund (IIE-SRF) and CUGS, Trinity College, for their academic and financial support. FR and LDLL respectively thank UniAndes-Colombia and CONDESAN for their academic and financial support (particularly within the Andean Forests Program and the Adaptation at Altitude project financed by SDC). We thank Jean Pierre Veillon, a true pioneer in the study of Venezuelan forests whose vision and energy led to the establishment and monitoring of a set of long-term plots nationwide. EV acknowledge funding from the Amazon Forest Inventory Network (RAINFOR) project for field work and data collection between 2004 and 2022 for San Eusebio plots. The RAINFOR forest monitoring network in Venezuela has been supported by the Natural Environment Research Council (http://www.nerc.ac.uk/) (grants NE/D005590/1andNE/ I028122/1), and the Gordon and Betty Moore Foundation (https://www.moore.org/).

References

Acevedo, M. F., Ataroff, M., Monteleone, S., & Estrada, C. (2001). Aberturas del dosel y espectro de la luz en el sotobosque de una selva nublada andina de Venezuela. *Universidad del Zulia, 9*(2), 165–183.

Acevedo, M. F., Ataroff, M., Monteleone, S., & Estrada, C. (2003). Heterogeneidad estructural y lumínica del sotobosque de una selva nublada andina de Venezuela. *Interciencia, 28*(7), 394–403.

Ataroff, M. (2001). Venezuela. In M. Kappelle & A. D. Brown (Eds.), *Bosques Nublados del Neotrópico. Domingo de Heredia* (pp. 397–442). Editorial IMBIO.

Ataroff, M. (2003). Selvas y bosques de montaña. In M. Aguilera, A. Azócar, & E. González-Jiménez (Eds.), *Biodiversidad en Venezuela* (pp. 762–810). Ediciones CONICIT-Fundación Polar.

Ataroff, M., & García-Núñez, C. (2013). Selvas y bosques nublados de Venezuela. In E. Medina, O. Huber, J. Nassar, & P. Navarro (Eds.), *Recorriendo el Paisaje Vegetal de Venezuela* (pp. 125–155). Ediciones Instituto Venezolano de Investigaciones Científicas (IVIC).

Ataroff, M., & Rada, F. (2000). Deforestation impact on water dynamics in a Venezuelan Andean cloud forest. *Ambio, 29*(7), 438–442. https://doi.org/10.1579/0044-7447-29.7.4400

Ataroff, M., & Sarmiento, L. (2003). Diversidad en los Andes de Venezuela. I. Mapa de unidades ecológicas del Estado Mérida. CD-ROM, ICAE, Mérida, Venezuela.

Baker, T. R., Phillips, O. L., Malhi, Y., Almeida, S., Arroyo, L., Di Fiore, A., Erwin, T., Higuchi, N., Killeen, T. J., Laurance, S. G., Laurance, W. F., Lewis, S. L., Monteagudo, A., Neill, D. A., Núñez Vargas, P., Pitman, N. C. A., Silva, J. N. M., & Vásquez Martínez, R. (2004). Increasing biomass in Amazonian forest plots. *Philosophical Transactions of the Royal Society B: Biological Sciences, 359*, 353–365.

Bellingham, P. J., & Tanner, E. V. J. (2000). The influence of topography on tree growth, mortality, and recruitment in a tropical montane forest. *Biotropica, 32*, 378–384. https://doi.org/10.1111/j.1744-7429.2000.tb00484.x

Bevilacqua, M., Cárdenas, L., & Medina, D. (2006). *Las áreas protegidas en Venezuela: diagnóstico de su condición 1993–2004* (p. 163). Fundación Empresas Polar.

Bonnesoeur, V., Locatelli, B., Guariguata, M., Ochoa-Tocachi, B. F., Vanacker, V., Mao, Z., Stokes, A., & Mathez-Stiefel, S. L. (2019). Impacts of forests and forestation on hydrological services in the Andes: A systematic review. *Forest Ecology and Management, 433*(15), 569–584. https://doi.org/10.1016/j.foreco.2018.11.033

Bono, G. (1996). Flora y Vegetación del Estado Táchira, Venezuela. Museo Reg. Scienze Naturali, Torino. Monografie XX. p. 951.

Bracho, J. (2014). *Respuestas ecofisiológicas de tres especies arbóreas a diferentes regímenes de luz en el bosque nublado San Eusebio* [Undergraduate Thesis]. Mérida (Venezuela): Facultad de Ciencias, Universidad de Los Andes, p.52.

Breshears, D. D., Adams, H. D., Eamus, D., McDowell, N. G., Law, D. J., Will, R. E., Williams, A. P., & Zou, C. B. (2013). The critical amplifying role of increasing atmospheric moisture demand on tree mortality and associated regional die-off. *Frontiers in Plant Science, 4*(266), 1–4. https://doi.org/10.3389/fpls.2013.00266

Brodribb, T. J., Powers, J., Cochard, H., & Choat, B. (2020). Hanging by a thread? Forests and drought. *Science, 368*, 261–266.

Bruijnzeel, L. A., Scatena, F. N., & Hamilton, L. S. (Eds.). (2010). *Tropical montane cloud forests. Science for conservation and management*. Cambridge University Press.

Bruijnzeel, L. A., Mulligan, M., & Scatena, F. N. (2011). Hydrometeorology of tropical montane cloud forests: Emerging patterns. *Hydrological Processes, 25*(3), 465–498. https://doi.org/10.1002/hyp.7974

Bueno, A., & Llambí, L. D. (2015). Facilitation and edge effects influence vegetation regeneration in old-fields at the tropical Andean forest-line. *Applied Vegetation Science, 18*(4), 613–623.

Cabrera, H. M. (1999). *Mecanismos de aclimatación a la luz en especies de la sucesión de bosques tropicales montanos* [Doctorate Thesis]. Mérida (Venezuela): Facultad de Ciencias, Universidad de Los Andes, p. 171.

Chacón-Moreno, E., Rodríguez-Morales, M., Paredes, D., Suárez del Moral, P., & Albarrán, A. (2021). Impacts of global change on the spatial dynamics of treeline in Venezuelan Andes. *Frontiers in Ecology and Evolution, 9*, 615223. https://doi.org/10.3389/fevo.2021.615223

Chave, J., Coomes, D., Jansen, S., Lewis, S. L., Swenson, N. G., & Zanne, A. E. (2009). Towards a worldwide wood economics spectrum. *Ecology Letters, 12*, 351–366. https://doi.org/10.1111/j.1461-0248.2009.01285.x

Chazdon, R. L. (2008). Beyond deforestation: Restoring forests and ecosystem services on degraded lands. *Science, 320*, 1458–1460.

Chazdon, R. L., & Fetcher, N. (1984). Photosynthetic light environments in a lowland tropical rain forest in Costa Rica. *Journal of Ecology, 72*, 553–564. https://doi.org/10.2307/2260066

Chazdon, R., Pearcy, R., Lee, D., & Fetcher, N. (1996). Photosynthetic responses of tropical forest plants to contrasting light environments. In S. S. Mulkey, R. L. Chazdon, & A. P. Smith (Eds.), *Tropical forest plant ecophysiology* (pp. 5–55). Chapman & Hall.

Chazdon, R. L., Falk, D. A., Banin, L. F., Wagner, M., Wilson, S., Grabowski, R. C., et al. (2021). The intervention continuum in restoration ecology: Rethinking the active–passive dichotomy. *Restoration Ecology*, e13535. https://doi.org/10.1111/rec.13535

Choat, B., Brodribb, T. J., Brodersen, C. R., Duursma, R. A., Lopez, R., & Medlyn, B. E. (2018). Triggers of tree mortality under drought. *Nature, 558*, 531–539. https://doi.org/10.1038/s41586-018-0240-x

Christmann, T., & Oliveras, M. I. (2021). A synthesis and future research directions for tropical mountain ecosystem restoration. *Scientific Reports, 11*, 23948. https://doi.org/10.1038/s41598-021-03205-y

Christmann, T., Palomeque, X., Armenteras, D., Wilson, S. J., Malhi, Y., & Oliveras, M. I. (2023). Disrupted montane forest recovery hinders biodiversity conservation in the tropical Andes. *Global Ecology and Biogeography, 32*, 793–808. https://doi.org/10.1111/geb.13666

Chua, S. C., & Potts, M. D. (2018). The role of plant functional traits in understanding forest recovery in wet tropical secondary forests. *Science of the Total Environment, 642*, 1252–1262. https://doi.org/10.1016/j.scitotenv.2018.05.397

Dávila, Y. (2009). *Efecto de borde y respuestas funcionales durante el establecimiento de Alchornea triplinervia y Myrcia acuminata en un fragmento de selva nublada andina* [Masters Thesis]. Mérida (Venezuela): Facultad de Ciencias, Universidad de Los Andes, p. 80.

Delaney, M., Brown, S., Lugo, A. E., Torres-Lezama, A., & Bello, N. (1997). The distribution of organic carbon in major components of forests located in five life zones of Venezuela. *Journal of Tropical Ecology, 13*(5), 697–708. https://doi.org/10.1017/S0266467400010877

Díaz, S., Kattge, J., Cornelissen, J. H. C., Wright, I. J., Lavorel, S., Dray, S., et al. (2016). The global spectrum of plant form and function. *Nature, 529*, 167–171. https://doi.org/10.1038/nature16489

Dulhoste, R. (2002). *Caracterización florístico-estructural y ecofisiologia de las especies dominantes de un bosque secundario en la Selva Nublada* [Masters Thesis]. Mérida (Venezuela): Facultad de Ciencias, Universidad de Los Andes, p. 87.

Endler, J. A. (1993). The color of light in forests and its implications. *Ecological Monographs, 63*, 1–27. https://doi.org/10.2307/2937121

Engwald, S. (1999). *Diversitat und Okologie der vaskularen Epiphyten in einem Berg-und einem Tieflandergenwald in Venezuela* [Doctorate Thesis]. Bonn (Deutschland): Universität Bonn, p. 390.

Estrada, J. (2003). *Análisis multivariante de la variación altitudinal de la composición florística en la Cordillera de Mérida*. Trabajo de Ascenso. Universidad de Los Andes. Facultad de Ciencias, Centro Jardín Botánico. Mérida, Venezuela.

Fadrique, B., Báez, S., Duque, A., Malizia, A., Blundo, C., Carilla, J., Osinaga-Acosta, O., Malizia, L., Silman, M., Farfán-Ríos, W., et al. (2018). Widespread but heterogeneous responses of Andean forests to climate change. *Nature, 564*, 207–212. https://doi.org/10.1038/s41586-018-0715-9

Fahey, T., Sherman, R., & Tanner, E. (2016). Tropical montane cloud forest: Environmental drivers of vegetation structure and ecosystem function. *Journal of Tropical Ecology, 32*(5), 355–367. https://doi.org/10.1017/S0266467415000176

Fassbender, H., & Grimm, U. (1981a). Ciclos biogeoquímicos en un ecosistema forestal de Los Andes Occidentales de Venezuela. II. Producción y descomposición de los residuos vegetales. *Turrialba, 31*, 39–47.

Fassbender, H., & Grimm, U. (1981b). Ciclos biogeoquímicos en un ecosistema forestal de Los Andes Occidentales de Venezuela. IV. Modelos y conclusiones. *Turrialba, 31*, 101–108.

Foster, P. (2001). The potential negative impacts of global climate change on tropical montane cloud forests. *Earth Science Reviews, 55*, 73–106. https://doi.org/10.1016/s0012-8252(01)00056-3

García Núñez, C., Azócar, A., & Rada, F. (1995). Photosynthetic acclimation to light in juveniles of two cloud forest tree species. *Trees, 10*, 114–124. https://doi.org/10.1007/bf00192192

Geist, H., & Lambin, E. (2002). Proximate causes and underlying driving forces of tropical deforestation. *Bioscience, 52*(2), 143–150.

Gentry, A. (1993). *Biodiversity and conservation of neotropical montane forests*. The New York Botanical Garden.

GFW-VEN. (2002). *The state of Venezuela's forest: A case study of the Guayana region* (p. 132). Global Forest Watch, Venezuelan Chapter-WRI-Fundación Polar.

González, W., Llambí, L. D., & Smith, J. K. (2011). Dinámica sucesional del componente arbóreo en la zona de transición entre el bosque y el páramo en los Andes Tropicales. *Ecotropicos, 24*(1), 60–79.

Grantham, H. S., Duncan, A., Evans, T. D., et al. (2020). Anthropogenic modification of forests means only 40% of remaining forests have high ecosystem integrity. *Nature Communications, 11*, 5978. https://doi.org/10.1038/s41467-020-19493-3

Grimm, U., & Fassbender, H. (1981a). Ciclos biogeoquímicos en un ecosistema forestal de Los Andes Occidentales de Venezuela. I. Inventario de las reservas orgánicas y minerales (N, P, K, Ca, Mg, Mn, Fe, Al, Na). *Turrialba, 31*, 27–37.

Grimm, U., & Fassbender, H. (1981b). Ciclos biogeoquímicos en un ecosistema forestal de Los Andes Occidentales de Venezuela. III. Ciclo hidrológico y translocación de elementos químicos con el agua. *Turrialba, 31*, 89–99.

Grubb, P. J. (1977). Control of forest growth and distribution on wet tropical mountains: With special reference to mineral nutrition. *Annual Review of Ecology and Systematics, 8*, 83–107. https://doi.org/10.1146/annurev.es.08.110177.000503

Gutiérrez, N., Gärtner, S., Gaviria, J., Meier, W., & Reif, A. (2012). Successional vegetation patterns in abandoned pastures of the lower montane cloud forest zone in the Venezuelan Andes. *Phytocoenologia, 42*(3–4), 101–132.

Hansen, M., Potapov, P., Moore, R., Hancher, M., Turubanova, S. A., Tyukavina, A., Thau, D., Stehman, S. V., Goetz, S. J., Loveland, T., et al. (2013). High-resolution global maps of 21st-century forest cover change. *Science, 2013*(342), 850–853.

Hansen, M. C., Wang, L., Song, X.-P., Tyukavina, A., Turubanova, S., Potapov, P. V., & Stehman, S. V. (2020). The fate of tropical forest fragments. *Science Advances, 6*(11), eaax8574. https://doi.org/10.1126/sciadv.aax857

Herold, M., Román-Cuesta, R. M., Mollicone, D., Hirata, Y., Van Laake, P., Asner, G. P., Souza, C., Skutsch, M., Avitabile, V., MacDicken, K., et al. (2011). Options for monitoring and estimating historical carbon emissions from forest degradation in the context of REDD+. *Carbon Balance and Management, 6*, 13. http://www.cbmjournal.com/content/6/1/13

Hetsch, W., & Hoheisel, H. (1976). Standorts –und Vegetationsgliederung in einem tropischen Nebelwald. *Allgemeine. Forst-und. Jagd-Zeitung, 147*, 200–209.

Hogan, K. P., & Machado, J. L. (2002). La luz solar: consecuencias biológicas y su medición. In G. Kattan & M. R. Guariguata (Eds.), *Ecología y Conservación de Bosques Neotropicales* (pp. 119–143). Libro Universitario Regional, EULAC-GTZ.

Hokche, O., Berry, P., & Huber, O. (2008). *Nuevo Catálogo de la Flora Vascular de Venezuela*. Fundación Instituto Botánico de Venezuela.

Homeier, J., Breckle, S. W., Gunter, S., Rollenbeck, R. T., & Leuschner, C. (2010). Tree diversity, forest structure and productivity along altitudinal and topographical gradients in a species rich Ecuadorian montane rain Forest. *Biotropica, 42*(2), 140–148. https://doi.org/10.1111/j.1744-7429.2009.00547.x

Hostettler, T. (2002). Tropical montane cloud forests: A challenge for conservation. *Bois et forêts des tropiques., 274*(4), 19–31.

Huber, O. (1986a). El clima. In O. Huber (Ed.), *La Selva Nublada de Rancho Grande, Parque Nacional Henri Pittier* (pp. 17–30). Fondo Editorial Acta Científica Venezolana.

Huber, O. (1986b). Las selvas nubladas de Rancho Grande: observaciones sobre su fisionomía, estructura y fenología. In O. Huber (Ed.), *La Selva Nublada de Rancho Grande, Parque Nacional Henri Pittier* (pp. 131–170). Caracas.

Huérfano, A., Fedón, I., & Julián, M. (2020). *Libro Rojo de la Flora Venezolana. Segunda edición*. Instituto Experimental Jardín Botánico Dr. Tobías Lasser. Universidad Central de Venezuela. Caracas, Venezuela.

Jaimes, J. C. (2015). *Dinámica de la aclimatación del aparato fotosintético a cambios en el ambiente de luz en individuos juveniles de especies arbóreas de la sucesión temprana en la Selva Nublada Andina* [Undergraduate Thesis]. Mérida (Venezuela): Facultad de Ciencias, Universidad de Los Andes, p. 54.

Josee, C., Francisco, C., Navarro, G., Barrena, V., Cabrera, E., Chacón-Moreno, E., Ferreira, W., Peralvo, M., Saito, M., & Tovar, A. (2009). Ecosistemas de los Andes del Norte y Centro. Bolivia, Colombia, Ecuador, Perú y Venezuela. Secretaría General de la Comunidad Andina, Programa Regional ECOBONA-Intercooperación, CONDESAN-Proyecto Páramo Andino, Programa BioAndes, NatureServe, IAvH, LTA-UNALM, ICAE-ULA, CDC-UNALM, RUMBOL SRL. Lima. ISBN: 978-9972-787-77-5.

Kitajima, K. (1994). Relative importance of photosynthetic traits and allocation patterns as correlates of seedling shade tolerance of 13 tropical trees. *Oecologia, 98*(3), 419–428.

Kitayam, K., & Aiba, S. I. (2002). Ecosystem structure and productivity of tropical rain forests along altitudinal gradients with contrasting soil phosphorus pools on Mount Kinabalu, Borneo. *Journal of Ecology, 90*(1), 37–51. https://doi.org/10.1046/j.0022-0477.2001.00634.x

Lamprecht, H. (1954). *Estudios Selviculturales en los Bosques del Valle de La Mucuy, Cerca de Mérida*. Facultad de Ciencias Forestales, Universidad de Los Andes.

Lamprecht, H. (1980). Silvicultura en los Trópicos. Deutsche Gesellschat fur Technische Zusammenarbeit (GTZ) GmbH.

Laurance, W. (2000). Do edge effects occur over large spatial scales? *Trends in Ecology & Evolution, 15*(4), 134–135. https://doi.org/10.1016/s0169-5347(00)01838-3

Leuschner, C., Moser, G., Bertsch, C., Röderstein, M., & Hertel, D. (2007). Large altitudinal increase in tree root/shoot ratio in tropical mountain forests of Ecuador. *Basic and Applied Ecology, 8*(3), 219–230. https://doi.org/10.1016/j.baae.2006.02.004

Lewis, S. L., Lloyd, J., Sitch, S., Mitchard, E. T. A., & Laurance, W. F. (2009). Changing ecology of tropical forests: Evidence and drivers. *Annual Review of Ecology, Evolution, and Systematics, 40*, 529–549.

Lieberman, M., Lieberman, D., Peralta, R., & Hartshorn, G. S. (1995). Canopy closure and the distribution of tropical forest tree species at La Selva, Costa Rica. *Journal of Tropical Ecology, 11*(2), 161–178. https://doi.org/10.1017/s0266467400008609

Linares, A. (2008). *Análisis florístico y estructural de la vegetación de una Selva Nublada Andina en un gradiente altitudinal en La Mucuy, Estado Mérida* [Doctorate Thesis]. Mérida (Venezuela): Facultad de Ciencias, Universidad de Los Andes, p. 139.

Lindorf, H. (2013). Volkmar Vareschi como docente en la Escuela de Biología de la Universidad Central de Venezuela. In E. Medina, O. Huber, & J. M. Nassar (Eds.), *Recorriendo el Paisaje Vegetal de Venezuela* (pp. 21–42). Ediciones IVIC, IVIC.

Llambí, L. D., Puentes, J., & García-Núñez, C. (2013). Spatial relations and population structure of a dominant tree along a treeline ecotone in the tropical Andes: Interactions at gradient and plant neighbourhood scales. *Plant Ecology and Diversity, 6-7*, 343–353.

Lohbeck, M., Poorter, L., Lebrija-Trejos, E., Martínez-Ramos, M., Meave, J. A., Paz, H., et al. (2013). Successional changes in functional composition contrast for dry and wet tropical forest. *Ecology, 94*, 1211–1216. https://doi.org/10.1890/12-1850.1

Lusk, C. H., Chazdon, R., & Hofmann, G. (2006). A bounded null model explains juvenile tree community structure along light availability gradients in a temperate rain forest. *Oikos, 112*(1), 131–137. https://doi.org/10.1111/j.0030-1299.2006.13922.x

Marquez, O. (1988). Génesis de una sequencia de suelos en el Bosque Experimental de San Eusebio, La Carbonera, Estado Mérida. *Rev Forest Venez., 32*, 133–150.

McMichael, C. N. H., Matthews-Bird, F., Farfan-Rios, W., & Feeley, K. J. (2017). Ancient human disturbances may be skewing our understanding of Amazonian forests. *Proceedings of the National Academy of Sciences, 114*, 522–527.

Medina, E. (1986). Aspectos ecofisiológicos de plantas de bosques nublados tropicales: el bosque nublado de Rancho Grande. In O. Huber (Ed.), *La Selva Nublada de Rancho Grande, Parque Nacional Henri Pittier* (pp. 189–196). Fondo Editorial Acta Científica Venezolana.

Michelangeli, F. (1998). *La Selva Nublada: Rancho Grande*. Armitano Editores.

Montilla, F. (2009). *La hojarasca y su papel en la erosión hídrica en tres ecosistemas y agroecosistemas de Los Andes venezolanos* [Masters Thesis]. Mérida (Venezuela): Facultad de Ciencias, Universidad de Los Andes, p. 129.

Moser, G., Hertel, D., & Leuschner, C. (2007). Altitudinal change in LAI and stand leaf biomass in tropical montane forests: A transect study in Ecuador and a pan-tropical meta-analysis. *Ecosystems, 10*(6), 924–935. https://doi.org/10.1007/s10021-007-9063-6

Muller-Landau, H. C. (2009). Sink in the African jungle. *Nature, 457*, 969.

Myers, N., Mittermeier, R. A., Mittermeier, C. G., da Fonseca, G. A. B., & Kent, J. (2000). Biodiversity hotspots for conservation priorities. *Nature, 403*(24), 853–858.

Nagy, L., Eller, C. B., Mercado, L., Cuesta, F., Llambí, L. D., et al. (2023). South American mountain ecosystems and global change – a case study for integrating theory and field observations for land surface modelling and ecosystem management. *Plant Ecology & Diversity, 16*(1–2), 1–27. https://doi.org/10.1080/17550874.2023.2196966

Naranjo, M. E. (2013). *Impacto de la fragmentación de la selva nublada y su reemplazo por pasturas sobre su estructura y ecohidrología en los Andes Venezolanos* [Doctorate Thesis]. Mérida (Venezuela): Facultad de Ciencias, Universidad de Los Andes, p. 276.

Oliveira, R. S., Eller, C. B., Bittencourt, P. R. L., & Mulligan, M. (2014). The hydroclimatic and ecophysiological basis of cloud forest distributions under current and projected climates. *Annals of Botany, 113*(6), 909–920. https://doi.org/10.1093/aob/mcu060

Oliveras, I., Roman-Cuesta, R. M., Urquiaga-Flores, E., Quintano-Loayza, J. A., Kala, J., Huaman, V., Lizarraga, N., Sans, G., Quispe, K., Lopez, E., et al. (2017). Fire effects and ecological recovery pathways of tropical montane cloud forests along a time chronosequence. *Global Change Biology, 24*(2), 758–772. https://doi.org/10.1111/gcb.13951

Pacheco, C., Aguado, I., & Mollicone, D. (2011). Las causas de la deforestación en Venezuela: un estudio retrospectivo. *Biollania, 10*, 281–292.

Phillips, O. L., & Lewis, S. L. (2014). Evaluating the tropical forest carbon sink. *Global Change Biology, 20*, 2039–2041. https://doi.org/10.1111/gcb.12423

Phillips, O. L., Baker, T. R., Arroyo, L., Higuchi, N., Killeen, T. J., Laurance, W. F., Lewis, S. L., Lloyd, J., Malhi, Y., Monteagudo, A., Neill, D. A., Nüñez Vargas, P., Silva, J. N. M., Terborgh, J., Vásquez Martínez, R., Alexiades, M., Almeida, S., Brown, S., Chave, J., Comiskey, J. A., Czimczik, C. I., Di Fiore, A., Erwin, T., Kuebler, C., Laurance, S. G., Nascimento, H. E. M., Olivier, J., Palacios, W., Patiño, S., Pitman, N. C. A., Quesada, C. A., Saldias, M., Torres Lezama, A., & Vinceti, B. (2004). Pattern and process in Amazon tree turnover, 1976-2001. *Philosophical Transactions of the Royal Society B: Biological Sciences, 359*, 381–407.

Quevedo, A. M., Jerez, M., Schwarzkopf, T., & García-Núñez, C. (2015). Distribution of juveniles of tree species along a canopy closure gradient in a tropical cloud forest of the Venezuelan Andes. *iForest, 9*, 363–369. https://doi.org/10.3832/ifor1568-008

Quevedo, A. M., Schwarzkopf, T., García, C., & Jerez, M. (2016). Ambiente de luz del sotobosque de una selva nublada andina: estructura del dosel y estacionalidad climática. *Revista de Biología Tropical, 64*(4), 1699–1707.

Quevedo, A. M., García-Núñez, C., Jerez, M., Jaimez, R., & Schwarzkopf, T. (2018). Leaf acclimation strategies to contrasting light conditions in saplings of different shade tolerance in a tropical cloud forest. *Functional Plant Biology, 45*(9), 968–982. https://doi.org/10.1071/FP17308

Quevedo-Rojas, A., & Jerez-Rico, M. (2020). Mixed forest plantations with native species for ecological restoration in cloud forests of the Venezuelan Andes. In A. C. Gonçalves (Ed.), *Silviculture*. https://doi.org/10.5772/intechopen.95006

Quiroz, L. (2010). *Definición de comunidades arbóreas asociadas a sitios específicos en una selva nublada de Los Andes venezolanos* [Masters Thesis]. Mérida (Venezuela): Facultad de Ciencias, Universidad de Los Andes, p. 144.

Raich, J. W., Russell, A. E., Kitayama, K., Parton, W. J., & Vitousek, P. M. (2006). Temperature influences carbon accumulation in moist tropical forests. *Ecology, 87*(1), 76–87. https://doi.org/10.1890/05-0023

Ramirez, L., Llambí, L. D., Schwarzkopf, T., Gamez, L. E., & Marquez, N. J. (2009). Vegetation structure along the forest-páramo transition belt in the Sierra Nevada de Mérida: Implications for understanding treeline dynamics. *Ecotropicos, 22*(2), 83–98.

Reich, P. B. (2014). The world-wide 'fast-slow' plant economics spectrum: A traits manifesto. *Journal of Ecology, 102*, 275–301. https://doi.org/10.1111/1365-2745.12211

Rodríguez, J. P., & Rojas-Sánchez, F. (1999). *Libro Rojo de la Fauna de Venezuela*. Provita y Fundación Empresas Polar, Caracas, Venezuela.

Rodríguez-Morales, M., Chacón-Moreno, E., & Ataroff, M. (2009). Transformación del paisaje de selvas de montaña en la Cuenca del río Capaz, Andes Venezolanos. *Ecotropicos, 22*, 64–82.

Roman-Cuesta, R. M., Carmona-Moreno, C., Lizcano, G., New, M., Silman, M., Knoke, T., Malhi, Y., Oliveras, I., Asbjornsen, H., & Vuille, M. (2014). Synchronous fire activity in the tropical high Andes: An indication of regional climate forcing. *Global Change Biology, 20*(6), 1929–1942. https://doi.org/10.1111/gcb.12538

Schneider, J., Gaviria, J., & Zizka, G. (2003). Inventario florístico de un bosque altimontano húmedo en el Valle de San Javier. Edo. Mérida, Venezuela. *Plántula, 3*(2), 65–81.

Schwartz, N. B., Feng, X., Muscarella, R., Swenson, N. G., Umaña, M. N., Zimmerman, J. K., & Uriarte, M. (2020). Topography and traits modulate tree performance and drought response in a tropical forest. *Frontiers in Forests and Global Change, 3*, 596256. https://doi.org/10.3389/ffgc.2020.596256

Schwarzkopf, T., Susan, J. R., Fahey, T. J., & Degloria, S. (2011). Are cloud forest tree structure and environment related in the Venezuelan Andes? *Austral Ecology, 36*(3), 280–289. https://doi.org/10.1111/j.1442-9993.2010.02160.x

Soh, M. C. K., Mitchell, N. J., Ridley, A. R., Butler, C. W., Puan, C. L., & Peh, K. S.-H. (2019). Impacts of habitat degradation on tropical montane biodiversity and ecosystem services: A systematic map for identifying future research priorities. *Frontiers in Forests and Global Change, 2*, 83. https://doi.org/10.3389/ffgc.2019.00083

Stadtmüller, T. (1987). *Los Bosques Nublados en el Trópico Húmedo*. Centro Agronómico Tropical de Investigación y Enseñanza (CATIE).

Steinhardt, U. (1979). Untersuchungen uber den Wasser- und Nahrstoffhaushalt eines Andinen Wolkenwaldes in Venezuela. *Göttinger Bodenkundliche Berichte., 56*, 1–185.

Strauss-Debenedetti, S., & Bazzaz, F. (1996). Photosynthetic characteristics of tropical trees along successional gradients. In S. S. Mulkey, R. L. Chazdon, & A. P. Smith (Eds.), *Tropical forest plant ecophysiology* (pp. 162–186). Chapman & Hall.

Torres-Lezama, A., Ramírez-Angulo, H., Vilanova, E., & Barros, R. (2008). Forest resources in Venezuela: Current status and prospects for sustainable management. *Bois et forêts des tropiques., 295*(1), 21–33.

Valcarcel, R. (1982). *Clasificación y mapeo de sitios en la Estación Experimental de San Eusebio basada en criterios físicos-hidrológicos del suelo* [Masters Thesis]. Mérida (Venezuela): Facultad de Ciencias Forestales y Ambientales, Universidad de Los Andes, p. 153.

Vancutsem, C., Achard, F., Pekel, J.-F., Vieilledent, G., Carboni, S., Simonetti, D., Gallego, J., Aragão, L. E. O. C., & Nasi, R. (2021). Long-term (1990–2019) monitoring of forest cover changes in the humid tropics. *Science Advances, 7: eabe 16*, 1–21.

Vareschi, V. (1992). *Ecología de la Vegetación Tropical*. Sociedad Venezolana de Ciencias Naturales.

Veillon, J. P. (1965). Variación altitudinal de la masa forestal de los bosques primarios en la vertiente nor-occidental de la Cordillera de Los Andes, Venezuela. *Turrialba, 15*, 216–224.

Veillon, J. P. (1985). El crecimiento de algunos bosques naturales de Venezuela en relación con los parámetros del medio ambiente. *Rev Forest Venez, 29*, 1–122.

Veillon, J. P. (1989). *Los Bosques Naturales de Venezuela. Parte I: el Medio Ambiente*. Instituto de Silvicultura, Universidad de Los Andes.

Vilanova, I. (1996). *Intercambio de gases y relaciones hídricas en plantas del sotobosque de la selva nublada de La Mucuy, Estado Mérida, Venezuela* [Undergraduate Thesis]. Mérida (Venezuela): Facultad de Ciencias, Universidad de Los Andes, p. 117.

Vilanova, E. (2019). *Patterns and processes in tropical forests: an analysis of forest structure and function from long-term permanent plots in Venezuela, Northern South America*. PhD Dissertation, University of Washington, Seattle, USA.

Vilanova, E., Ramírez-Angulo, H., Torres-Lezama, A., Aymard, G., Gámez, L., Durán, C., Hernández, L., Herrera, R., van der Heijden, G., Phillips, O. L., et al. (2018). Environmental drivers of forest structure and stem turnover across Venezuelan tropical forests. *PLoS One, 13*(6), e0198489. https://doi.org/10.1371/journal.pone.0198489

Werden, L. K., Zarges, S., Holl, K. D., Oliver, C. L., Oviedo-Brenes, F., Rosales, J. A., & Zahawi, R. A. (2022). Assisted restoration interventions drive functional recovery of tropical wet forest tree communities. *Frontiers in Forests and Global Change, 5*, 935011. https://doi.org/10.3389/ffgc.2022.935011

Werner, F. A., & Homeier, J. (2015). Is tropical montane forest heterogeneity promoted by a resource-driven feedback cycle? Evidence from nutrient relations, herbivory and litter decomposition along a topographical gradient. *Functional Ecology, 29*(3), 430–440. https://doi.org/10.1111/1365-2435.12351

Willis, K. J., & MacDonald, G. M. (2011). Long-term ecological records and their relevance to climate change predictions. *Annual Review of Ecology, Evolution, and Systematics, 42*, 267–287. https://doi.org/10.1146/annurev-ecolsys-102209-144704

Wilson, S. J., & Rhemtulla, J. M. (2018). Small montane cloud forest fragments are important for conserving tree diversity in the Ecuadorian Andes. *Biotropica, 50*(4), 586–597. https://doi.org/10.1111/btp.12542

Wright, S. J. (2005). Tropical forests in a changing environment. *Trends in Ecology & Evolution, 20*, 553–560.

Wright, I. J., Reich, P. B., Westoby, M., Ackerly, D. D., Baruch, Z., Bongers, F., et al. (2004). The worldwide leaf economics spectrum. *Nature, 428*, 821–827. https://doi.org/10.1038/nature02403

Wright, S. J., Kitajima, K., Kraft, N. J. B., Reich, P. B., Wright, I. J., Bunker, D. E., et al. (2010). Functional traits and the growth–mortality trade-off in tropical trees. *Ecology, 91*(12), 3664–3674. https://doi.org/10.1890/09-2335.1

Forest Biomass and Species Diversity for Conservation Status Assessment in Subtropical Andean Forests

Cecilia Blundo, Yohana G. Jimenez, Sergio J. Ceballos, Agustina Malizia, Julieta Carilla, Romina D. Fernandez, Oriana Osinaga Acosta, H. Ricardo Grau, Silvia Pacheco, Sebastián A. Albanesi, Roberto J. Cáceres, Carla J. Cárdenas, Alejandro Brown, and Lucio R. Malizia

1 General Characteristics of Subtropical Andean Forests

Subtropical Andean forests represent the southernmost extension of the Andean forests, which spans across South America from Venezuela to Argentina (Kappelle & Brown, 2001). In this extreme of the distribution, from southern Bolivia to northwestern Argentina (18–28 °S), forests show some differences respect to the tropical Andes, particularly in forest structure and species composition (Kappelle & Brown, 2001; Malizia et al., 2020), as well as in forest dynamics (Baez et al., 2015; Duque et al., 2021). In general, subtropical Andean forests have less stem density and lower basal area per unit area, and a distinct floristic composition with more representation of temperate species compared to tropical Andean forests (Brown et al., 2001;

C. Blundo (✉) · Y. G. Jimenez · S. J. Ceballos · A. Malizia · J. Carilla · R. D. Fernandez
O. Osinaga Acosta · H. R. Grau
Instituto de Ecología Regional, CONICET, Universidad Nacional de Tucumán, Yerba buena, Tucumán, Argentina

S. Pacheco · R. J. Cáceres · C. J. Cárdenas · A. Brown
Fundación ProYungas (FPY), Yerba buena, Tucumán, Argentina

S. A. Albanesi
Fundación ProYungas (FPY), Yerba buena, Tucumán, Argentina

Instituto de Biodiversidad Neotropical, CONICET, Universidad Nacional de Tucumán, Yerba buena, Tucumán, Argentina

L. R. Malizia
Fundación ProYungas (FPY), Yerba buena, Tucumán, Argentina

Instituto de Biodiversidad Neotropical, CONICET, Universidad Nacional de Tucumán, Yerba buena, Tucumán, Argentina

Centro de Estudios Territoriales Ambientales y Sociales, Facultad de Ciencias Agrarias, Universidad Nacional de Jujuy, San Salvador de Jujuy, Jujuy, Argentina

N. Clerici (ed.), *Conservation of Andean Forests*,
https://doi.org/10.1007/978-3-031-80805-0_7

Malizia et al., 2020; González-Caro et al., 2023). In terms of carbon storage, Duque et al. (2021) showed that forest productivity decreases toward the subtropical Andes.

In Argentina, the subtropical Andean forests are of great importance in terms of biodiversity and provision of environmental goods and services. These forests occupy an area of about 40.000 km^2 and support a human population of more than two million people mostly settled in the eastern foothills. Although there is not a precise quantification, the number of species of the flora and fauna within this Andean region potentially represent up to 40% of the country's species richness, in less than 2% of the national territory (Brown et al., 2006). These mountain forests are key in the provision of water for human consumption and irrigation of ca. 400.000 ha of crops (Brown et al., 2006), in addition to generating important timber and non-timber forest resources (Malizia et al., 2009). Currently, natural and semi-natural areas represent about 80% of subtropical Andean forests, that is 20% was transformed into crops or urbanization, and this occurs mainly in the lowlands (Nanni et al., 2020). Protected areas represent 9% of natural areas and are under national or provincial jurisdiction (e.g., national parks and provincial reserves), and some of them are managed by institutions (e.g. Universidad Nacional de Tucumán) and private landowners.

The broad geographical gradient along 2000 m of elevation (400–2300 m a.s.l.), and 750 km of latitudinal extension (22–28 °S), harbors a great heterogeneity of environmental conditions. Climate is monsoonal with a marked dry season and occasionally frost in winter, with annual rainfall between 800 and 2000 mm, and mean annual temperature between 14 and 22 °C (Bianchi & Yañez, 1992). In addition, water input through fog interception may be as important as direct rainfall above ~1000 m elevation (Hunzinger, 1997). These geographical gradients define the distribution of forests along discontinued mountain ranges, with a clear segregation between northern and southern forests (Brown et al., 2001), and differentiation among altitudinal vegetation zones, namely: premontane forests (400–900 m a.s.l.), lower montane forests (900–1500 m a.s.l.), and upper montane forests (1500–2300 m a.s.l) (Cabrera & Willink, 1980). Flora and fauna are relatively well known along both latitudinal and elevation gradients.

The forest structure displays variation in response to climatic and topographic heterogeneity, with decreasing species richness toward southern forests not only in tree species (Morales et al., 1995; Brown et al., 2001), but also in lianas (Malizia et al., 2015), and epiphytes (Ceballos, 2023). Along the elevation gradient, stem density and basal area increase, while canopy height and wood density of tree species decrease with elevation (Carilla & Grau, 2011; Malizia et al., 2012; Blundo et al., 2021). Tree species composition changes from deciduous and wind-dispersed species in premontane forests toward semi-deciduous or evergreen species and predominance of zoochory in lower and upper montane forests (Blundo et al., 2012; Malizia et al., 2012). Successional forests in abandoned productive lands may be dominated by native species, with a trend to converge towards typical mature forests (Grau et al., 1997; Carilla & Grau, 2011). However, sometimes successional forests become invaded by non-native species, with structural and functional characteristics different from native forests (Grau et al., 2008; Ceballos et al., 2020; Fernandez et al., 2020; Jimenez et al., 2023).

The distribution of animal species, such as, mammals, birds or amphibians follow the forest structure patterns across geographical gradients of subtropical Andean. Richness of mammal species decreases with latitude (Ojeda et al., 2008), as well as birds (Bellis et al., 2009) and amphibians (Lavilla et al., 2000). Focusing on mammal diversity, latitudinal decline of species richness has been associated with geographical barriers (i.e., discontinued mountain ranges) and certain traits of species (Ojeda et al., 2008). Along the elevation gradient, species richness of medium and large size mammal's decreases (Di Bitetti et al., 2013), somewhat previously reported for small mammals and bats (Ojeda & Mares, 1989). Domestic livestock grazing has notorious effects on mammal community assemblage (Nanni, 2015), affecting the abundance of different species. Despite forest fragmentation in lowlands due to land-use changes, it has been documented that medium and large mammal species make use of the forest corridors in the region (Albanesi et al., 2019).

In this chapter, we analyze patterns of above ground biomass (AGB) and tree species diversity from forest plots data to show differences along the latitudinal gradient, and among premontane, lower and upper montane forests. Then, we combine forest plot data with satellite imagery to map patterns of AGB and tree diversity across subtropical Andean forests of Argentina. The combined analysis of these patterns allowed us to identify areas with high conservation value at the regional scale (e.g., high AGB stocks, high tree diversity, or both). Furthermore, we analyze relationships between forest structure, in terms of AGB and tree diversity, and mammal species richness at the landscape scale, using camera trap surveys. We conclude discussing these results and their practical implications for land management and conservation planning, as well as for ecological and biodiversity research.

2 Measuring AGB and Biodiversity and Combination of Field Data and Satellite Images

For more than 30 years, we have managed a forest plot monitoring system with permanent plots established across subtropical Andean forests of northwestern Argentina (Fig. 1a). The *Subtropical Network of Permanent Plots* (Red Subtropical de Parcelas Permanentes, RedSPP; https://ier.conicet.gov.ar/red-subtropical-de-parcelas-permanentes-redspp-2/) has contributed to describe the main ecological patterns across latitudinal and elevation gradients, and re-censuses data allowing us to study forest dynamics. The RedSPP has 78 permanent plots (71.6 hectares) that are systematically measured every 5–10 years, following standardized protocols for Andean forests (Osinaga et al., 2014). These permanent plots generally have 1 hectare in size, and all trees with tree diameter ≥10 cm at breast (1.30 m height) are measured, marked with numbered aluminum tag and identified to species level. For each plot, we estimated AGB using the pantropical allometric equation developed

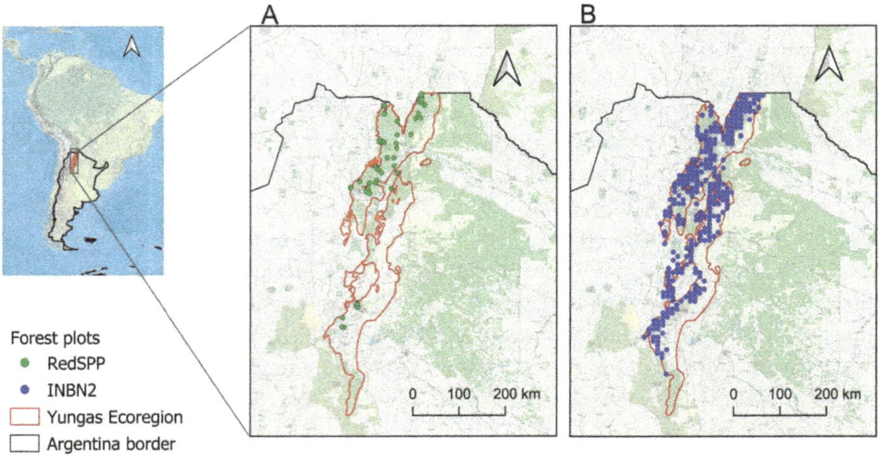

Fig. 1 Distribution of permanent forest plot monitoring systems across subtropical Andean forests of northwestern Argentina, (**a**) forest plots of RedSPP (Subtropical Network of Permanent Plots), and (**b**) forest plots of INBN2 (Second National Inventory of Native Forests)

by Chave et al. (2014), which included tree diameter and height, and wood density of species. Species wood density was extracted from local databases (Easdale et al., 2007; INTI-CITEMA, 2007; https://www.inti.gob.ar/publicaciones/descargac/365), and when local data were unavailable we used global data (Chave et al., 2006). The RedSPP database includes a total of 161 tree species (represented by 33,614 trees) out of the 226 tree species cited for the subtropical Andean forests (Digilio & Legname, 1966; Legname, 1982). We used these permanent plots data to describe patterns of AGB, tree species richness and species accumulation curves at the regional scale. We performed rarefaction curves of species richness controlled by the amount of individuals because there are some plots of different sizes.

Then, we linked forest plot data with satellite images to map AGB and tree diversity at the regional scale. To include forest areas not sampled by RedSPP, we added forest data from the *Second National Inventory of Native Forests* (INBN2; Dirección Nacional de Bosques, 2021; Fig. 1b). The INBN2 included 296 plots 0.1-ha in size established in a 10-km grid. Plot establishment follows a protocol which considers the general guidelines suggested by FAO for Latin America (Secretaría de Ambiente y Desarrollo Sustentable de la Nación, 2019). To harmonize the two databases (371 plots total), we consider one-tenths the size in 1-ha plots of RedSPP to match with the size of INBN2 plots. We estimated AGB as described above, and we calculated the Shannon Index with the relative abundance of tree species for all plots.

We processed and analyzed remote sensing data (Landsat 8 and MODIS imagery, and Radar data from Sentinel-1) using Google Earth Engine (GEE), selecting predictor variables based on Essential Biodiversity Variables (EBVs), as suggested by Vihervaara et al. (2017). We utilized Landsat 8 surface reflectance images for both the wet (October to March) and dry (June to August) periods of 2019 and 2020. The selected spectral bands included B1 (coastal aerosol), B2 (blue), B3 (green), B4 (red), B5 (near-infrared), and B6 (short-wave infrared). Additionally, we calculated the Normalized Difference Vegetation Index (NDVI). For each period, we extracted the median values of these bands and the NDVI to represent seasonal conditions. Cloud cover was filtered to include only images with less than 20% cloud coverage, and further cloud masking was applied using a custom function. We incorporated MODIS combined Leaf Area Index (LAI) and Fraction of Photosynthetically Active Radiation (FPAR) products for each season. The median values for the LAI and FPAR were computed for the wet and dry periods. Radar imagery from the Sentinel-1 GRD dataset was included to capture structural information of the vegetation. We filtered the data for VV and VH polarizations for the wet season. We calculated the mean, median, minimum, and maximum values for the VV polarization, and the mean values for the VH polarization. Additional variables included were latitude, tree probability (from Google Dynamic World), evapotranspiration data (from TerraClimate), elevation and slope (from the Shuttle Radar Topographic Mission). The analysis followed a methodology for species distribution modeling in GEE (Crego et al., 2022). With this information, we performed a correlation analysis to identify and remove highly correlated variables, ensuring the robustness of our predictive models. For biodiversity, the final set of selected bands included: NDVI, elevation (dem), slope, B4 (wet period), B6 (wet period), B4 (dry period), B5 (dry period), latitude, tree probability, radar median, radar minimum, radar maximum, and LAI (dry period). For biomass, the final set of selected bands included: elevation, NDVI (dry period), B2 (wet period), B3 (wet period), B4 (wet period), B5 (wet period), B6 (wet period), tree probability, radar median, radar minimum, radar maximum, and FPAR (dry period).

We conducted 10 classifications analyzing the relationship between biomass, the Shannon Index and their respectively predictor variables using a machine learning algorithm (Random Forest, Breiman, 2001). With the final maps we sampled a 10-km systematic grid to obtain AGB and Shannon index at each sampling point (N = 592) to analyze the relationship between forest biomass and tree diversity in subtropical Andean forests. We evaluated the fit of different models (linear and polynomial) for AGB as a function of the tree diversity.

Regarding mammalian biodiversity, the ProYungas Foundation, a non-governmental organization for conservation and sustainable development of subtropical forests (https://proyungas.org.ar/), has been carrying out studies of mammals with camera traps in subtropical Andean forests. Several surveys on private properties, mainly located in premontane forests (~600 m a.s.l), have evaluated mammal's diversity comparing continuous forests and areas close to crops, such as, sugarcane and citrus (e.g., Di Bitetti et al., 2013; Albanesi et al., 2016, 2019). With the objective of evaluating the relationship between AGB and tree diversity (as descriptors of forest structure) and mammal species richness, we used the geographical location of camera traps to extract values of AGB and tree Shannon index from the maps described above. Thus, we combine data at the landscape scale (~8.000 hectares) for mammal species richness, AGB and tree Shannon index in 368 points separated by at least 1.5 km. We adjusted a generalized linear model (GLM) with

Poisson distribution for mammal species richness as a function of AGB and tree Shannon index. All analyzes were performed in R (R Core Team, 2023).

3 AGB Stocks and Tree Species Diversity: Results from the RedSPP Permanent Plots

Patterns of forest biomass decrease along latitudinal and elevation gradients (Table 1). AGB stock for northern forests is 189 ± 8 Mg ha^{-1}, decreasing toward southern forests to 147 ± 11 Mg ha^{-1}. AGB also decreases from premontane forests toward upper montane forests both in the north (from 221 to 151 Mg ha^{-1}) and the south regions (from 163 to 149 Mg ha^{-1}). Climate variation does not seem to explain these patterns directly (Blundo et al., 2021); however, temperature and rainfall explain changes in forest structure and species composition (Malizia et al., 2012; Blundo et al., 2012). For example, there is less basal area per unit area with dominance of hardwood species in premontane forests (e.g., *Anadenanthera colubrina*, *Myracrodruon urundeuva*, *Phyllostylon rhamnoides*), and greater basal area per unit area dominated by softwoods species in upper montane forest (e.g., *Alnus acuminata*, *Podocarpus parlatorei*, *Juglans australis*). Furthermore, the range of variation in biomass among forest plots may be large (Table 1), because most of the plots are established in mature forests, but there are some plots in successional forests (e.g., post-agriculture in some southern plots) or in forests with history of timber extraction (e.g., some plots in northern forests).

Tree species richness decreases with latitude, a pattern previously documented (Morales et al., 1995), with 148 tree species registered in northern forests and 67 tree species registered in southern forests (Table 1, Fig. 2a). Tree species richness

Table 1 Summary of forest aboveground biomass (AGB; Mg ha^{-1}), total tree species registered and tree species richness (ha^{-1}) in the permanent plots of RedSPP. Elevation range: premontane forests: 400–900 m a.s.l, lower montane forests: 900–1500 m a.s.l, and upper montane forests: 1500–2300 m a.s.l.

	N plots (ha)	Mean AGB ± sd (Mg ha^{-1})	Total tree species (n)
Subtropical Andes (Argentina)	75 (69.4)	174 ± 7	161
Northern forests (22–24 °S)	47 (47)	189 ± 8	148
Southern forests (26–28 °S)	28 (22.4)	147 ± 11	67
Northern forests		Mean AGB (range)	Richness (ha^{-1})
Premontane forests	20 (20)	221 (114–312)	25–49
Lower montane forests	12 (12)	183 (126–233)	31–53
Upper montane forests	15 (15)	151 (82–247)	4–35
Southern forests			
Premontane forests	6 (5)	163 (131–188)	8–24
Lower montane forests	11 (13.7)	137 (31–229)	9–31
Upper montane forests	11 (3.7)	149 (40–249)	4–15

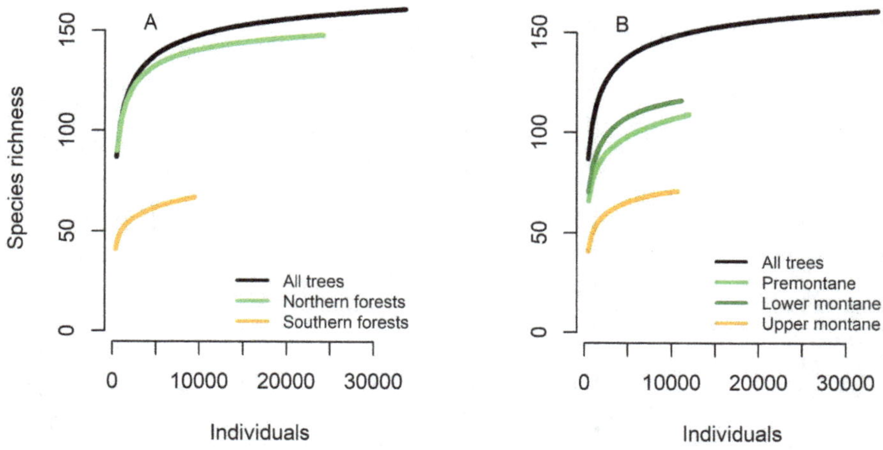

Fig. 2 Rarefaction curves of species richness by individual's sample: (**a**) differentiating northern and southern forests, and (**b**) differentiating forest types along elevation

shows a nonlinear pattern with elevation gradient (R-squared = 0.29, P < 0.001), with higher species richness at middle elevation (~1200 m a.s.l) and decreasing toward upper montane forests (Table 1). Total tree species richness registered by altitudinal vegetation zone were: 109 in premontane forests, 116 in lower montane forests, and 71 tree species in plots of upper montane forests (Fig. 2b). Few species are present along all elevations; for example, *Parapiptadenia excelsa*, *Ocotea porphyria* and *Allophylus edulis* (Malizia et al., 2012). High species turnover occurs among forest types and between northern and southern forests, with only 24 out the 161 tree species present across the complete latitudinal extension of subtropical Andean forests.

4 AGB and Tree Diversity Relationship for Regional Conservation Status Assessment

Final maps showed biomass values ranging between 18 and 222 Mg ha^{-1}, with a mean squared error of 80, and Shannon Index varied between 0.4 (in the crop and grassland interface zone) and 1.8, with a mean squared error of 0.49 (Fig. 3). We found a strong positive correlation (Pearson' coefficient = 0.75, P < 0.0001) between AGB and Shannon index in the 592 sampled points from these maps. Although biomass tends to increase with tree species diversity, forests with the highest biomass (150–200 Mg ha^{-1}) did not show maximum Shannon index, while areas with the highest Shannon index (2.1–2.3) showed a mean AGB of 120 Mg ha^{-1} (Fig. 4). These discrepancies imply that the superimposition of patterns of biomass and tree species diversity could generate alternative situations in which high conservation areas result from different combinations of high biomass and high species diversity.

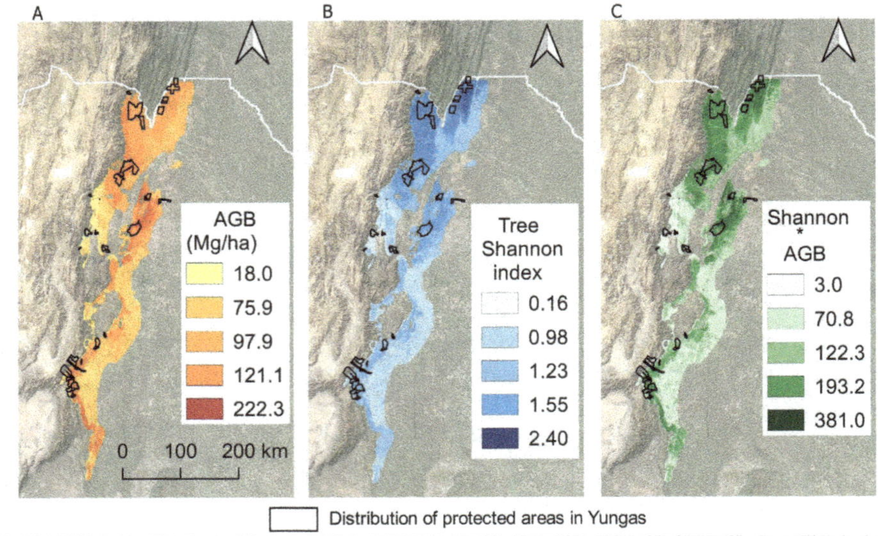

Fig. 3 Maps of (**a**) above ground biomass (AGB), and (**b**) tree diversity estimated with Shannon index for subtropical Andean forests of Northwestern Argentina. In (**c**) map resulting from the superimposition of AGB and tree Shannon index maps. The black polygons in the maps indicate the location and extension of protected areas

Fig. 4 Relationship between forest diversity measured as tree Shannon index and above ground biomass (AGB) in points sampled from maps shown in Fig. 3. AGB per hectare on *axis y*. A polynomial model ($y \sim x + x^2$), showed a R-squared = 0.59

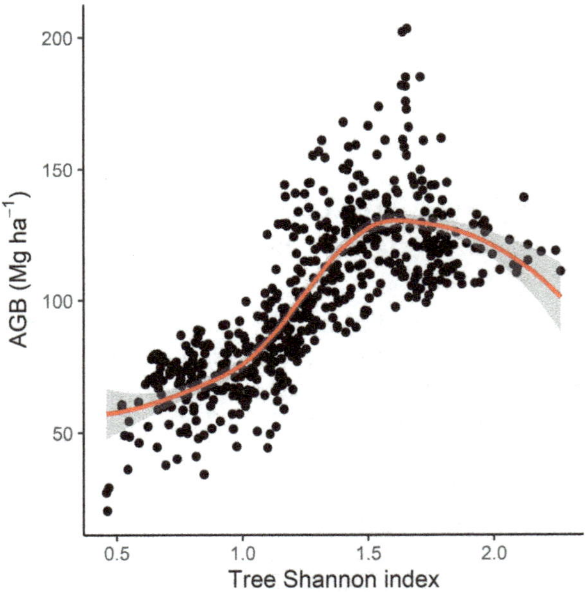

For example, forests with moderate selective logging have less stocks of biomass than mature forests but present high tree diversity due to successional processes promoting the coexistence of pioneer and late-succession species (Blundo & Malizia, 2009). On the other side, forests close to urban areas have higher stock of biomass than mature forests despite being dominated by non-native tree species diminishing stand-level biodiversity (Jimenez et al., 2022, Malizia et al., *in review*). In addition, here we only analyzed tree diversity, however Duque et al. (2014) have suggested that other life forms should also be considered (e.g., epiphytes, lianas), particularly along mountain forests because they increase biodiversity although biomass decreases with elevation.

The effect of biodiversity on the productivity of ecosystems is essential to understand the impact of local species loss on the functioning of natural ecosystems. The relationship productivity-biodiversity may be used for accurate valuation of biodiversity and successful integration of biological conservation and socioeconomic development in forest ecosystems (Liang et al., 2016). There are increasing efforts to describe the structure of forests communities, and quantify relationships between plant species diversity and biomass, as a surrogate for productivity in long-lived trees (e.g., Vilà et al., 2007; Vance-Chalcraft et al., 2010; Zhang et al., 2012). In general, significant relationships between tree diversity and productivity were found, but the magnitude and direction are not consistent and seem to depend on the surrounding environment.

5 Forest Structure and Mammal Species Richness at the Landscape Scale

Over more than 10 years, different camera traps surveys carried out by Foundation ProYungas have generated 3.040 photographs of mammals, with records for 32 species of medium (24 species; ≤15 kg average) and large (eight species; >15 kg average) size mammals, which represent approximately 75% of mammal species of medium and large size cited for these forests (Jayat et al., 2009; Jayat & Ortiz, 2010). Results from these surveys highlight that among the most frequent large mammals stand out gray brocket deer (*Mazama gouazoubira*), collared peccary (*Pecari tajacu*) and lowland tapir (*Tapirus terrestris*), and with fewer registrations appear puma (*Puma concolor*) and jaguarundi (*P. yagouaroundi*), and the emblematic jaguar (*Panthera onca*). Among the most common medium-sized mammals we find tayra (*Eira barbara*), agouti (*Dasyprocta sp.*) and tapeti (*Sylvilagus brasiliensis*). Mammal species richness tends to be greater in areas of continuous forest, where a single camera trap may record between 10 and 12 species (Albanesi et al., 2019).

We found that the relationships between mammal species richness and biomass and tree diversity (as forest structure descriptors), were not significant (Fig. 5). According to the maps of AGB and tree Shannon index developed in this study, the

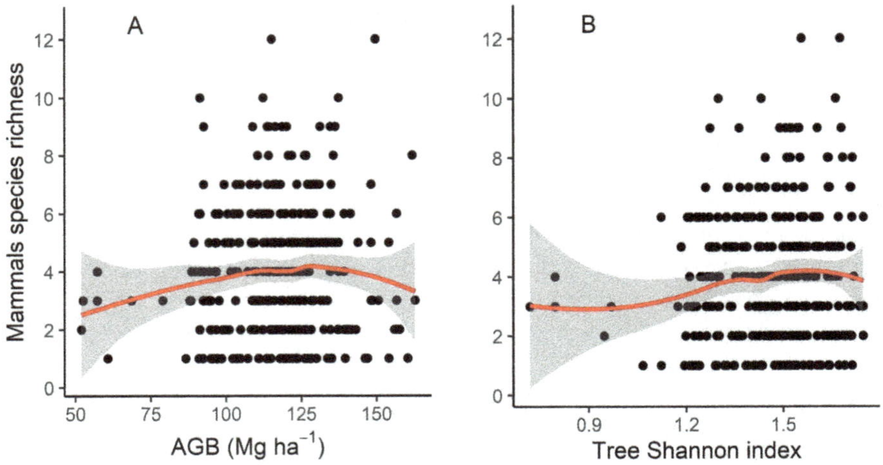

Fig. 5 Relationships among mammal species richness and forest structure in terms of (**a**) above ground biomass (AGB), and (**b**) tree diversity (Shannon Index). Albanesi et al. (2019) explained in detail the methodology of mammal surveys with camera-traps

sites where camera traps were located show a relatively high variation in the AGB stock (between 52 and 163 Mg ha^{-1}), and in tree Shannon index (between 0.7 and 1.7). Probably, direct measurements of AGB and tree diversity at the camera trap sites instead of estimated values, could help to identify patterns more clearly. However, the sites where the highest values of mammal species richness were recorded (between 8 and 12 species per camera trap), present a relatively good forest structure, with 100–125 tons of biomass per hectare and Shannon indices greater than 1.5 (Fig. 5).

There have been documented positive correlations between mammal species richness, carbon stocks and net primary productivity across different forest regions (Siikamäki & Newbold, 2012; Gebert et al., 2019). Large and medium size mammals play a crucial role in controlling ecosystem processes, such as nutrient cycling and energy flow, by turning over high amounts of biomass (Veldhuis et al., 2018). Decades ago, Carey and Johnson (1995) showed higher mammal diversity and abundance in forests with a good and continuous structure (e.g., large diameters of trees, closer canopy cover). Another study reported the same pattern in the Atlantic forest, with high mammal diversity in forests with higher carbon stocks (Galetti et al., 2017). Our results show some evidence supporting the idea of meeting more species of mammals in areas of continuous forests with moderate to high values of biomass and tree diversity, which are considered indicators of relatively good forest structure.

6 Perspective of Forest Management and Land-Use Planning in Subtropical Andean Forests

Forests are undergoing changes in their structure and species composition in response to global environmental changes. Climate and land-use changes are the main components of these changes that modify forest structure, demography and species composition (Vitousek et al., 1997; Lewis et al., 2004). Mapping forest structure at regional spatial scale is important to understand the provision of goods and ecosystem services, and to generate information that contributes to decision-making regarding forest management and conservation. The integration of information generated with permanent plots and remote sensing data represents an opportunity to extrapolate local measures of structure and dynamics to broad regional scales (Chambers et al., 2007). However, given the spatial variation that exists in forests with high biodiversity (e.g., species composition changes over short distances), the limitation to perform regional maps is the availability of forest monitoring data (Baker et al., 2021). In subtropical Andean forests, long-term monitoring of large plots (e.g., 1-ha plots of the RedSPP) has facilitated the knowledge of the growth and mortality rates of species, forest carbon balance, and the impact of climate change on forest composition, among other ecological issues. Increasing knowledge in all these aspects is essential for efficient management of forest areas (e.g. knowledge about growth of timber species). In addition, data from the national inventory allows us to systematically cover all types of forests at the regional scale, increasing the probability of including forests in different conservation status, often sub-sampled by monitoring systems for scientific purposes.

This is the first time that it has been possible to unify forest data from the two largest monitoring initiatives in subtropical Andean forests of northwestern Argentina, such as the RedSPP and INBN2, designed for different objectives and with some methodological differences but similar core data. Additionally, we have achieved maps of AGB and tree diversity with acceptable margins of error that can surely be improved. We found a positive relationship between AGB and tree Shannon diversity index, quite evident when looking at the maps, but also supported statistically. This is a starting point in the generation of a tool that could help decision makers in planning the use of the territory in these mountain forests. In addition, although we do not find strong evidence of the relationship between mammal species richness and AGB and tree diversity, forests with a continuous canopy and moderate to high AGB and tree diversity, tend to promote higher mammal species richness.

Modern land use change started early in the southern subtropical Andean forests, since sugarcane cultivation began in the mid-nineteenth century (Reboratti, 2014). This has resulted in a greater amount of area transformed with premontane forests severely reduced (Brown & Malizia, 2004). In contrast, montane forests still retain a greater proportion of their original area and in many places have been expanding (e.g., Gutierres & Grau, 2014; Nanni et al., 2015; Jimenez et al., 2021), in addition to harboring greater biodiversity. This situation is reflected in the maps that we generate in this study, from which we estimated about 1.377.000 hectares of forest with AGB stocks between 120 and 220 Mg ha^{-1} and tree Shannon index between 1.6 and 2.4 (around 20–30 tree species per hectare). The protected areas in this sector, include about 246.000 hectares under national and provincial jurisdiction. In this context, about 18% of forests with higher AGB and tree diversity in the northern subtropical Andes are under some level of legal protection. The creation of new protected areas could be one of the options to foster forest conservation. However, the combination of little pressure for transformation of montane forests into modern agriculture, and a land tenure structure characterized by large properties ranging from the foothills to montane areas, allows the implementation of alternative strategies that combine intensive agriculture in flat terrains and privately owned conservation in the slopes. For example, the "Paisajes Productivos Protegidos" of Fundación ProYungas (Productive Protected Landscapes) is an initiative that promotes the integrated management of productive and natural areas at the landscape scale (Brown, 2021), and adds a substantial number of hectares to the forest conservation system in an innovative scheme; more than 2.5 million of hectares in the North of Argentina, including subtropical Andean forests, Chaco and Atlantic Forests, and private areas of forests in Paraguay, Chile and Bolivia.

The Yungas ecoregions have experienced anthropogenic pressure in the past, leading to the deforestation of the lower belt in the southern part as a consequence of agricultural expansion. Currently, some areas are still under pressure from urban development, but the proportion affected is low due to a significant portion of the region being under protection. Additionally, socioeconomic changes have led to a reduction of rural population and in livestock activities, with possible recovery of native fauna. However, other processes of change are emerging, in particular expansion of exotic species (e.g., *Ligustrum lucidum*, *Gleditzia triacanthos*, *Morus alba*, etc.) often spreading from peri-urban areas (Grau et al., 2008; Jimenez et al., 2022). This is generating new patterns of diversity that may become a focus of future research on the relationships between biomass, diversity and conservation in the region.

In summary, the subtropical Andean forests of Argentina represent a highly valuable ecosystem. Decades of field-based permanent plots, now coupled with remote sensing data, and a growing database of camera-trap records of mammals, allowed a detailed characterization of biomass and biodiversity and put numbers at their strong positive relationship. As novel processes of change emerge, this baseline information will allow real time monitoring as a tool for landscape and territorial planning.

Acknowledgments We especially thank the work team of the *Dirección Nacional de Bosques* for providing us supporting with the inventory data. We thank E. Aráoz for reading the manuscript critically.

Literature Cited

Albanesi, S. A., Jayat, J. P., & Brown, A. D. (2016). Patrones de actividad de mamíferos de medio y gran porte en el pedemonte de yungas del noroeste argentino. *Mastozoología Neotropical, 23*(2), 335–358.

Albanesi, S. A., Alberti, P., Jayat, J. P., & Brown, A. D. (2019). Mamíferos de mediano y gran porte en corredores boscosos del piedemonte de yungas del noroeste argentino. *Mastozoología Neotropical, 26*(2), 220–232.

Baez, S., Malizia, A., Carilla, J., Blundo, C., Aguilar, M., Aguirre, N., et al. (2015). Large-scale patterns of turnover and basal area change in Andean forests. *PLoS One, 10*(5), e0126594.

Baker, T. R., Vicuña Miñano, E., Banda-R, K., del Castillo Torres, D., Farfan-Rios, W., Lawson, I. T., et al. (2021). From plots to policy: How to ensure long-term forest plot data supports environmental management in intact tropical forest landscapes. *Plants, People, Planet, 3*(3), 229–237.

Bellis, L. M., Rivera, L., Politi, N., Martín, E., Perasso, M. L., Cornell, F., & Renison, D. (2009). Latitudinal patterns of bird richness, diversity and abundance in Polylepis australis mountain forest of Argentina. *Bird Conservation International, 19*(3), 265–276.

Bianchi, A. R., & Yañez, C. E. (1992). *Las precipitaciones en el noroeste Argentino: Salta*. INTA Estación Experimental Agropecuaria.

Blundo, C., & Malizia, L. R. (2009). Impacto del aprovechamiento forestal en la estructura y diversidad de la Selva Pedemontana. In *Selva pedemontana de las Yungas, historia natural ecología y manejo de un ecosistema en peligro* (pp. 387–406). ProYungas.

Blundo, C., Malizia, L. R., Blake, J. G., & Brown, A. D. (2012). Tree species distribution in Andean forests: Influence of regional and local factors. *Journal of Tropical Ecology, 28*(1), 83–95.

Blundo, C., Malizia, A., Malizia, L. R., & Lichstein, J. W. (2021). Forest biomass stocks and dynamics across the subtropical Andes. *Biotropica, 53*(1), 170–178.

Breiman, L. (2001). Random forests. *Machine Learning, 45*, 5–32. https://doi.org/10.1023/A:1010933404324

Brown, A. D. (2021). *Ambiente y Producción II* (Ediciones del Subtropico ed., p. 173). Tucumán.

Brown, A. D., & Malizia, L. R. (2004). Las selvas pedemontanas de las Yungas. *Ciencia hoy, 14*(83), 52–63.

Brown, A. D., Grau, H. R., Malizia, L. R., & Grau, A. (2001). Los Bosques Nublados de la Argentina. In M. Kappelle & A. D. Brown (Eds.), *Bosques Nublados de Latinoamerica* (pp. 623–659). Editorial INBio.

Brown, A. D., Martinez Ortiz, U., Acerbi, M., & Corcuera, J. F. (2006). *La situación ambiental argentina 2005*.

Cabrera, A. L., & Willink, A. (1980). *Biogeografía de América Latina*. Segunda edición corregida. Colección de Monografías Científicas de la Secretaría General de la Organización de los Estados Americanos, Programa Regional de Desarrollo Científico y Tecnológico.

Carey, A. B., & Johnson, M. L. (1995). Small mammals in managed, naturally young, and old-growth forests. *Ecological Applications, 5*(2), 336–352.

Carilla, J., & Grau, R. (2011). Tendencias sucesionales de los bosques montanos subtropicales del noroeste argentino. *Bosque (Valdivia), 32*(2), 97–111.

Ceballos, S. J. (2023). Vascular epiphytes in Argentinian Yungas: Distribution, diversity, and ecology. *The Botanical Review, 89*(1), 91–113.

Ceballos, S. J., Jiménez, Y. G., & Fernandez, R. D. (2020). Estructura de los bosques de Gleditsia triacanthos en función de la edad (valle de La Sala, Tucumán, Argentina). *Ecología Austral, 30*, 251–259.

Chambers, J. Q., Asner, G. P., Morton, D. C., Anderson, L. O., Saatchi, S. S., Espírito-Santo, F. D., et al. (2007). Regional ecosystem structure and function: Ecological insights from remote sensing of tropical forests. *Trends in Ecology & Evolution, 22*(8), 414–423.

Chave, J., Muller-Landau, H. C., Baker, T. R., Easdale, T. A., Steege, H. T., & Webb, C. O. (2006). Regional and phylogenetic variation of wood density across 2456 neotropical tree species. *Ecological Applications, 16*(6), 2356–2367.

Chave, J., Réjou-Méchain, M., Búrquez, A., Chidumayo, E., Colgan, M. S., Delitti, W. B., et al. (2014). Improved allometric models to estimate the aboveground biomass of tropical trees. *Global Change Biology, 20*(10), 3177–3190.

Crego, R. D., Stabach, J. A., & Connette, G. (2022). Implementation of species distribution models in Google Earth Engine. *Diversity and Distributions, 28*(5), 904–916.

Di Bitetti, M. S., Albanesi, S. A., Foguet, M. J., De Angelo, C., & Brown, A. D. (2013). The effect of anthropic pressures and elevation on the large and medium-sized terrestrial mammals of the subtropical mountain forests (Yungas) of NW Argentina. *Mammalian Biology, 78*(1), 21–27.

Digilio, A. P. L., & Legname, P. R. (1966). *The indigenous trees of the province of Tucuman, Argentina*. UNT, Instituto Miguel Lillo.

Dirección Nacional de Bosques. (2021). *Datos del Segundo Inventario Nacional de Bosques Nativos de la República Argentina*. Ministerio de Ambiente y Desarrollo Sostenible de la Nación.

Duque, A., Feeley, K. J., Cabrera, E., Callejas, R., & Idarraga, A. (2014). The dangers of carbon-centric conservation for biodiversity: A case study in the Andes. *Tropical Conservation Science, 7*(2), 178–191.

Duque, A., Peña, M. A., Cuesta, F., González-Caro, S., Kennedy, P., Phillips, O. L., et al. (2021). Mature Andean forests as globally important carbon sinks and future carbon refuges. *Nature Communications, 12*(1), 2138.

Easdale, T. A., Healey, J. R., Grau, H. R., & Malizia, A. (2007). Tree life histories in a montane subtropical forest: Species differ independently by shade-tolerance, turnover rate and substrate preference. *Journal of Ecology, 95*, 1234–1249.

Fernandez, R. D., Ceballos, S. J., Aragón, R., Malizia, A., Montti, L., Whitworth-Hulse, J. I., et al. (2020). A global review of Ligustrum lucidum (OLEACEAE) invasion. *The Botanical Review, 86*, 93–118.

Galetti, M., Brocardo, C. R., Begotti, R. A., Hortenci, L., Rocha-Mendes, F., Bernardo, C. S. S., et al. (2017). Defaunation and biomass collapse of mammals in the largest Atlantic forest remnant. *Animal Conservation, 20*(3), 270–281.

Gebert, F., Njovu, H. K., Treydte, A. C., Steffan-Dewenter, I., & Peters, M. K. (2019). Primary productivity and habitat protection predict elevational species richness and community biomass of large mammals on Mt. Kilimanjaro. *Journal of Animal Ecology, 88*(12), 1860–1872.

González-Caro, S., Tello, J. S., Myers, J. A., Feeley, K., Blundo, C., Calderón-Loor, M., et al. (2023). Historical assembly of Andean tree communities. *Plants, 12*(20), 3546.

Grau, H. R., Arturi, M. F., Brown, A. D., & Aceñolaza, P. G. (1997). Floristic and structural patterns along a chronosequence of secondary forest succession in Argentinean subtropical montane forests. *Forest Ecology and Management, 95*(2), 161–171.

Grau, H. R., Hernández, M. E., Gutierrez, J., Gasparri, N. I., Casavecchia, M. C., Flores-Ivaldi, E. E., & Paolini, L. (2008). A peri-urban neotropical forest transition and its consequences for environmental services. *Ecology and Society, 13*(1), 35.

Gutierres, A. J., & Grau, H. R. (2014). Assessment of swaps and persistence in land cover changes in a subtropical peri urban region, NW Argentina. *Landscape and Urban Planning, 127*, 83–93.

Hunzinger, H. (1997). Hydrology of montane forests in the sierra de San Javier, Tucumán, Argentina. *Mountain Research and Development, 17*, 299–308.

Jayat, J. P., & Ortiz, P. E. (2010). Mamíferos del pedemonte de Yungas de la Alta Cuenca del Río Bermejo en Argentina: una línea de base de diversidad. *Mastozoología Neotropical, 17*(1), 69–86.

Jayat, J. P., Ortiz, P. E., & Miotti, M. D. (2009). Mamíferos de la selva pedemontana del noroeste argentino. Ecología, historia natural y conservación de la selva pedemontana de las Yungas australes (AD Brown, PG Blendinger, T. Lomáscolo P. García Bes Ediciones del Subtrópico, Tucumán.

Jimenez, Y. G., Aráoz, E., Grau, H. R., & Paolini, L. (2021). Linking forest transition, plant invasion and forest succession theories: Socioeconomic drivers and composition of new subtropical andean forests. *Landscape Ecology, 36*, 1161–1176.

Jimenez, Y. G., Ceballos, S. J., Aráoz, E., Blundo, C., Carilla, J., Malizia, A., & Grau, H. R. (2022). Spatial, temporal and ecological patterns of Peri-urban Forest transitions. An example from subtropical Argentina. *Frontiers in Forests and Global Change, 5*, 761957.

Jimenez, Y. G., Grau, H. R., Fernandez, R. D., Ceballos, S. J., Blundo, C., Malizia, A., et al. (2023). The role of species composition in montane reforested areas to offset carbon emissions. *Remote Sensing Applications: Society and Environment, 32*, 101024.

Kappelle, M., & Brown, A. D. (2001). *Bosques nublados del neotrópico* (pp. 301–370). Instituto Nacional de la Biodiversidad.

Lavilla, E. O., Vaira, M., Ponssa, M. L., & Ferrari, L. (2000). Batracofauna de las Yungas Andinas de Argentina: una síntesis. *Cuadernos de herpetología, 14*, 5–26.

Legname, P. R. (1982). *Arboles indígenas del noroeste argentino (Salta, Jujuy, Santiago del Estero y Catamarca)*.

Lewis, S. L., Malhi, Y., & Phillips, O. L. (2004). Fingerprinting the impacts of global change on tropical forests. *Philosophical Transactions of the Royal Society of London. Series B: Biological Sciences, 359*(1443), 437–462.

Liang, J., Crowther, T. W., Picard, N., Wiser, S., Zhou, M., Alberti, G., et al. (2016). Positive biodiversity-productivity relationship predominant in global forests. *Science, 354*(6309), aaf8957.

Malizia, L. R., Pacheco, S., & Loiselle, B. (2009). Árboles de valor forestal en las Yungas de la Alta Cuenca del río Bermejo. In *Selva pedemontana de las Yungas, historia natural ecología y manejo de un ecosistema en peligro* (pp. 105–120).

Malizia, L., Pacheco, S., Blundo, C., & Brown, A. D. (2012). Caracterización altitudinal, uso y conservación de las Yungas Subtropicales de Argentina. *Ecosistemas: Revista Cietifica y Tecnica de Ecologia y Medio Ambiente, 21*(1–2), 53–73.

Malizia, A., Campanello, P. I., Villagra, M., & Ceballos, S. (2015). Geographical, taxonomical and ecological aspects of Lianas in subtropical forests of Argentina. In N. Parthasarathy (Ed.), *Biodiversity of Lianas* (Sustainable development and biodiversity) (Vol. 5). Springer.

Malizia, A., Blundo, C., Carilla, J., Osinaga Acosta, O., Cuesta, F., Duque, A., et al. (2020). Elevation and latitude drives structure and tree species composition in Andean forests: Results from a large-scale plot network. *PLoS One, 15*(4), e0231553.

Morales, J. M., Siromba, M., & Brown, A. D. (1995). Riqueza de árboles en las Yunas argentinas. In A. D. Brown & H. R. Grau (Eds.), *Investigación, conservación y desarrollo en selvas subtropicales de montaña* (pp. 163–174). Laboratorio de Investigaciones Ecológicas de las Yungas.

Nanni, A. S. (2015). Dissimilar responses of the Gray brocket deer (Mazama gouazoubira), Crab-eating fox (Cerdocyon thous) and Pampas fox (Lycalopex gymnocercus) to livestock frequency in subtropical forests of NW Argentina. *Mammalian Biology, 80*(4), 260–264.

Nanni, A. S., Gasparri, N. I., & Grau, H. R. (2015). Redistribution of forest biomass in the heterogeneous environment of subtropical Andes undergoing agriculture adjustment. *Applied Geography, 62*, 107–114.

Nanni, A. S., Piquer Rodríguez, M., Rodriguez, M. D., Núñez Regueiro, M. M., Periago, M. E., Aguiar, S., et al. (2020). Presiones sobre la conservación asociadas al uso de la tierra en las ecorregiones terrestres de la Argentina. *Ecología Austral, 30*, 304–320.

Ojeda, R. A., & Mares, M. A. (1989). *A biogeographic analysis of the mammals of Salta Province, Argentina*. Special Publication 27. The Museum.

Ojeda, R. A., Barquez, R. M., Stadler, J., & Brandl, R. (2008). Decline of mammal species diversity along the Yungas forest of Argentina. *Biotropica, 40*(4), 515–521.

Osinaga, O., Báez, S., Cuesta, F., Malizia, A., Carrilla, J., Aguirre, N., & Malizia, L. (2014). *Monitoreo de diversidad vegetal y carbono en bosques andinos-Protocolo extendido*. Protocolo 2-Versión 1. CONDESAN/IER-UNT/COSUDE.

R Core Team. (2023). *R: A language and environment for statistical computing*. R Foundation for Statistical Computing.

Reboratti, C. (2014). El noroeste entre la globalización y la marginación. *Geograficando, 10*.

Secretaría de Gobierno de Ambiente y Desarrollo Sustentable de la Nación. (2019). *Segundo Inventario Nacional de Bosques Nativos: manual de campo*. Secretaría de Gobierno de Ambiente y Desarrollo Sustentable de la Nación.

Siikamäki, J., & Newbold, S. C. (2012). Potential biodiversity benefits from international programs to reduce carbon emissions from deforestation. *Ambio, 41*, 78–89.

Vance-Chalcraft, H. D., Willig, M. R., Cox, S. B., Lugo, A. E., & Scatena, F. N. (2010). Relationship between aboveground biomass and multiple measures of biodiversity in subtropical forest of Puerto Rico. *Biotropica, 42*(3), 290–299.

Veldhuis, M. P., Gommers, M. I., Olff, H., & Berg, M. P. (2018). Spatial redistribution of nutrients by large herbivores and dung beetles in a savanna ecosystem. *Journal of Ecology, 106*(1), 422–433.

Vihervaara, P., Auvinen, A. P., Mononen, L., Törmä, M., Ahlroth, P., Anttila, S., et al. (2017). How essential biodiversity variables and remote sensing can help national biodiversity monitoring. *Global Ecology and Conservation, 10*, 43–59.

Vilà, M., Vayreda, J., Comas, L., Ibáñez, J. J., Mata, T., & Obón, B. (2007). Species richness and wood production: A positive association in Mediterranean forests. *Ecology Letters, 10*(3), 241–250.

Vitousek, P. M., D'antonio, C. M., Loope, L. L., Rejmanek, M., & Westbrooks, R. (1997). Introduced species: A significant component of human-caused global change. *New Zealand Journal of Ecology, 21*, 1–16.

Zhang, Y., Chen, H. Y., & Reich, P. B. (2012). Forest productivity increases with evenness, species richness and trait variation: A global meta-analysis. *Journal of Ecology, 100*(3), 742–749.

Forest Connectivity Loss Surpasses Deforestation in the Colombian Andean Forests

Camilo A. Correa-Ayram, Oscar Godínez-Gómez, and Paulo J. Murillo-Sandoval

1 Introduction

In many tropical regions, imposing mountains and volcanoes rising from the tropical rainforest lowlands is prominent. Mountain environments can be vertically divided into altitudinal belts or zones that mark the distribution of known mountain ecosystem types called "*Orobiomes*" (Walter, 1973). Colombia is distinguished by the majesty of the Andean Mountain range, which unfolds in three parallel cordilleras: Western, Central and Eastern. The Western Cordillera, which has a lower altitude and narrower, reaches a maximum elevation of 4100 m, lacking snow-capped peaks but with a pronounced influence from the Pacific Ocean on its western side. On the other hand, the Central Cordillera has the most significant altitude, exceeding 3500 m with peaks rising beyond 5300 m. The Eastern Cordillera is the longest, with an average elevation of 2500 m. Its eastern side limits from north to south with the savannas of the Orinoco region and the tropical humid lowland forests of the Amazon region (Van der Hammen, 2000; Andrade, 2022).

The forests that extend over the Andes Cordillera at an altitudinal range from 1000 to 3500 m. are known as "Andean forests". The Andean forests in Colombia

C. A. Correa-Ayram (✉)
Pontificia Universidad Javeriana, Facultad de Estudios Ambientales y Rurales, Departamento de Ecología y Territorio, Bogotá, Colombia
e-mail: correa.c@javeriana.edu.co

O. Godínez-Gómez
School of Natural Resources and Environment, University of Florida, Gainesville, FL, USA
e-mail: oscargodinezgome@ufl.edu

P. J. Murillo-Sandoval
Departamento de Topografía, Facultad de Ciencias del Hábitat, Diseño e Infraestructura, Universidad del Tolima, Ibagué, Colombia
e-mail: pjmurillos@ut.edu.co

N. Clerici (ed.), *Conservation of Andean Forests*,
https://doi.org/10.1007/978-3-031-80805-0_8

are characterized by their high diversity of flora and fauna, as well as their high number of endemic species (Van der Hammen, 1989; Myers et al., 2000; Armenteras et al., 2007; Rahbek et al., 2019). This biodiversity results from several factors present in the Colombian territory, including climate, geology, geographical complexity, and the influence of two oceans (Raven et al., 2020; Renjifo et al., 2020). In particular, the Andean forests in Colombia have high rainfall and frequent fog presence, which is crucial to the distribution of Andean Forest ecosystems representing primary water sources (Bruijnzeel, 2001; Gomez-Peralta et al., 2008). Rain enters the forest as precipitation, intercepted by the vegetation through gravity (Cavelier, 1991). This interception process causes tiny droplets to be propelled by wind against tree trunks, branches, and leaves of the vegetation and coalesce into larger droplets that eventually drain to the ground. Fog interception is more significant on slopes exposed to the wind (windward slopes) than on sheltered ones (leeward slopes). This phenomenon increases with altitude on the same mountain and the angle inclination of the leaves (Cavelier, 1991). Fog plays a crucial role in maintaining high moisture levels in the soil and atmosphere, attenuating seasonal fluctuations derived from changes in rainfall patterns.

Unfortunately, Andean forests are highly vulnerable mountain ecosystems due to deforestation and other drivers of change. The imbalance of their ecological integrity directly compromises ecological and evolutionary processes and their ecosystem functions (Balthazar et al., 2015; Nicholson et al., 2021; Suescún et al., 2023). Although the current extent of the Andean forest is well-defined, it has significantly reduced. For example, the Andean region in Colombia spans 325,000 km^2, with 66% of it covered by Andean forests, which amounts to approximately 215,325 km^2 (Etter et al., 2017). According to the Mapbiomas land cover dataset in 2023 and a comparison with the colombian ecosystems map data from Etter et al. (2017), at least 45% of its original area has been transformed. During the last few decades, agricultural expansion, the introduction of pastures, road infrastructure development, mining, and other human pressures derived from human population growth and institutional frameworks have been the main drivers of the deforestation of Andean forests in Colombia (Rodríguez Eraso et al., 2013).

Forest connectivity refers to the degree to which forests are interconnected or linked. Although forest fragmentation has been historically documented (Armenteras et al., 2003; Calbi et al., 2020; Calderón-Caro & Benavides, 2022), its impact on forest connectivity processes in the Andean region remains poorly understood. Correa Ayram et al. (2020) considered that, along with tropical dry forests, Andean forests are the Colombian ecosystems that suffered the most increase over time in human pressure. For example, the Red List of Ecosystems identifies additional ecosystems in the Andean region as critically endangered, including the subhumid tropical Andean forests, the medium-dense forests of the high Andean cordilleran high plains, and the medium-dense forests of the subhumid high Andean hills and mountains (IUCN-CEM, 2022). The Andes in Colombia are a territory of global importance for biodiversity conservation and ecosystem services. However, the Andean region has a high population density and is mainly transformed by agricultural expansion and induced grasslands for livestock. It also has a high density of linear roads and energy infrastructure. These ecosystems are also under pressure caused by processes of agricultural frontier expansion with introduced grasses such as Kikuyu grass (*Pennisetum clandestinum*) (Etter et al., 2017). These conditions threaten forest connectivity in the Andean region, limiting vital long-term ecological processes, species migration, and gene flow.

Andean forests provide essential natural contributions to people (Mathez-Stiefel et al., 2017; Móstiga et al., 2023), such as water supply, erosion control, and carbon storage, and they are biodiversity hotspots harboring several endemic species. Maintaining connectivity between forest patches allows the dispersal of plants and movement of animals between subpopulations as the climate changes, reducing the risk of extinction and increasing adaptability to climate change. Loss of forest connectivity threatens to isolate subpopulations, inhibit species migrations to favorable habitats, alter pollination and seed dispersal, and degrade genetic diversity. The fundamental role of Andean forests in terms of connectivity ranges from acting as corridors between the Neotropical and Nearctic environments as a crucial evolutionary process (Hoorn et al., 2010; Clerici et al., 2019) to balancing hydrological connectivity in the Andean-Amazon region and Andean-Llanos region (Anderson et al., 2018), processes that are fundamental to maintaining Colombian water resources.

In this chapter, we address forest connectivity as a characteristic of the landscape that enables, to a higher or lesser extent, the movement and dispersal of species, genetic exchange, and other ecological flows across the landscape (Taylor et al., 1993; Saura et al., 2011). This analysis focuses on Andean forest ecosystems between 1000 and 3500 m. We used the Andean region of Colombia as the study area boundary. Based on annual land cover data provided by the Mapbiomas project (Souza & Azevedo, 2017), we analyzed the status of forest connectivity in the Colombian Andean forests between 1985 and 2022 and identified hotspots of landscape connectivity loss. Quantifying declines in connectivity during this period is essential to understand the extent of fragmentation, identify priority areas for conservation, and develop strategies to restore landscape connectivity (Murillo-Sandoval et al., 2022; Linero-Triana et al., 2023). These findings can inform environmental policies and land use planning initiatives to balance development with strategies to maintain and restore the connectivity of threatened Andean forest ecosystems.

2 Study Area

We centered our study in the Andean region of Colombia, which extends for 325,000 km^2. This region fully covers the mountain ecosystems (*orobiomes*) of the Evergreen Tropical Rainforest Zonobiome (Walter, 1973). However, taking the Ecosystems Map of Colombia by Etter et al. (2017) as a reference, we focused only on the Andean Forest ecosystems distributed with an original extension of 215,325 km^2. Other mountain ecosystems classified within the zonal Orobiomes of the Evergreen Tropical Rainforest Zonobiome, such as the high Andean xerophytic shrublands and the inter-Andean valley xerophytic shrublands were excluded. Páramo ecosystem and bogs were also excluded (Fig. 1).

Fig. 1 Study area. Current and original Andean forests indicated the potential forest in the Andean region

3 Methodology

Our strategy to analyze Colombia's Andean-Amazon connectivity loss is divided into two parts. First, we used the annual land cover maps from the MapBiomas Colombia initiative produced from the pixel-by-pixel classification of Landsat satellite images between 1985 and 2022 (Mapbiomas, 2023). These maps have an overall accuracy of 82%. The maps are publicly available and can be found on the project's website (https://colombia.mapbiomas.org/). To understand the distribution of the Andean forest, we reclassified the annual land cover maps into two categories: Transformed (all anthropic land cover classes) and non-transformed (all forest categories). Additionally, we used the Ecosystem Map of Colombia as a reference to define the original limits of the Andean forest (Etter et al., 2017). To clearly define the original forest, we clustered all categories within the boundary of the Andean region of Colombia, which corresponded to the Orobiomes of the Evergreen Tropical Rain Forest Zonobiome (Walter, 1973). The Orobiomes correspond to pronounced mountainous environments within the Zonobiomes and are subdivided into altitude strips. A total of eleven ecosystems are part of the Andean Colombian forest and they were used for the analysis. Second, once we defined the distribution of areas transformed and non-transformed, we determined the change in landscape connectivity from 1985 to 2022 at 5-years intervals. As a proxy to quantify forest connectivity, we used the Equivalent Connected Area (ECA) index. The ECA index represents the area of a hypothetical continuous and fully connected habitat patch that provides the same connectivity probability as the existing network of habitat patches in each area (Saura et al., 2011). We further applied a set of different median dispersal distances of focal species in the region ranging from 1 to 70 km to cover habitat accessibility for multiple species.

3.1 Landscape Connectivity Changes at Different Scales

The Equivalent Connected Area (ECA) index was calculated for seven periods aggregated in 5-year time steps: 1985–1990, 1990–1995, 1995–2000, 2000–2005, 2005–2010, 2010–2015, and 2015–2022. The ECA quantified *habitat,* understood as remaining non-transformed forest areas, and *connected habitat* non-transformed but still connected areas given a dispersal distance (Saura et al., 2011). We calculated the ECA index at three scales: (1) using the entire extension of the Andean region of Colombia (2) using the ecosystems, and (3) using a hexagonal grid of 7.8 km^2.

The ECA index was normalized between 0 and 100 (ECAnorm), indicating the percentage of reachable habitat within each unit of analysis (i.e., Andean region of Colombia, each ecosystem, and each hexagon), and showing how much of the habitat is connected for species with different median dispersal distances (Santini et al., 2016). We calculated ECA*norm* at different median dispersal distances to cover habitat accessibility for multiple species (1–5 km-10–30 km-50–70 km) based on Castillo et al. (2020). We selected 10 km as the central value for the median dispersal distance to discuss our results (Martensen et al., 2017). We also calculated the percentage change of the ECA index, also known as *dECA*, which is the difference between the ECA value over a period of time, expressed as a percentage of the initial value. In our case, we calculated dECA for each temporal period and an additional comparison between the whole temporal period between 1985 and 2022 (Murillo-Sandoval et al., 2022). As a measure of comparison, we included the dA indicator as the percentage of habitat variation (A = total amount of habitat area in the landscape). This comparison identifies that negative changes in *habitat* (dA < 0) are linked to dA whereas *connected habitat (dECA)* is linked to loss of connectivity. We can determine whether the loss of connected area (ECA) for a particular dispersal range exceeds the loss of habitat area (A), or vice versa. Some examples might help to understand these relationships. A scenario in which non-transformed areas are spatially isolated from the previously existing ones (considered as dispersal distance). In that case, a hypothetical expansion of habitat area may result in limited connectivity benefits (dECA < dA). In contrast, when non-transformed lands serve as connecting elements among other non-transformed areas, an increase in connectivity is more likely than that in transformed lands (dECA > dA). A neutral scenario occurs when the entire non-transformed area constitutes a continuous before and after area extent, resulting in dECA = dA (Saura et al., 2011). The negative values (dECA < 0) obtained were identified as connectivity losses, assuming greater intensity of loss as the values approach dECA = −100. We used the Makurhini R package (https://github.com/connectscape/Makurhini) for these analyses (Godínez-Gómez & Correa-Ayram, 2020).

4 Results

4.1 Andean Region Changes

The Andean region suffered significant losses in total Andean forest coverage and landscape connectivity. We found that the loss of connectivity in the Andean forests is much greater than the forest loss attributed to their area. By 2022, 43% of the original forest had been lost by 2022, while forest connectivity had decreased by 85%. In other words, only 15% of the forest is well connected. Between 1985 and 2022, there has been an estimated 7.7% reduction in Andean forest extent, equivalent to over one million hectares transformed in the last 37 years. Even more critical is the connectivity outcome; between 1985 and 2022, almost 15% of the well-connected area of this ecosystem was lost in less than four decades (Fig. 2). The analysis of changes in dECA and dA value during each period shows a rapid decline in Andean forest area and connected forest. The transformation of the Andean forest

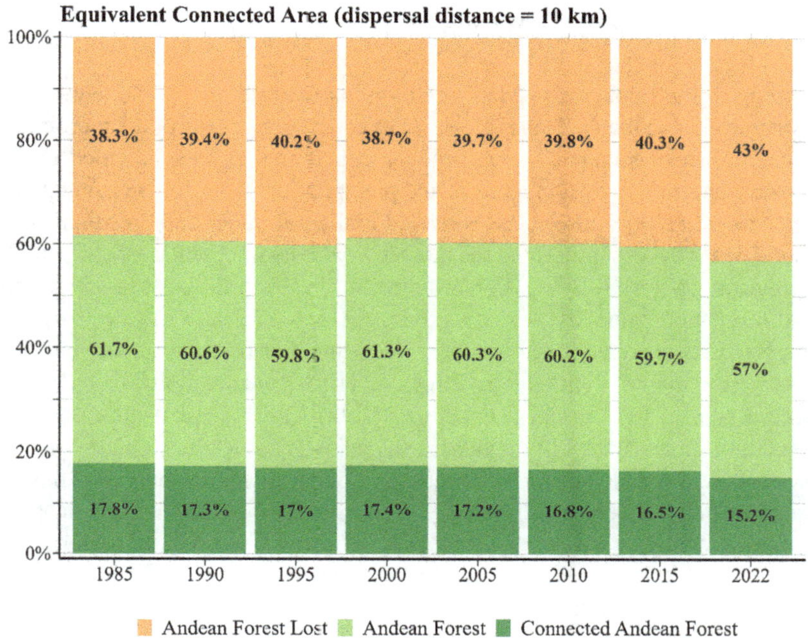

Fig. 2 Distribution (%) of deforested Andean forest (orange) and remnant Andean forest (light green). It is possible to observe the percentage of potentially connected remnant Andean forest using a dispersion distance of 10 km (dark green)

between 1985 and 2000 occurred at an accelerated rate, with losses of nearly 2% in the first 10 years (1985–1995) and a slight deceleration in 1995–2000. However, between 1985 and 1995, the connectivity loss occurred even more slowly in relative terms, with a reduction of almost 0.8% in connected Andean forest areas (the loss of Andean forest cover exceeded the loss of connectivity). There was a net gain in forest cover of 1.5% between 1995 and 2000, following losses close to 2% in previous years. Interestingly, there is an increase of 0.4% in the surface area of connected forests in the same period, reversing the loss of connectivity of the Andean forest for the first time. This result may exemplify the ease of losing connectivity contrasted with the difficulty of its recovery.

After a partial recovery by 2000, the Andean forest returned to its negative trend, with constant losses in the following two decades, dramatically accelerating the loss of the Andean forest area since 2015 with a reduction of 4.3% by 2022. However, due to its ecological implications, the most alarming data is the increasing loss of connectivity. While between 2000 and 2010, connectivity loss fluctuated by ~0.6% per quinquennium, from 2010 to 2022, the loss of connectivity reached almost 1.6%.

4.2 Forest Ecosystems Change

According to the Ecosystems Map of Colombia (Etter et al., 2017), eleven forest ecosystems exist in the Andean region. These ecosystems correspond to pronounced mountainous environments, which produce orographic units subdivided into altitude strips determined by temperature and precipitation changes. We aggregated the eleven ecosystems into three classes related to their altitude: high, median, and low forests. The high forest presented the most dramatic loss in habitat reachability (dA) and connectivity loss (dECA). For instance, in the High dense forest in humidity sub-Andean mountains (B19a) ecosystem, dA = −11.9 and dECA = −64. The medium forests presented the most significant number of ecosystems, with seven, but only one the Medium dense forest in High-Andean mountains ecosystem showed a gain in habitat and connectivity (B22). Other ecosystems in this class, such as B20, show dECA>dA, suggesting a positive rise in connectivity. Low forest ecosystems (B23) and forest, herbaceous, and wetlands ecosystems (P1) presented a negative performance in connectivity. In both classes, dECA<dA indicates that habitat area increase does not contribute to connectivity (see Figs. 3 and 4).

The peripheral areas and the forest around geographic valleys are places with marked loss in connectivity. In the Pacific, the western part of the Western Mountain

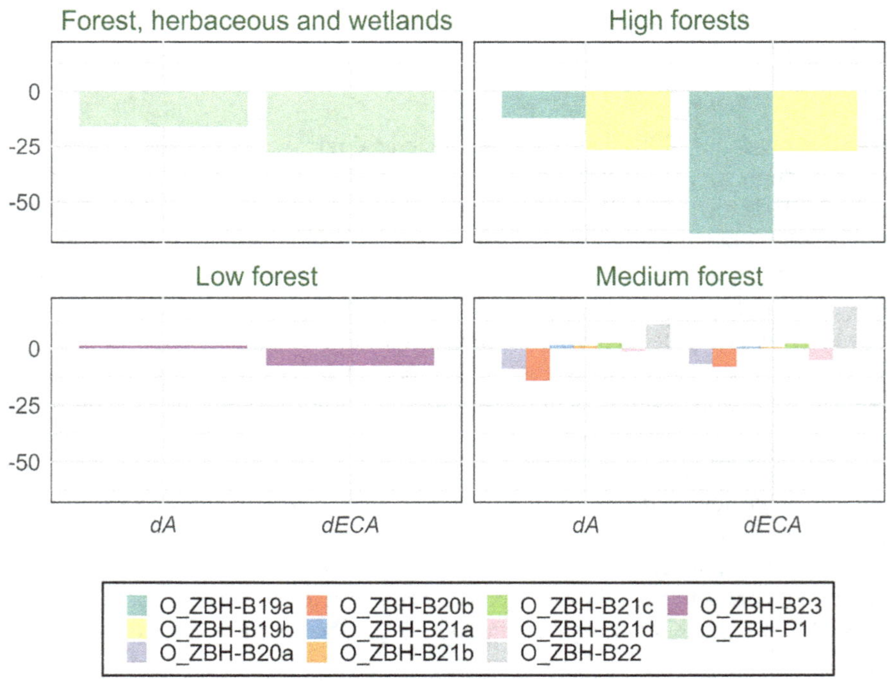

Fig. 3 dA and dECA values for each ecosystem in the Andean region (see ecosystem names in Table 1).

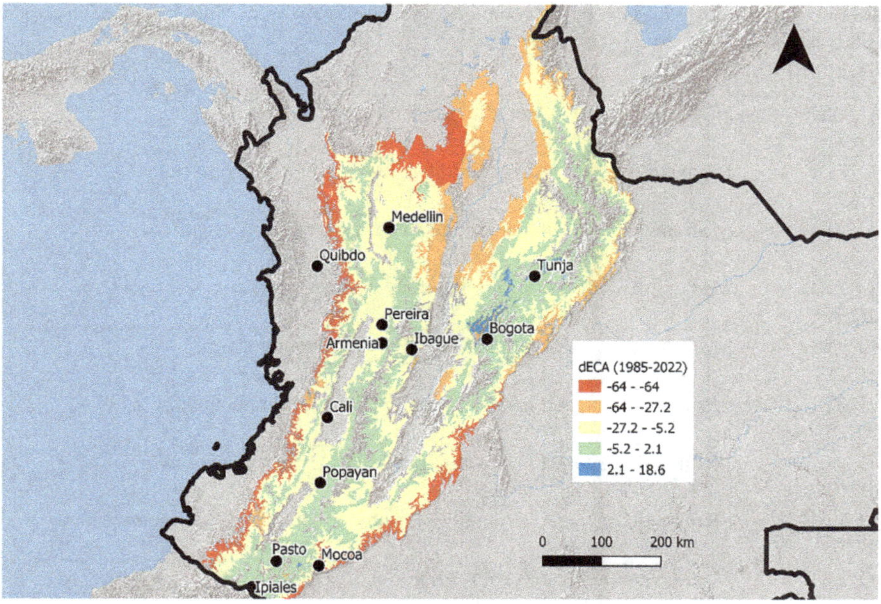

Fig. 4 dECA between 1985 and 2022 at ecosystem scale.

showed diminished connectivity following the longest strip presented in Colombia. Similarly, in the south of the East Mountain, which shares forests with the Amazon, there is a strip of connectivity loss. The most notable area with connectivity loss is in the north Andes.

4.3 *Local Hotspots of Connectivity Loss*

We found high values of connectivity loss (dECA > −70%) distributed throughout the Colombian Andean region (Fig. 5a). We identified three regions with high rates of connectivity loss: the Southern Andes, the Eje Cafetero, and the Cundiboyacense Plateau (Fig. 5b). In the first hotspot in the Eje Cafetero, the loss of connectivity is significant (Fig. 5b1). This region is in the center-west of Colombia, on the Western and Central Mountain ranges. It encompasses much of the Cauca Valley region and the volcanic massif, which still harbors necessary remaining patches of cloudy forest that, along with the paramos and snow-capped peaks, make up the water support of the country's central region. Likewise, this hotspot of Andean forest connectivity loss extends from the Quindío department towards Risaralda, reaching the southeast of Antioquia. The second hotspot is the Southern Andes (Fig. 5b2). This hotspot extends over the region of Nudo de los Pastos, the Popayán Plateau, and the western slope of the Colombian Massif, considered Colombia's most important freshwater-producing ecoregion. Finally, the Cundiboyacense Plateau hotspot (Fig. 5b3)

Table 1 List of ecosystems linked to Orobiomes of the Evergreen Tropical Rain Forest Zonobiome (Etter et al., 2017)

Code	Ecosystem name
O_ZBH-B19a	High dense forest in humidity sub-Andean mountains
O_ZBH-B19b	High dense forest in very humidity sub-Andean mountains
O_ZBH-B20a	Medium dense forest in humidity Andean mountains
O_ZBH-B20b	Medium dense forest in very humidity Andean mountains -Eje Cafetero
O_ZBH-B21a	Medium dense forest in humidity High-Andean mountains
O_ZBH-B21b	Medium dense forest in very humidity High-Andean mountains
O_ZBH-B21c	Medium dense forest in sub-humidity High-Andean mountains
O_ZBH-B21d	Medium dense forest in humidy and sub-humidity High-Andean mountains
O_ZBH-B22	Medium dense forest in High-Andean mountains
O_ZBH-B23	Low dense forest and herbaceous in dry Andean mountains of intra-Andean valleys
O_ZBH-P1	Forest, herbaceous and wetlands in high-Andean mountains

comprises a zone of high and flat lands located in the eastern Andean mountain range, between the Colombian departments of Cundinamarca, Boyacá, and the capital district of Bogotá. It plays an essential regulatory hydraulic function in aquifer recharge and flood mitigation, which is very critical for the country's most populated area. In this area, the Bogotá savanna, the Ubaté valley, and the Tunja and Sogamoso valleys are the subregions most affected by the loss of connectivity of Andean forests. The sub-humid forests of the Cundiboyacense Plateau are one of the most critical ecosystems in the country, pressured by the expansion of the agricultural frontier and mainly replaced by pastures induced by livestock (Etter et al., 2017).

5 Discussion

The Andean region has experienced significant losses in total Andean forest cover and connectivity. We find that connectivity loss is greater than the documented area deforested. Despite a 43% reduction in the original forest cover by 2022, forest connectivity has experienced a concerning 85% decrease. The loss of critical habitat patches hampers the movement capacity of species with limited dispersal ability (Herrera et al., 2017), limiting the resilience and metapopulation persistence capacity. This effect on connectivity compromises the conservation of emblematic

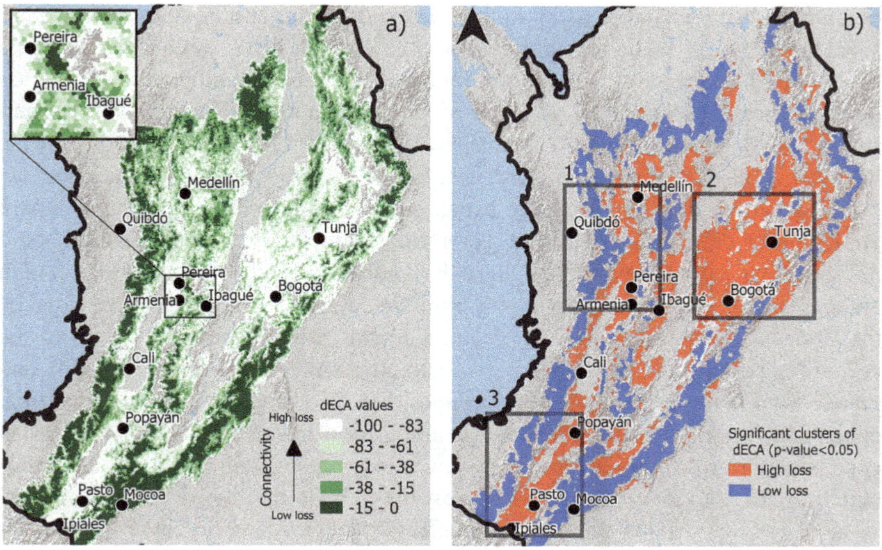

Fig. 5 Spatial distribution of connectivity loss of the Andean forest at hexagon level with a subset in 3 medium-size cities (**a**). Hotspots of loss in dECA, using the original area of Andean forest to 2022 (**b**). The three hotspots are (1) Eje Cafetero, (2) the Cundiboyacense plateau, and (3) the Southern Andean

terrestrial species such as the Andean bear (*Tremarctos ornatus*) and the cougar (*Puma concolor*) (Hohbein & Nibbelink, 2021). It has also affected the dispersal of birds, preferring high altitudes, such as the four species of crested eagles (Genus: *Spizaetus*) found in Colombia or the majestic Andean guans (*Penelope montagnii*) (WCS, 2023). The Andean forest connection is essential for species dispersal, supporting genetic exchange and evolutionary processes through the diversity of Neotropical forest ecosystems. However, the documented extreme connectivity loss prevents species' natural migration along altitudinal and latitudinal gradients, resulting in local and macro-regional isolation (Ocampo-Peñuela et al., 2022).

The loss of connectivity over the past 500 years is linked to the rural population density dynamics in the Andean region (Etter & Villa, 2000; Etter & van Wyngaarden, 2000). The Andes Mountain range hosts the most densely populated areas of Colombia. This high demographic concentration, driven by historical, economic, and road connectivity factors, strongly pressures the Andean forest. The growing demand for housing and expanding agricultural and livestock production areas to supply large cities has led to losing connectivity of natural habitats. Unplanned urbanization processes are rapidly advancing on the mountain foothills of the Andes, turning continuous forests into isolated green patches and disrupting landscape connectivity essential for the genetic flow of flora and fauna populations.

Our results at the ecosystem level also reveal critical patterns. High-dense forests are currently more disconnected. This is alarming given nature's essential contributions to people, such as water supply. Additionally, forests in the foothills seem to be highly affected by increasing human demand for agriculture and urban expansion. For instance, the ecosystem between the Andean Mountain and the Amazon basin is highly disconnected, and the foothills in the Choco biogeographic region appear to present a similar trend. While a border effect may affect connectivity metrics calculation, foothills show a consistent long strip of connectivity loss. At the hexagon level (7.8 km^2), a not surprising result is the location of significant loss hotspots surrounding highly dense urban areas. Cundinamarca plateau, Southern Andes, and the Eje Cafetero are the regions most affected (Fig. 5). Colombia should develop a national strategy to safeguard the Andean forests, especially those in high and medium altitudes. Considering the multiple environmental services provided by the Andean forests, such as water regulation and carbon capture, we warn of the severe impact this loss of connectivity could have on regional climate stability and human well-being in the following decades. We recommend urgent conservation, ecological restoration, and territorial planning measures. Strategies such as urban expansion towards foothills must be reviewed to guarantee that new roads and infrastructure development do not affect forest loss. Given that the resilience and survival of species depend on intact biological corridors, it is urgent to implement plans either by protecting continuous forests or reforesting key areas for animal dispersion. If not remedied, the fate of the Andean forests would be compromised.

Remote sensing is a key element not fully exploited to create landscape connectivity indicators. Dense historical remote sensing time series and improvements in spectral and spatial resolutions are currently available to develop a better understanding of connectivity at different scales. But some limitations remain. In this study, the land cover maps were derived from Landsat, which is suitable for a regional scale but less effective in dealing with small-scale changes. This prevents capturing fine-scale patterns, such as scattered trees, living fences, and small forest patches, which act as stepping-stones to facilitate ecological flows of Andean forest species in landscapes dominated by agriculture or other land use. Also, some problems using optical data are linked to effective classification, especially in very cloudy regions such as the Andean Mountains. While the scale and some errors can affect metric calculations, the results are consistent with well-documented human drivers of change. Future work should include the ecological characteristics of umbrellas or threatened species in the Andean forests and how they influence landscape connectivity. The landscape matrix should consider indicators such as habitat requirements, dispersal range, biotic interactions, and species adaptation capacity to anthropogenic disturbances. The influence of the anthropogenic matrix could be included in the calculation of the dECA index by incorporating a resistance or friction layer to species mobility (e.g., dECAmatrix in Saura et al., 2011). A commonly used proxy for landscape resistance could be a measure of the magnitude of human pressure on the landscape, which has been used in some studies focused on the Andean-tropical ecoregion in Mexico and the United States (Castillo et al., 2020; Correa Ayram et al., 2017; Belote et al., 2016). Incorporating ecological factors and the pressures exerted on them would allow for a more precise assessment of functional connectivity for the species of interest in Andean forests.

Similarly, future research on ecological restoration priorities in Andean forests might complement the approach presented here to identify sites where meaningful forest connections are being lost more accurately. This analysis enables the determination of more effective interventions in terms of connectivity rather than solely focusing on the restored area. This approach would positively impact restoration projects, allowing a more efficient allocation of budgetary resources by prioritizing the recovery of landscape functional connectivity. Consequently, more significant optimization of efforts and a better cost-benefit ratio would be achieved in ecological restoration initiatives in these ecologically important ecosystems.

6 Conclusions

The drastic loss of connectivity in the Andean forests of Colombia is concerning. The loss of connectivity far exceeds documented deforestation, an indicator of human large-scale impacts affecting the movement of species and gradually isolating remaining forested patches. The accelerated loss of connectivity is locally isolating flora and fauna populations, hindering their natural migration and jeopardizing their long-term resilience and persistence. The main drivers of change are well-known, and they are linked to highly concentrated population density, unplanned urbanization processes, and increasing land demand for agriculture and livestock farming. If this trend is not urgently reversed through decisive conservation measures, the future of Andean forest will be severely compromised. This affect regional climate stability, the supply of essential natural benefits, and human well-being. It is imperative to integrate the ecological factors of key species and anthropogenic influence to optimize the recovery of functional connectivity in restoration initiatives.

References

Anderson, E. P., Jenkins, C. N., Heilpern, S., Maldonado-Ocampo, J. A., Carvajal-Vallejos, F. M., Encalada, A. C., et al. (2018). Fragmentation of Andes-to-Amazon connectivity by hydropower dams. *Science Advances, 4*(1), eaao1642.

Andrade, M. G. (2022). Los Bosques Andinos e Interandinos. In *Colombia país de bosques* (pp. 58–65). Alpha Editorial SA.

Armenteras, D., Gast, F., & Villareal, H. (2003). Andean forest fragmentation and the representativeness of protected natural areas in the eastern Andes, Colombia. *Biological Conservation, 113*(2), 245–256. https://doi.org/10.1016/S0006-3207(02)00359-2

Armenteras, P., Cadena, V., Camilo, E., & Moreno, S. (2007). *Evaluación del estado de los bosques de niebla y de la meta 2010 en Colombia.*

Balthazar, V., Vanacker, V., Molina, A., & Lambin, E. F. (2015). Impacts of forest cover change on ecosystem services in high Andean mountains. *Ecological Indicators, 48*, 63–75.

Belote, R. T., Dietz, M. S., McRae, B. H., Theobald, D. M., McClure, M. L., Irwin, G. H., McKinley, P. S., Gage, J. A., & Aplet, G. H. (2016). Identifying corridors among large protected areas in the United States. *PLoS One, 11*(4), e0154223. https://doi.org/10.1371/journal.pone.0154223

Bruijnzeel, L. A. (2001). Hydrology of tropical montane cloud forests: A reassessment. In *Land use and water resources research* (Vol. 1, pp. 1–1).

Calbi, M., Clerici, N., Borsch, T., & Brokamp, G. (2020). Reconstructing long term high Andean forest dynamics using historical aerial imagery: A case study in Colombia. *Forests, 11*(8), 1–20. https://doi.org/10.3390/F11080788

Calderón-Caro, J., & Benavides, A. M. (2022). Deforestation and fragmentation in the most biodiverse areas in the Western Cordillera of Antioquia (Colombia); [Deforestación y fragmentación en las áreas más biodiversas de la Cordillera Occidental de Antioquia (Colombia)]. *Biota Colombiana, 23*(1). https://doi.org/10.21068/2539200X.942

Castillo, L. S., Correa Ayram, C. A., Matallana Tobón, C. L., Corzo, G., Areiza, A., González, M. R., Serrano, F., Chalán Briceño, L., Sánchez Puertas, F., More, A., Franco, O., Bloomfield, H., Aguilera Orrury, V. L., Rivadeneira Canedo, C., Morón-Zambrano, V., Yerena, E., Papadakis, J., Cárdenas, J. J., Golden Kroner, R. E., & Godínez-Gómez, O. (2020). Connectivity of protected areas: Effect of human pressure and subnational contributions in the ecoregions of tropical Andean countries. *Land, 9*(8), 239. https://doi.org/10.3390/land9080239

Cavelier, J. (1991). *El Ciclo del Agua en Bosques Montanos*. Bosques de niebla de Colombia. Banco de Occidente, Uribe Hurtado editores, Santafé de Bogotá (pp. 119–149).

Clerici, N., Salazar, C., Pardo-Díaz, C., Jiggins, C. D., Richardson, J. E., & Linares, M. (2019). Peace in Colombia is a critical moment for Neotropical connectivity and conservation: Save the northern Andes–Amazon biodiversity bridge. *Conservation Letters, 12*(1), e12594.

Correa Ayram, C. A., Mendoza, M. E., Etter, A., & Pérez Salicrup, D. R. (2017). Anthropogenic impact on habitat connectivity: A multidimensional human footprint index evaluated in a highly biodiverse landscape of Mexico. *Ecological Indicators, 72*, 895–909. https://doi.org/10.1016/j.ecolind.2016.09.007

Correa Ayram, C. A., Etter, A., Díaz-Timoté, J., Rodríguez Buriticá, S., Ramírez, W., & Corzo, G. (2020). Spatiotemporal evaluation of the human footprint in Colombia: Four decades of anthropic impact in highly biodiverse ecosystems. *Ecological Indicators, 117*(June), 106630. https://doi.org/10.1016/j.ecolind.2020.106630

Etter, A., & van Wyngaarden, W. (2000). Patterns of landscape transformation in Colombia, with emphasis in the Andean region. *Ambio: A Journal of the Human Environment, 29*(7), 432–439.

Etter, A., & Villa, L. A. (2000). Andean forests and farming systems in part of the Eastern Cordillera (Colombia). *Mountain Research and Development, 20*(3), 236–246. https://doi.org/10.1659/0276-4741(2000)020[0236:AFAFSI]2.0.CO

Etter, A., Andrade, A., Saavedra, K., Amaya, P., Arevalo, P., Cortés, J., & Soler, D. (2017). Lista roja de ecosistemas de Colombia (Vers. 2.0). In *Internacional-Colombia*. Pontificia Universidad Javeriana y Conservación Internacional-Colombia.

Godínez-Gómez, O., & Correa-Ayram, C. (2020). *Connectscape/Makurhini: Analyzing landscape connectivity (v1.0.0) (1.0.0)*. Zenodo. https://doi.org/10.5281/zenodo.3771605

Gomez-Peralta, D., Oberbauer, S. F., McClain, M. E., & Philippi, T. E. (2008). Rainfall and cloud-water interception in tropical montane forests in the eastern Andes of Central Peru. *Forest Ecology and Management, 255*(3–4), 1315–1325.

Herrera, L. P., Sabatino, M. C., Jaimes, F. R., & Saura, S. (2017). Landscape connectivity and the role of small habitat patches as stepping stones: an assessment of the grassland biome in South America. *Biodiversity and Conservation, 26*(14), 3465–3479. https://doi.org/10.1007/s10531-017-1416-7

Hohbein, R. R., & Nibbelink, N. P. (2021). Omnidirectional connectivity for the Andean bear (Tremarctos ornatus) across the Colombian Andes. *Landscape Ecology, 36*, 3169–3185.

Hoorn, C., Wesselingh, F. P., ter Steege, H., Bermudez, M. A., Mora, A., Sevink, J., Sanmartín, I., Sanchez-Meseguer, A., Anderson, C. L., Figueiredo, J. P., Jaramillo, C., Riff, D., Negri, F. R.,

Hooghiemstra, H., Lundberg, J., Stadler, T., Särkinen, T., & Antonelli, A. (2010). Amazonia through time: Andean uplift, climate change, landscape evolution, and biodiversity. *Science, 330*(6006), 927 LP–927931. https://doi.org/10.1126/science.1194585

IUCN-CEM. (2022). *The IUCN Red List of Ecosystems*. Version 2022-1. http://iucnrle.org. Downloaded on 06-03-2024.

Linero-Triana, D., Correa-Ayram, C. A., & Velásquez-Tibatá, J. (2023). Prioritizing ecological connectivity among protected areas in Colombia using a functional approach for birds. *Global Ecology and Conservation, 48*, e02713. https://doi.org/10.1016/j.gecco.2023.e02713

MapBiomas Colombia Project. (2023). *Collection 1 of annual land cover and land use maps*. Accessed on 2024 via link: https://plataforma.colombia.mapbiomas.org/

Mathez-Stiefel, S. L., Peralvo, M., Báez, S., Rist, S., Buytaert, W., Cuesta, F., et al. (2017). Research priorities for the conservation and sustainable governance of Andean forest landscapes. *Mountain Research and Development, 37*(3), 323–339.

Martensen, A. C., Saura, S., & Fortin, M. J. (2017). Spatio-temporal connectivity: Assessing the amount of reachable habitat in dynamic landscapes. *Methods in Ecology and Evolution, 8*(10), 1253–1264. https://doi.org/10.1111/2041-210X.12799

Móstiga, M., Armenteras, D., Vayreda, J., & Retana, J. (2023). Nature's contributions to people (NCPs) and biodiversity hotspots: A step towards multifunctionality of conservation areas in Peru. *Perspectives in Ecology and Conservation, 21*, 329–339.

Murillo-Sandoval, P. J., Clerici, N., & Correa-Ayram, C. (2022). Rapid loss in landscape connectivity after the peace agreement in the Andes-Amazon region. *Global Ecology and Conservation, 38*, e02205. https://doi.org/10.1016/j.gecco.2022.e02205

Myers, N., Mittermeier, R. A., Mittermeier, C. G., Da Fonseca, G. A., & Kent, J. (2000). Biodiversity hotspots for conservation priorities. *Nature, 403*(6772), 853–858.

Nicholson, E., Watermeyer, K. E., Rowland, J. A., Sato, C. F., Stevenson, S. L., Andrade, A., et al. (2021). Scientific foundations for an ecosystem goal, milestones and indicators for the post-2020 global biodiversity framework. *Nature Ecology & Evolution, 5*(10), 1338–1349.

Ocampo-Peñuela, N., Suárez-Castro, A. F., Díaz-Timoté, J. J., Gómez-Valencia, B., Olaya-Rodríguez, M. H., Sánchez-Clavijo, L. M., & Correa-Ayram, C. A. (2022). Increased exposure of Colombian birds to rapidly expanding human footprint. *Environmental Research Letters, 17*(11), 114050.

Rahbek, C., Borregaard, M. K., Colwell, R. K., Dalsgaard, B. O., Holt, B. G., Morueta-Holme, N., et al. (2019). Humboldt's enigma: What causes global patterns of mountain biodiversity? *Science, 365*(6458), 1108–1113.

Raven, P. H., Gereau, R. E., Phillipson, P. B., Chatelain, C., Jenkins, C. N., & Ulloa Ulloa, C. (2020). The distribution of biodiversity richness in the tropics. Science. *Advances, 6*(37), eabc6228.

Renjifo, L. M., Amaya-Villarreal, A. M., & Butchart, S. H. (2020). Tracking extinction risk trends and patterns in a mega-diverse country: A Red List Index for birds in Colombia. *PLoS One, 15*(1), e0227381.

Rodríguez Eraso, N., Armenteras-Pascual, D., & Alumbreros, J. R. (2013). Land use and land cover change in the Colombian Andes: Dynamics and future scenarios. *Journal of Land Use Science, 8*(2), 154–174.

Santini, L., Saura, S., & Rondinini, C. (2016). Connectivity of the global network of protected areas. *Diversity and Distributions, 22*(2), 199–211. https://doi.org/10.1111/ddi.12390

Saura, S., Estreguil, C., Mouton, C., & Rodríguez-Freire, M. (2011). Network analysis to assess landscape connectivity trends: Application to European forests (1990–2000). *Ecological Indicators, 11*(2), 407–416. https://doi.org/10.1016/j.ecolind.2010.06.011

Souza, C., & Azevedo, T. (2017). *MapBiomas general handbook* (pp. 1–23). MapBiomas.

Suescún, D., León, J. D., Villegas, J. C., & Correa-Londoño, G. A. (2023). Nutrient loss to erosion responds to rain characteristics under transformed landscapes in the Río Grande basin, Colombian Andes. *Ecohydrology, 16*(3), e2519.

Taylor, P. D., Fahrig, L., Henein, K., & Merriam, G. (1993). Connectivity is a vital element of landscape structure. *Oikos, 68*, 571–573.

Van der Hammen, T. (1989). History of the montane forests of the northern Andes. In F. Ehrendorfer (Ed.), *Woody plants — Evolution and distribution since the tertiary*. Springer. https://doi.org/10.1007/978-3-7091-3972-1_6

Van der Hammen, T. (2000). Aspectos de historia y ecología de la biodiversidad norandina y amazónica. *Revista de la Academia Colombiana de Ciencias, 24*(91), 231–245.

Walter, H. (1973). *Vegetation of the earth in relation and ecological Systems of the Geo-Biosphere* (2nd ed.). Springer.

Wildlife Conservation Society (WCS). (2023). *Enhancing Forest connectivity in the Colombian Andes*. Expansion of Cuchilla del San Juan Regional Integrated Management District. WCSNewsroom. https://newsroom.wcs.org/

Phenology and Potential Primary Productivity Response to Climate Change in the Andean Mediterranean Forests: A Remote Sensing Analysis

Marcelo D. Miranda, Cynnamon Dobbs, and Valentina Bravo

1 Introduction

Recent studies have emphasized an increased vulnerability of forest ecosystems to climate change resulting in more widespread and severe drought-induced decline episodes around the world (Lindner et al., 2010; Hartmann et al., 2015; Miranda et al., 2020). Medi terranean regions have shown an increase in aridity, as a combination of precipitation decline, increased temperatures, and heat waves, with consequences for forest declines, especially on the diversity and survival of Mediterranean forests (Peñuelas et al., 2017).

Within the Andean Mediterranean region of Chile, aridity has increased because of a higher frequency of heatwave events along with a 10-year mega-drought (Boisier et al., 2016; Garreaud et al., 2020), resulting in vegetation primary productivity loss (Glade et al., 2016; Mccullum & Torres-Pérez, 2020) and recent massive tree mortality (Miranda et al., 2023), which is expected to be more intense in the future (Matskovsky et al., 2021).

M. D. Miranda (✉)
Faculty of Agricultural and Natural Systems, Pontificia Universidad Católica de Chile, Santiago, Chile

Center of Applied Ecology and Sustainability (CAPES), Pontificia Universidad Católica de Chile, Santiago, Chile
e-mail: mmirands@uc.cl

C. Dobbs
Department of Natural Resources and the Environment, University of Connecticut, Storrs, CT, USA

V. Bravo
Faculty of Agricultural and Natural Systems, Pontificia Universidad Católica de Chile, Santiago, Chile

N. Clerici (ed.), *Conservation of Andean Forests*,
https://doi.org/10.1007/978-3-031-80805-0_9

The Andean Mediterranean forest has adaptation strategies for dry periods, given that it developed during the last glacial maximum 20,000 years ago (Rundel et al., 2016). Trees and shrubs exhibit leathery leaves adapted to high levels of radiation and evapotranspiration, lignotubers and epicormic buds with nutritional reserves that allow to withstand long dry periods and stress from fire events, and extensive and deep roots system (Armesto et al., 1996; Blondel et al., 2010). In other cases, trees have facultative deciduous foliage that adapts to drought and high-altitude cold conditions (Armesto et al., 1996, Blondel et al., 2010). Nevertheless, given the unprecedented speed of climate change, there is no guarantee that all species in these ecosystems will be able to evolve and adapt to these unprecedented changes (Fuentes-Castillo et al., 2020).

Satellite-borne sensors provide quantitative and spatially continuous data that improves availability to characterize and monitor ecosystem structures and more recently, functional attributes of ecosystems at different spatial and temporal resolutions (Paruelo, 2008; van Leeuwen et al., 2013). Numerous studies have shown that vegetation indices data obtained from satellite images, such as the Normalized Difference Vegetation Index (NDVI) and the Enhanced Vegetation Index (EVI), are directly related to ecosystem functionality, particularly aboveground net primary productivity (Glenn et al., 2008). They are based on optical measurements of the vegetation canopy, principally in the red and infra-red wavelength, being sensitive to temporal changes in leaf chlorophyll, leaf area, canopy cover, and structure (Gamon et al., 1995; Fensholt et al., 2004; Forkel et al., 2013).

Trend detection with time series of vegetation indices can help to identify and quantify changes in ecosystem properties from local to global scale (Forkel et al., 2013). Methods based on time series decomposition and remote sensing data have shown to be effective in quantifying forest damage caused by forest fires, insect infestations, or tree dieback associated with drought events (Byer & Jin, 2017; Housman et al., 2018; Khodaee et al., 2020).

Satellite phenology, which involves spectral satellite data to track seasonal variations in land surfaces, provides consistent estimates of primary productivity, as well as temporal events associated with the onset, peak, end, and duration for large areas using vegetation indices (Zhang et al., 2009; Ivits et al., 2013). Studies of primary productivity and its phenology based on satellite vegetation indices enable continuous monitoring thereof, allowing rapid updates as technological advances (Ferchichi et al., 2022).

Information that allow to understand spatial and temporal variability of vegetation productivity, especially in those Andean Mediterranean ecosystems located in the southern hemisphere, result essential to evaluate future loss of ecosystem goods and services and to evaluate increase in vulnerability to natural disasters. These two aspects are critical due to their social and ecological consequences.

The aim of this chapter is to present a temporal analysis of the potential primary productivity and its associated phenology in the last two decades for Mediterranean forests located in the Andean range of Chile. This region is important since it has been subjected to a profound drought and several heatwave events during the last

10 years, affecting its structure, composition, functions and the provision of ecosystem services to society.

2 Andean Mediterranean Forests

According to the National Vegetational and Land Use Cadaster of Chile, the country is covered by 18,030,375 hectares of natural forests (CONAF, 2021). The Andean forests studied in this work are part of the ecological region of Mediterranean Forests, which in Chile spans approximately between latitudes 32° and 37° South. Specifically, to the north, it includes the sub-region of Mediterranean forests and the temperate evergreen forests with the presence of mainly broad-leaved species (Donoso, 1993). The Mediterranean region holds global significance as a biodiversity hotspot and a carbon sink due to its species richness and vegetation structure (Ameray et al., 2021). Globally, these areas have been among the most altered in the past 200 years due to land conversion for agriculture, recurring forest fires, and the development of the forest industry associated with exotic species plantations.

From a climatic standpoint, there is a clear north-south gradient in precipitation and temperatures, characterized by a Mediterranean climate condition (Csb) with a warm, dry summer in the northern section, and a temperate climate condition (Cfb) without a dry season and a warm summer covering the central and southern sectors of the study region (Peel et al., 2007). Precipitation ranges between 222 and 1393 mm between latitudes 32.16° and 36.60° South, respectively (Santibáñez Quezada, 2017). Regarding temperatures, maxima in January reach 29.7 °C and a minimum in July of 5.2 °C in the northern section, and 28.9 and 3.4 °C in the same periods in the southern section (36.6° South Latitude).

According to the potential natural vegetation classification of Chile (Luebert & Pliscoff, 2017), the studied mountain area hosts seven potential vegetational formations (Table 1). Among them, deciduous forests associated with species of the genus Nothofagus predominate in terms of potential surface (54.23%), followed by sclerophyll forests (28.36%), thorny forests (5.65%), shrubland (4.35%), and coniferous forests (3.09%).

Table 1 List and surface of the potential natural forest types according to Luebert y Pliscoff (2017)

Forest type	Area (ha)
Deciduous	973,178
Sclerophyll	508,984
Thorny	101,427
Coniferous	55,387
Evergreen	12,268
Scrubland	78,094
Thorny scrubs	37,068
Total	1,766,406

According to Donoso (1981), Gajardo (1995), and Luebert and Pliscoff (2017), it is important to note that the sclerophyll forest type is represented by the presence of at least one of the following species: Quillay (*Quillaja saponaria*), Litre (*Lithraea caustica*), Peumo (*Cryptocarya alba*), Espino (*Acacia caven*), Maitén (*Maytenus boaria*), Algarrobo (*Prosopis chilensis*), Belloto (*Beilschmiedia miersii*), Boldo (*Peumus boldus*), Bollén (*Kageneckia oblonga*), and Molle (*Schinus latifolius*). The deciduous-evergreen forest type dominated by *N. oblicua-N. alpina-N.dombeyi* is divided into three subtypes, characterized by the level of development and degradation of the forest (secondary renewal and pure secondary forest, original remnants, and degraded forests), where species of the genus Nothofagus dominate. On the other hand, the evergreen forest type is subdivided into five subtypes characterized by their floristic composition (Ñadis, coastal olivillo (*Aextoxicon punctatum*), and canelo (*Drimys winteri*) renewals) and the synecology of the communities that compose them (evergreen with emergent intolerants and tolerant evergreen).

According to the structural classification of vegetation by forest types (Donoso, 1993), the mountainous area contains 9 out of the 12 forest types native to Chile (Table 2) (Fig. 1). Of the total area, 54.79% is covered by combinations of species associated with the genus Nothofagus, forming deciduous and deciduous-evergreen forests, 40.88% by sclerophyll forests with leathery-leaved species, and 3.98% by forests composed of coniferous species. The evergreen forest type associated with laurifolious species is poorly represented in this area (0.14%) because its distribution begins in this sector and extends to 43.9° South Latitude.

From a structural perspective, deciduous and evergreen forests dominated by Nothofagus species have a coverage degree that generally exceeds 50%, with

Table 2 Forest type and sub-types are present in the study region according to Donoso (1993) and CONAF (2021)

Forests				Stand
Type	Sub-type	Area (ha)	Area (%)	Mean area (ha)
Conifers	*A. Araucana*	**64,996.69**	**3.98**	**56.03**
		27,631.18	1.69	71.21
	A. chilensis	37,365.51	2.29	48.40
Deciduous	*N. pumilio*	**331,442.10**	**20.30**	**49.97**
		145,049.76	8.88	39.53
	N. oblicua-N. glauca	186,392.34	11.41	62.89
Deciduous-Evergreen		**563,325.05**	**34.50**	**49.07**
	N. dombeyi-N. alpina-L. philippiana	46,437.23	2.84	94.96
	N. oblicua-N. alpina-N.dombeyi	516,887.81	31.65	47.02
Evergreen		**673,260.84**	**41.23**	**45.73**
	Sclerophyll *Q. saponaria-L. caustica- C. alba*	667,627.76	40.88	45.96
	Sclerophyll *J. chilensis-Q. saponaria*	3331.82	0.20	43.84
	N. dombeyi-L. philippiana-E. coccineum	2301.26	0.14	19.02
	Total	**1,633,024.68**	**100.00**	**48.03**

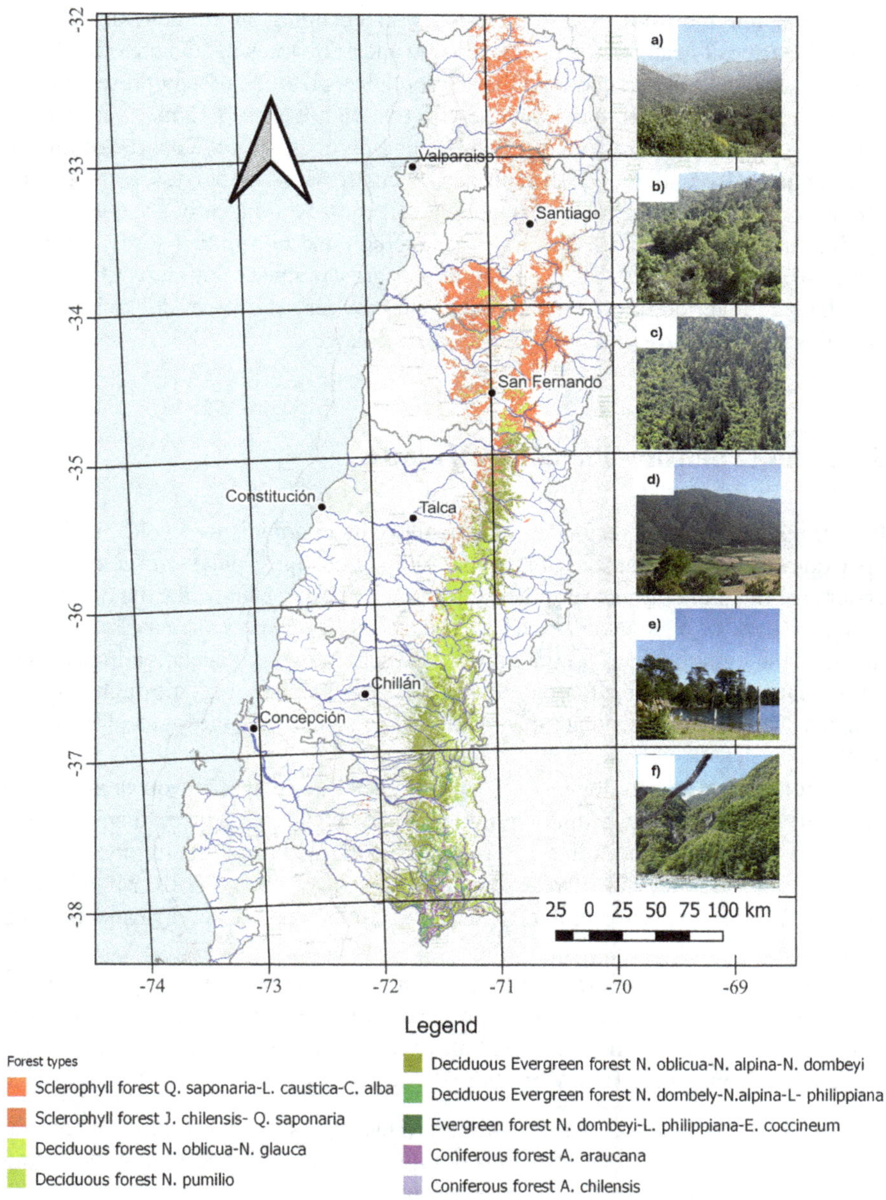

Fig. 1 Distribution of Andean Mediterranean forests according to the Forest Types Classification of Chile (Donoso, 1993). It is noteworthy that this classification is used in the Forestry Law for processes of control and incentive in the use of Chile's natural and planted forests. Pictures representing examples of: (**a**) evergreen forest sclerophyll with presence of J. chilensis, (**b**) deciduous forest N. oblicua-N. glauca, (**c**) coniferous forest A. chilensis, (**d**) deciduous forest N. oblicua-N. alpina, (**e**) coniferous forest A. Araucana, (**f**) evergreen forest N. dombeyi-L. philippiana-E. coccineum

individuals ranging from 15 to 40 m in height. Depending on the stage of development, their density can fluctuate between 600 and 1800 trees/ha. On the other hand, sclerophyll forests associated with the sclerophyll forest type mostly present coverages of less than 50% and individual heights that do not exceed 15 m. Their density can vary between 300 and 800 trees/ha, depending mainly on the degree of human intervention they undergo. The main ecosystem services associated with Andean Mediterranean forests are water production, soil conservation, carbon sequestration, production of litter as nutrients, firewood for fuel, and non-timber forest products (Smith-Ramírez et al., 2023). The main threats are associated with forest fires, land use change for agriculture, pest outbreaks, and browning processes due to increased droughts and heatwaves from ongoing climate change.

3 Remote Sensing Phenology Trends

Phenology, the study of recurring biological events in organisms' cycles, including their correlation with seasonal or annual climate changes, plays a crucial role in understanding plant primary productivity. It encompasses events like the onset, cessation, or peak of photosynthetic or growth activity over the year. Satellite sensors facilitate the monitoring of primary productivity by utilizing vegetation indices like NDVI and EVI, allowing for the characterization of phenology through satellite imagery. This capability enables the tracking of vegetation changes at local, regional, or global scales.

Remote sensing phenology, observed from satellite images, involves analyzing time series representing primary productivity behavior over more than 2 years. Different satellite sensors offer varying spatial and temporal resolutions, such as 1.1 km, 500 m, 250 m, and 10 m, associated with sensors like AVHRR, MODIS, and Sentinel 2 satellites, with temporal series extending up to 46, 21, and 6 years respectively.

From time series representing vegetation index values, several phenological events associated with vegetation primary productivity can be identified (Jönsson & Eklundh, 2002, 2004). These events, linked to both temporal and productivity values, include some of the following (Fig. 2):

1. Start of season (SOS): The date when productivity value increases by a certain percentage (x%) from the previous minimum value observed before the spring maximum. In this case x% corresponds to 30%.
2. End of season (EOS): The date when productivity reaches a certain percentage (y%) of the minimum value after the spring maximum. In this case y% corresponds to 30%.
3. Peak time (PET): The date when the maximum vegetation index value occurs within the growth season.
4. Seasonal integral (LIN): The sum of vegetation index values between SOS and EOS.

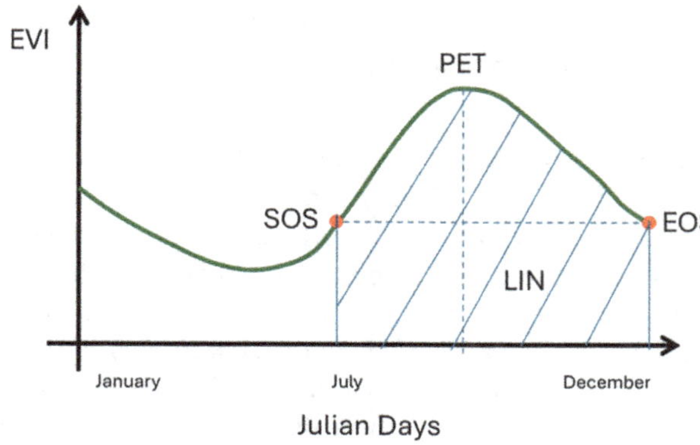

Fig. 2 Different phenological variables associated with a temporal series of EVI vegetation index values, considering a 1-year period. SOS Start of season, EOS End of season, PET Peak time, LIN Large integral

These metrics capture short- and long-term changes in ecosystem activity and are sensitive to intra- and inter-annual climate variations. However, understanding the internal ecosystem mechanisms producing these events remains uncertain when solely based on remote sensing data.

The quantification of phenological metrics involves fitting a function that relates the behavior of the vegetation index to time. The initial step in this adjustment involves analyzing the integrity of the time series, identifying anomalous and missing values, which can be imputed following the methodology proposed by Moritz and Bartz-Beielstein (2017). Subsequently, a model is fitted to the cleaned series that relates vegetation index values to time. Various functions can be employed for this purpose, including adaptive Savitzky-Golay, asymmetric Gaussian, double logistic, or non-parametric fits with varying degrees of complexity (Jönsson & Eklundh, 2002). Unlike global functions, such as those based on Fourier transformations, local functions are suitable at capturing changes in seasonal synchronization between different years, a characteristic that renders them suitable for studying vegetation dynamics and phenology.

The spatial resolution chosen for calculating phenological variables depends on the selected sensor. For MODIS, it is 6.25 hectares (250 × 250 meters), corresponding to the resolution of the satellite product MOD13Q1v061 "MOD13Q1.061 Terra Vegetation Indices 16-Day Global 250 m" (Didan, 2015). Additionally, the temporal coverage spans 20 years from 2002 to 2022. The temporal granularity is 16 days for EVI vegetation index data series and 12 months for the derived phenological variables.

The image downloading was conducted through the Google Earth Engine platform using the JavaScript language. Image analysis and statistical calculations were performed in a computer server using the Terra spatial analysis package available for the R language (R Core Team, 2023). In similar manner, phenological analysis was performed using the software Timesat 3.3 (Eklundh & Jönsson, 2016).

4 Productivity and Phenology in Andean Mediterranean Forests

The potential primary productivity observed through the MODIS sensor exhibits a noticeable north-south gradient across the study region (Fig. 3). The Enhanced Vegetation Index (EVI) serves as a proxy for vegetation vigor and vitality, where values exceeding 0.1 are indicative of live vegetation, values between 0 and 0.1 are associated with bare soil, and values less than 0 indicate the presence of water bodies, ice, or urban areas (Zhengxing et al., 2003).

Specifically, within the latitudes of 32°–34.7° and 36.2°–38.3°, the mean annual EVI values are notably lower than 0.22. In the northern section, these lower EVI values predominantly correspond to sclerophyll forests thriving under a regime of low precipitation (222–550 mm/year) and enduring annual drought periods of 9 and 7 months (Santibáñez Quezada, 2017). Conversely, in the southern section, the diminished EVI values are linked to deciduous forests, primarily dominated by *N. pumilio*, situated at higher elevations near the tree line limit.

Conversely, the highest EVI values (annual average >0.34) are observed between 35° and 38°, indicative of deciduous and deciduous-evergreen forests dominated by *N. oblicua, N. alpina, and N. dombeyi*. These species flourish below the 2000 m a.s.l., occupying areas characterized by favorable terrain physiography, humidity levels, and soil conditions. Additionally, an intermediate scenario is depicted by deciduous forests dominated by *N. oblicua-N. glauca* in the foothills of the Andes Mountain range (34.5°–36.8°), where EVI values range between 0.28 and 0.34, correlating with an average annual precipitation of around 700 mm/year.

Concerning the inter and intra-annual variability of the Enhanced Vegetation Index (EVI) (Fig. 4), distinct amplitude values are evident across the study region, reflecting the variability in vegetation productivity and its correlation with climatic factors. Higher amplitude values are predominantly observed in deciduous and shrubland forests, attributed to the natural during winter (EVI amplitude = 0.4). Conversely, lower amplitudes are characteristic of coniferous and evergreen forests (EVI amplitude = 0.10), indicative of their year-round foliage retention and extended periods of productivity. These differential responses associated with various vegetation types have been consistently observed in prior studies utilizing both global (van Leeuwen et al., 2013) and locally acquired sensor data within this study region (Glade et al., 2016).

Regarding the phenological variables, Start of Season (SOS) (Fig. 5a, Table 3) exhibits a distinct pattern of onset activity characterized by a pronounced latitudinal gradient. Sclerophyll forests situated on north-facing aspects in southern hemisphere initiate their activity between August 15 and September 15, while forest stands in more favorable environmental conditions (south-facing aspects) commence their SOS during the first 2 weeks of October (Julian days 257–295).

Conversely, conifer forests demonstrate their onset activity during the first 2 weeks of November (Julian days 304–321), accompanied by a lower standard

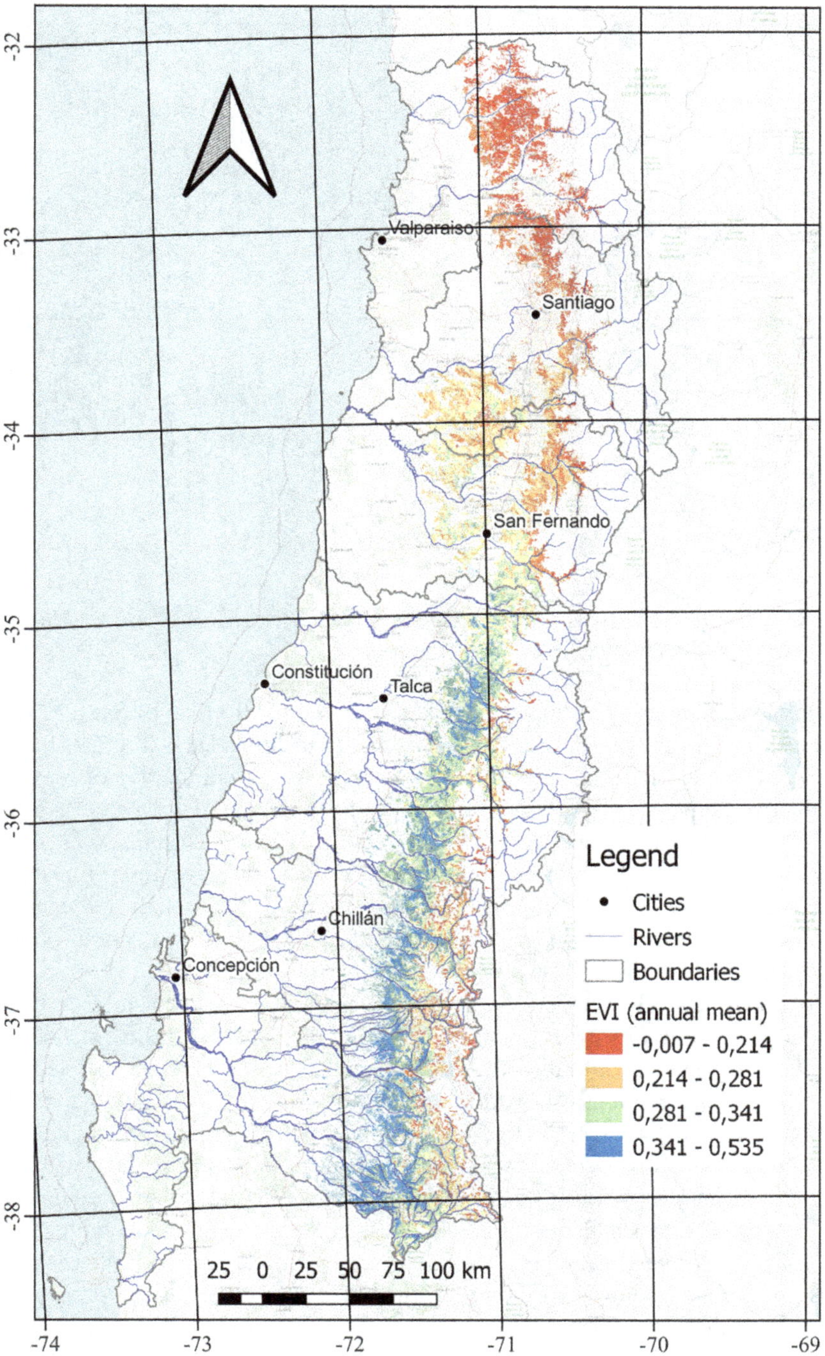

Fig. 3 Enhanced Vegetation Index (EVI) annual average between years 2002 and 2022

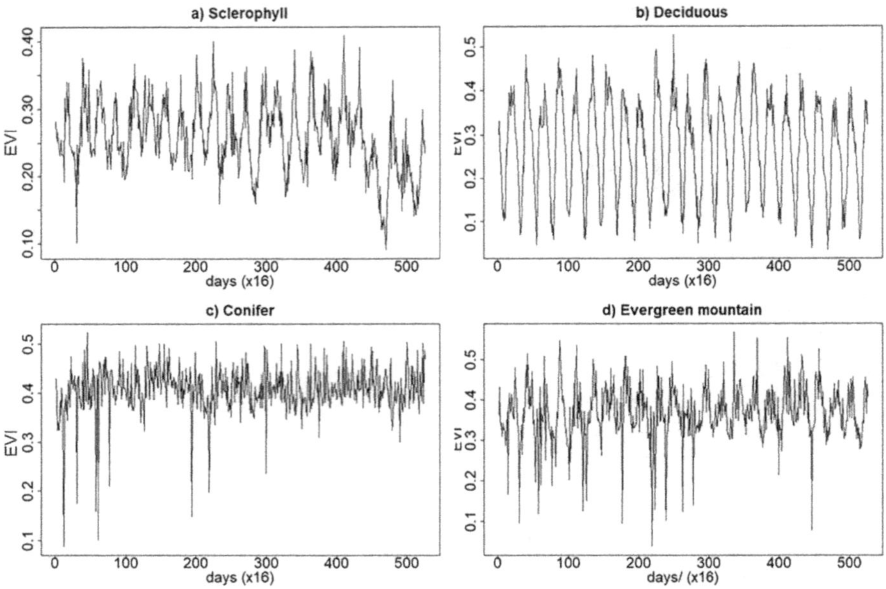

Fig. 4 Example of EVI signals for different forest types: (**a**) Sclerophyll, (**b**) Deciduous, (**c**) Conifer and (**d**) Evergreen Mountain

deviation. A similar latitudinal gradient is observed for the End of Season (EOS), resulting in a mean seasonal activity occurring at 215 Julian Days (Fig. 5b, Table 3).

Furthermore, the Peak Time (PET) variable, representing the day of maximum productive activity, occurs approximately 2.85 months after the Start of Season (SOS) for evergreen sclerophyll and evergreen *N. dombeyi-L. philippiana-E. coccineum* forests, 3.27 months for deciduous and deciduous-evergreen forests, and 3.36 months for coniferous forests (Fig. 5c, Table 3). Sclerophyll and evergreen forests exhibit notable variability, primarily attributable to the structural diversity they exhibit across their distribution (Bannister & Donoso, 2013).

Similarly to EVI, we found a pronounced latitudinal gradient in accumulated productivity (LIN) distinguishing the behavior of forests located north and south of the study region (Fig. 5c, Table 3). Accumulated LIN values fluctuate between 0.69 and 10.32. Sclerophyll forests exhibit two distinct groups of stands, with mean LIN values of 2.36 and 4.15, respectively, corresponding to drier and more humid environmental conditions. Deciduous forests have a mean LIN value of 5.81, with lower values associated with forest stands at higher altitudes dominated by *N. oblicua and N. pumilio*. Forest stands at lower altitudes, characterized by a combination of *N. oblicua, N. alpina, and N. glauca*, exhibit LIN values exceeding 6.66. In the case of conifer forests, lower LIN values are linked to forest stands dominated by *A. chilensis* distributed throughout the study region. Conversely, *A. Araucana* exhibits the highest LIN values but is specifically distributed in the southern part of the study region.

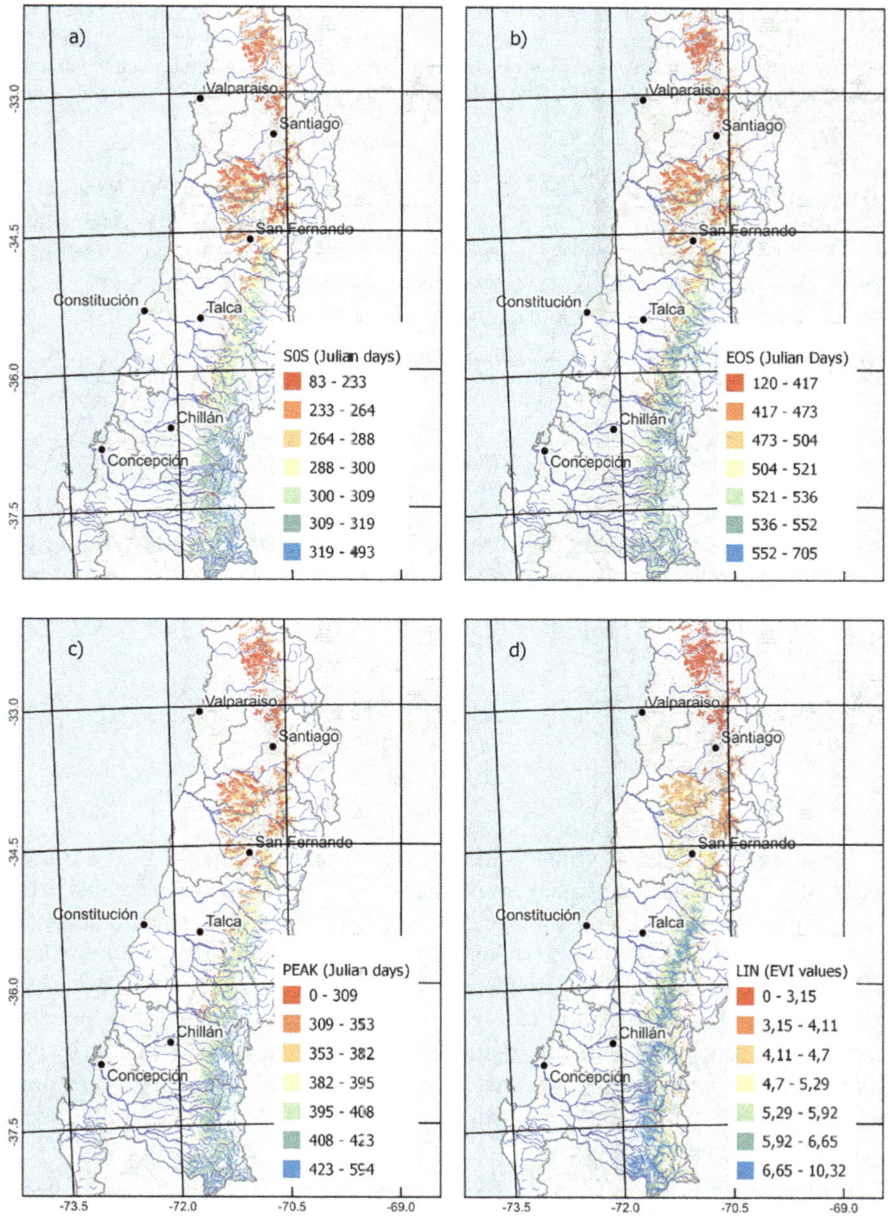

Fig. 5 Phenological variables: (**a**) Start of Season (SOS), (**b**) End of Season (EOS), (**c**) Peak Time (PET) and (**d**) accumulated integral (LIN). The first three variables are measured in Julian days (1–365) and the last in EVI values

Table 3 Mean and standard deviation of phenological variables start of season (SOS), end of season (EOS), day of maximum activity (PET), and accumulated productivity integral (LIN) for different types of forests associated with the study region. The values of the first three variables correspond to Julian days (1–365). The LIN variable is represented in EVI vegetation index values (−1, 1)

Forests		EVI		SOS		EOS		PEAK		LIN	
Type	Sub type	Mean	SD	Mean	SD	Mean	SD	Mean	SD	Mean	SD
Coniferous	*A. araucana*	0.274	0.063	321	13	547	16	423	12	5.568	1.047
	A. chilensis	0.283	0.075	303	22	531	30	403	26	5.406	1.281
Deciduous	*N. pumilio*	0.232	0.086	327	16	541	16	422	12	5.022	1.296
	N. oblicua-N. glauca	0.307	0.060	298	21	529	32	397	27	5.781	1.098
Deciduous-Evergreen	*N. dombeyi-N. alpina L. philippiana*	0.328	0.066	308	13	532	18	409	14	6.330	1.080
	N. oblicua-N. alpina-N. dombeyi	0.329	0.076	306	19	528	33	403	26	6.087	1.268
Evergreen	Sclerophyll *Q. saponaria-L. caustica-C. alba*	0.228	0.069	250	36	448	61	335	46	3.773	1.168
	Sclerophyll J. chilensis-Q. saponaria	0.278	0.025	238	37	438	52	327	45	4.572	0.422
	N. dombeyi-L. philippiana-E. coccineum	0.378	0.045	301	25	493	48	387	35	6.211	1.093

Regarding the trend of primary productivity, represented by the LIN variable, Fig. 6 illustrates significant values resulting from the linear regression applied to the 20-year observation series. The image depicting LIN trends on a standard deviation scale highlights three general behaviors: (a) sectors with a negative slope (indicating a decrease in accumulative productivity), primarily concentrated between parallels 32° and 36°S, (b) areas with a positive slope (reflecting an increase in accumulated productivity), mainly associated with vegetation located at the boundary of tree vegetation distribution between parallels 32° and 38°S and (c) zero slope areas, representing no significant changes, predominantly located toward the south of the study region.

Regarding the LIN trend at the level of forest types, it's notable that the evergreen forest exhibits the highest values of negative trends compared to its median value with the negative and positive global standard deviation (Fig. 7c, h, i). According to Miranda et al. (2023), the sclerophyll forests located between 32° and 34.5° Latitude South have been among the most affected ecosystems by the last 10 years of drought and heatwaves occurring across the Mediterranean biome.

On the other hand, LIN positive trends (Fig. 7a, d) are present in conifer forests dominated by *A. araucana* and evergreen forests dominated by *N. pumilio*. In this case, both forest types are characterized by their distribution in higher elevations surrounding the limit line for tree distribution.

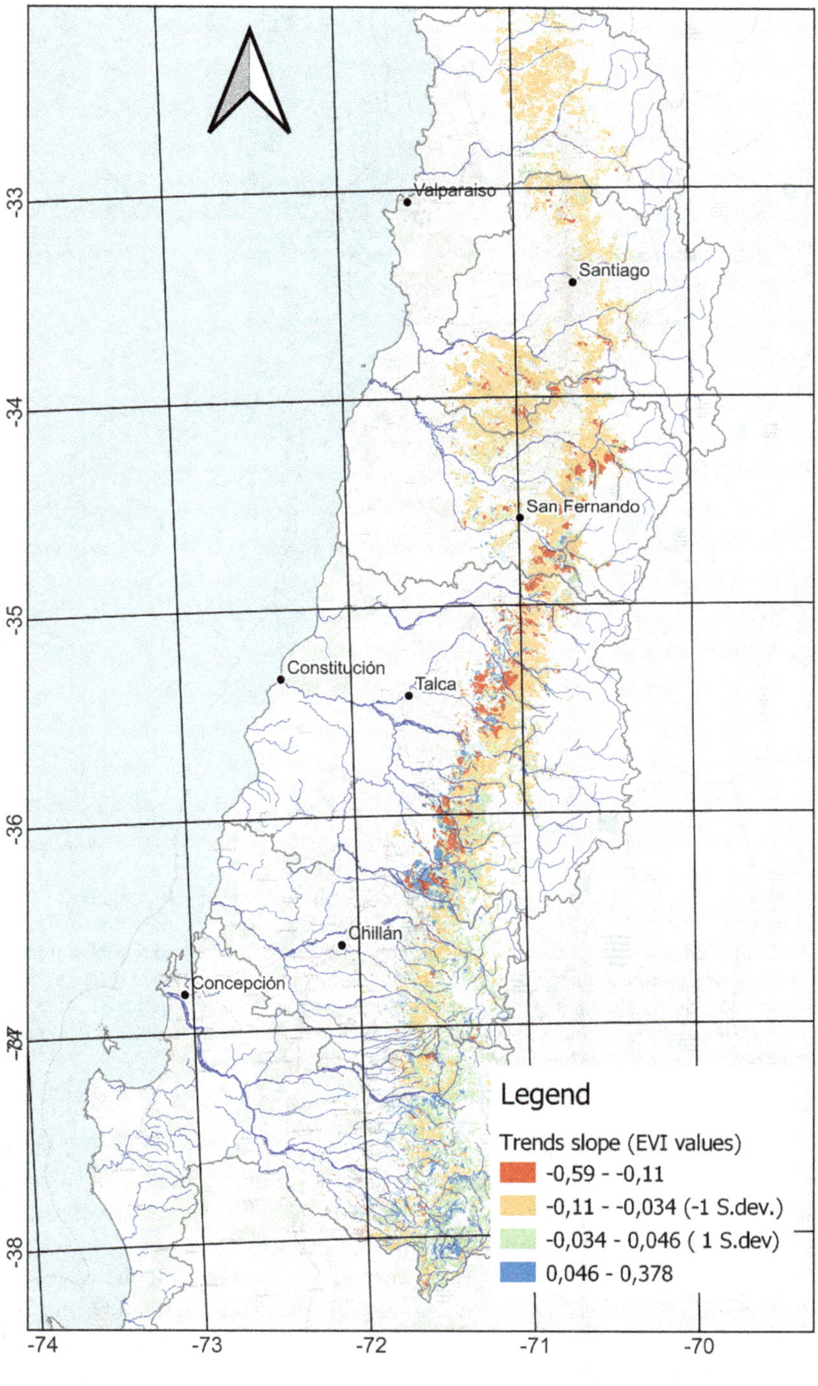

Fig. 6 Trend of primary productivity represented by the phenological variable LIN. Red-colored pixels represent a negative trend, while green-colored pixels represent a positive trend for the LIN variable for a 20-year observation period

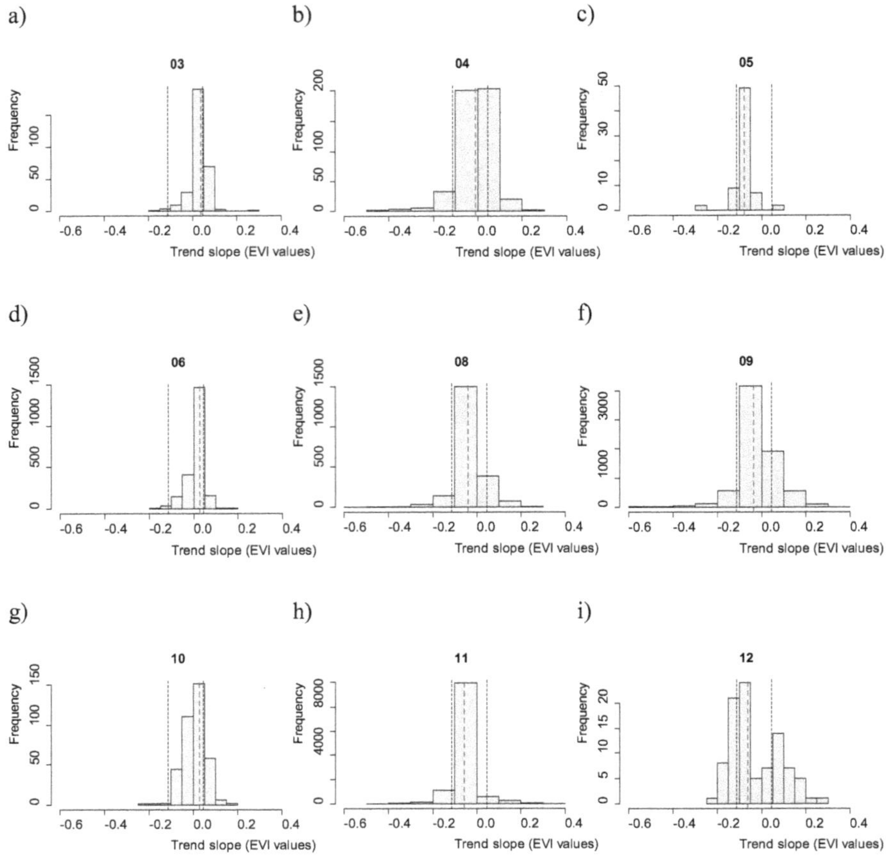

Fig. 7 Histograms of LIN trend by forest types: (**a**) coniferous A. Araucana, (**b**) coniferous A. chilensis, (**c**) evergreen forest J. chilensis-Q. saponaria, (**d**) deciduous forest N. pumilio, (**e**) deciduous forest N. oblicua-N- glauca, (**f**) deciduous evergreen forest N. oblicua-N. alpina-N.dombeyi, (**g**) deciduous-evergreen forest N. dombeyi-N. alpina-L. philippiana, (**h**) evergreen sclerophyll forest Q. saponaria-L. caustica-C. alba, (**i**) evergreen forest N. dombeyi-L. philippiana-E. coccineum. Black dotted lines represent one standard deviation in term of LIN trend slope at global scale, red dotted lines represents the median value at level of forest types

It's important to note that a series of isolated pixels with positive and negative slopes are randomly distributed throughout the study region. Their trend values may be due to anthropogenic processes independent of the climate, such as forest management, land use change, among others.

These results align with observations in recent years regarding the effect of the last heatwaves and prolonged drought, which began in 2010 (Garreaud et al., 2017). The peaks reached in 2019–2020 generated a widespread browning of vegetation between the Coquimbo and O'Higgins regions (Miranda et al., 2020, 2023). It should also be noted that productivity declines may also be associated with the effects of forest fires, especially those occurring in 2017 and 2022.

5 Conclusion

Across the study region, there exists a clear north-south gradient in primary productivity, as evidenced by MODIS sensor observations. This gradient reflects variations in vegetation types, climatological and environmental conditions associated with the Andean range.

EVI values serve as indicators of vegetation health and type, with higher values indicating lush vegetation and lower values associated with barren land or water bodies. Different vegetation types, such as sclerophyll forests in the north and deciduous forests in the south, exhibit clear distinct EVI patterns.

The study region experiences significant seasonal variability in vegetation dynamics, with deciduous and shrubland forests showing greater fluctuations in EVI values compared to evergreen forests. This variability could be linked to climatic factors and vegetation phenology.

Phenological variables such as Start of Season (SOS), End of Season (EOS), and Peak Time (PET) demonstrate distinct patterns across different forest types. Evergreen sclerophyll forests exhibit notable variability in PET, suggesting diverse responses to environmental cues.

Accumulated productivity values (LIN) reveal pronounced gradients across the study area, reflecting variations in environmental factors such as altitude and moisture availability. Different forest types exhibit characteristic LIN values, indicative of their ecological niches.

Analysis of primary productivity trends over a 20-year period highlights spatial heterogeneity, with sectors showing both positive and negative trends. Evergreen forests display the highest negative trends, likely due to climate-induced stressors, while coniferous (*A. Araucana*) and deciduous forests (*N. pumilio*) exhibit positive trends associated with elevation, due to the proximity of the distribution line of vegetation, there is a change in the climate trend, where temperatures increase while environmental humidity is maintained.

It's important to consider the influence of human activities such as forest management and land use change on primary productivity trends. These factors may contribute to isolated trends observed in certain areas, independent of climatic drivers.

Recent droughts and heatwaves have had significant impacts on vegetation health, leading to widespread browning and decreased productivity in certain regions. Forest fires further exacerbate these effects, emphasizing the vulnerability of ecosystems to climatic extremes.

Finally, phenological variables, by accounting for the dynamics of forest ecosystems, can be used as indicators of their adaptive response to climate change, especially droughts and heatwaves. To advance in this line, it is necessary to have a more detailed understanding of the behavior of forest subtypes and, above all, to incorporate other effects of change such as forest fires, illegal logging for fuel or timber, among other human pressures.

References

Ameray, A., Bergeron, Y., Valeria, O., Montoro Girona, M., & Cavard, X. (2021). Forest carbon management: A review of silvicultural practices and management strategies across boreal, temperate and tropical forests. *Current Forestry Reports, 7*(4), 245–266. https://doi.org/10.1007/s40725-021-00151-w

Armesto, J., Kalin Arroyo, M., & Villagrán, C. (Eds.). (1996). *Ecología de los bosques nativos de Chile*. Universitaria.

Bannister, J. R., & Donoso, P. J. (2013). Forest typification to characterize the structure and composition of old-growth evergreen forests on Chiloe Island, North Patagonia (Chile). *Forests, 4*(4), 1087–1105. https://doi.org/10.3390/f4041087

Blondel, J., Aronson, J., Bodiou, J.-Y., & Boeuf, G. (2010). *The Mediterranean Region: Biological diversity through time and space* (2nd ed.). Oxford University Press.

Boisier, J. P., Rondanelli, R., Garreaud, R. D., & Muñoz, F. (2016). Anthropogenic and natural contributions to the Southeast Pacific precipitation decline and recent megadrought in central Chile. *Geophysical Research Letters, 43*(1), 413–421. https://doi.org/10.1002/2015GL067265

Byer, S., & Jin, Y. (2017). Detecting drought-induced tree mortality in Sierra Nevada forests with time series of satellite data. *Remote Sensing, 9*(9), 929. https://doi.org/10.3390/rs9090929

CONAF. (2021). *Catastro de los Recursos Vegetacionales Nativos de Chile: Actualizaciones al año 2020* (p. 79) www.conaf.cl

Didan, K. (2015). MOD13Q1 MODIS/Terra vegetation indices 16-Day L3 Global 250m SIN Grid V006 [Data set]. In *NASA EOSDIS Land Processes Distributed Active Archive Center*.

Donoso, C. (1981). *Tipos forestales de los bosques nativos de Chile*. Investigación y Desarrollo Forestal, Corporación Nacional Forestal.

Donoso, C. (1993). *Bosques templados de Chile y Argentina: variacion, estructura y dinamica* (Tercera ed.). Editorial Universitaria.

Eklundh, L., & Jönsson, P. (2016). TIMESAT for processing time-series data from satellite sensors for land surface monitoring. In Y. Ban (Ed.), *Multitemporal remote sensing* (pp. 177–194). Springer International Publishing.

Fensholt, R., Sandholt, I., & Schultz, M. (2004). Evaluation of MODIS LAI, fAPAR and the relation between fAPAR and NVDI in a semi-arid environment usingin situ measurement. *Remote Sensing of Environment, 91*, 490–507.

Ferchichi, A., Abbes, A. B., Barra, V., & Farah, I. R. (2022). Forecasting vegetation indices from spatio-temporal remotely sensed data using deep learning-based approaches: A systematic literature review. *Ecological Informatics, 68*, 101552. https://doi.org/10.1016/j.ecoinf.2022.101552

Forkel, M., Carvalhais, N., Verbesselt, J., Mahecha, M. D., Neigh, C. S. R., & Reichstein, M. (2013). Trend Change detection in NDVI time series: Effects of inter-annual variability and methodology. *Remote Sensing, 5*(5), 2113–2144. https://doi.org/10.3390/rs5052113

Fuentes-Castillo, T., Hernández, H. J., & Pliscoff, P. (2020). Hotspots and ecoregion vulnerability driven by climate change velocity in Southern South America. *Regional Environmental Change, 20*(1), 27. https://doi.org/10.1007/s10113-020-01595-9

Gajardo, R. (1995). *La vegetación natural de Chile: clasificación y distribución geográfica* (p. 165). Editorial Universitaria.

Gamon, J. A., Field, C. B., Goulden, M. L., Griffin, K. L., Hartley, A. E., Joel, G., Peñuelas, J., & Valentini, R. (1995). Relationships between NDVI, canopy structure, and photosynthesis in three Californian vegetation types. *Ecological Applications, 5*(1), 28–41.

Garreaud, R. D., Alvarez-Garreton, C., Barichivich, J., Pablo Boisier, J., Christie, D., Galleguillos, M., LeQuesne, C., McPhee, J., & Zambrano-Bigiarini, M. (2017). The 2010–2015 megadrought in central Chile: Impacts on regional hydroclimate and vegetation. *Hydrology and Earth System Sciences, 21*(12), 6307–6327. https://doi.org/10.5194/hess-21-6307-2017

Garreaud, R. D., Boisier, J. P., Rondanelli, R., Montecinos, A., Sepúlveda, H. H., & Veloso-Aguila, D. (2020). The central Chile mega drought (2010–2018): A climate dynamics perspective. *International Journal of Climatology, 40*(1), 421–439. https://doi.org/10.1002/joc.6219

Glade, F. E., Miranda, M. D., Meza, F. J., & van Leeuwen, W. J. D. (2016). Productivity and phenological responses of natural vegetation to present and future inter-annual climate variability across semi-arid river basins in Chile. *Environmental Monitoring and Assessment, 188*(12), 676. https://doi.org/10.1007/s10661-016-5675-7

Glenn, E. P., Huete, A. R., Nagler, P. L., & Nelson, S. G. (2008). Relationship between remotely-sensed vegetation indices, canopy attributes and plant physiological processes: What vegetation indices can and cannot tell us about the landscape. *Sensors, 8*, 2136–2160. www.mdpi.org/sensors

Hartmann, H., Adams, H. D., Anderegg, W. R. L., Jansen, S., & Zeppel, M. J. B. (2015). Research frontiers in drought-induced tree mortality: Crossing scales and disciplines. *New Phytologist, 205*(3), 965–969. https://doi.org/10.1111/nph.13246

Housman, I. W., Chastain, R. A., & Finco, M. V. (2018). An evaluation of forest health insect and disease survey data and satellite-based remote sensing forest change detection methods: Case studies in the United States. *Remote Sensing, 10*(8), 1184. https://doi.org/10.3390/rs10081184

Ivits, E., Cherlet, M., Mehl, W., & Sommer, S. (2013). Ecosystem functional units characterized by satellite observed phenology and productivity gradients: A case study for Europe. *Ecological Indicators, 27*, 17–28. https://doi.org/10.1016/j.ecolind.2012.11.010

Jönsson, P., & Eklundh, L. (2002). Seasonality extraction by function fitting to time-series of satellite sensor data. *IEEE Transactions on Geoscience and Remote Sensing, 40*(8), 1824–1832. https://doi.org/10.1109/TGRS.2002.802519

Jönsson, P., & Eklundh, L. (2004). TIMESAT—A program for analyzing time-series of satellite sensor data. *Computers & Geosciences, 30*(8), 833–845. https://doi.org/10.1016/j.cageo.2004.05.006

Khodaee, M., Hwang, T., Kim, J. H., Norman, S. P., Robeson, S. M., & Song, C. (2020). Monitoring forest infestation and fire disturbance in the southern appalachian using a time series analysis of landsat imagery. *Remote Sensing, 12*(15), 2412. https://doi.org/10.3390/RS12152412

Lindner, M., Maroschek, M., Netherer, S., Kremer, A., Barbati, A., Garcia-Gonzalo, J., Seidl, R., Delzon, S., Corona, P., Kolström, M., Lexer, M. J., & Marchetti, M. (2010). Climate change impacts, adaptive capacity, and vulnerability of European forest ecosystems. *Forest Ecology and Management, 259*(4), 698–709. https://doi.org/10.1016/j.foreco.2009.09.023

Luebert, F., & Pliscoff, P. (2017). *Sinopsis bioclimática y vegetacional de Chile* (Vol. 107). Editorial Universitaria.

Matskovsky, V., Venegas-gonz, A., Roig, F. A., Le Quesne, C., Klock, K., Canales, C., & Mu, A. A. (2021). Tree growth decline as a response to projected climate change in the 21st century in Mediterranean mountain forests of Chile. *Global and Planetary Change, 198*, 103406. https://doi.org/10.1016/j.gloplacha.2020.103406

Mccullum, A., & Torres-Pérez, J. (2020). *National Aeronautics and Space Administration tracking vegetation phenology with remote sensing*. https://arset.gsfc.nasa.gov/land/we

Miranda, A., Lara, A., Altamirano, A., Di Bella, C., González, M. E., & Julio Camarero, J. (2020). Forest browning trends in response to drought in a highly threatened mediterranean landscape of South America. *Ecological Indicators, 115*(August 2019), 106401. https://doi.org/10.1016/j.ecolind.2020.106401

Miranda, A., Syphard, A. D., Berdugo, M., Carrasco, J., Gómez-González, S., Ovalle, J. F., Delpiano, C. A., Vargas, S., Squeo, F. A., Miranda, M. D., Dobbs, C., Mentler, R., Lara, A., & Garreaud, R. (2023). Widespread synchronous decline of Mediterranean-type forest driven by accelerated aridity. *Nature Plants, 9*, 1810–1817. https://doi.org/10.1038/s41477-023-01541-7

Moritz, S., & Bartz-Beielstein, T. (2017). imputeTS: Time series missing value imputation in R. *R Journal, 9*(1), 207–218. https://doi.org/10.32614/rj-2017-009

Paruelo, J. M. (2008). La caracterización funcional de ecosistemas mediante sensores remotos. *Ecosistemas: Revista Cietifica y Tecnica de Ecologia y Medio Ambiente, 17*(3), 4–22.

Peel, M. C., Finlayson, B. L., & Mcmahon, T. A. (2007). Updated world map of the Köppen-Geiger climate classification. *Hydrology and Earth System Sciences, 11*, 1633–1644.

Peñuelas, J., Sardans, J., Filella, I., Estiarte, M., Llusià, J., Ogaya, R., Carnicer, J., Bartrons, M., Rivas-Ubach, A., Grau, O., Peguero, G., Margalef, O., Pla-Rabés, S., Stefanescu, C., Asensio, D., Preece, C., Liu, L., Verger, A., Barbeta, A., et al. (2017). Impacts of global change on Mediterranean forests and their services. *Forests, 8*(12), 1–37. https://doi.org/10.3390/f8120463

R Core Team. (2023). *R: A language and environment for statistical computing*. R Foundation for Statistical Computing. https://www.R-project.org

Rundel, P. W., Arroyo, M. T. K., Cowling, R. M., Keeley, J. E., Lamont, B. B., & Vargas, P. (2016). Mediterranean biomes: Evolution of their vegetation, floras, and climate. *Annual Review of Ecology, Evolution, and Systematics, 47*(December), 383–407. https://doi.org/10.1146/annurev-ecolsys-121415-032330

Santibáñez Quezada, F. (2017). *Atlas agroclimático de Chile. Estado actual y tendencias del clima. Tomo III: Regiones de Valparaíso, Metropolitana, O'Higgins y Maule*. Universidad de Chile. Facultad de Ciencias Agronómicas.

Smith-Ramírez, C., Grez, A., Galleguillos, M., Cerda, C., Ocampo-Melgar, A., Miranda, M. D., Muñoz, A. A., Rendón-Funes, A., Díaz, I., Cifuentes, C., Alaniz, A., Seguel, O., Ovalle, J., Montenegro, G., Saldes-Cortés, A., Martínez-Harms, M. J., Armesto, J. J., & Vita, A. (2023). Ecosystem services of Chilean sclerophyllous forests and shrublands on the verge of collapse: A review. *Journal of Arid Environments, 211*(May 2022), 104927. https://doi.org/10.1016/j.jaridenv.2022.104927

van Leeuwen, W., Hartfield, K., Miranda, M., & Meza, F. (2013). Trends and ENSO/AAO driven variability in NDVI derived productivity and phenology alongside the Andes Mountains. *Remote Sensing, 5*(3), 1177–1203. https://doi.org/10.3390/rs5031177

Zhang, Y., Xu, M., Chen, H., & Adams, J. (2009). Global pattern of NPP to GPP ratio derived from MODIS data: Effects of ecosystem type, geographical location and climate. *Global Ecology and Biogeography, 18*(3), 280–290. https://doi.org/10.1111/j.1466-8238.2008.00442.x

Zhengxing, W., Chuang, L., & Alfredo, H. (2003). From AVHRR-NDVI to MODIS-EVI: Advances in vegetation index research. *Acta Ecologica Sinica, 23*(5), 979–987.

Non-native Plants in the Andes Ecoregions: Current Patterns and Future Perspectives

Romina D. Fernandez, Yohana G. Jimenez, M. Virginia Gonzalez, and H. Ricardo Grau

1 Background of Andean Region

The Andes, the world's longest continental mountain range, extend over 7000 kilometers along the western edge of South America spanning through seven countries: Venezuela, Colombia, Ecuador, Peru, Bolivia, Chile, and Argentina (Tovar et al., 2022). This impressive mountain system is characterized by a remarkable diversity of landscapes and ecosystems, shaped by the complex interactions of geology, climate, land use and livelihood activities (Fjeldså, 2007; Tovar et al., 2022; Pérez-Escobar et al., 2022). The unique geography of the Andes, marked by its towering peaks and ecological diversity, underscores its significance as a key natural feature in the South American continent. The Andean ecosystems are recognized as a global biodiversity hotspot due to the numerous endemic species adapted to the elevational zones and repeated biogeographic events of migrations and isolation (Särkinen et al., 2012; González-Caro et al., 2020; Pérez-Escobar et al., 2022). Beyond their ecological richness, Andean ecosystems hold immense cultural significance, shaping the identities and practices of communities reliant on the mountains for sustainability and spirituality (Radcliffe & Laurie, 2006; De la Cadena, 2010; Carilla et al., 2023a). The Andes play a crucial role in regulating regional and global climate patterns. These ecosystems offer other several services, such as soil fertility, water

Supplementary Information The online version contains supplementary material available at https://doi.org/10.1007/978-3-031-80805-0_10. The videos can be accessed individually by clicking the DOI link in the accompanying figure caption or by scanning this link with the SN More Media App.

R. D. Fernandez (✉) · Y. G. Jimenez · M. V. Gonzalez · H. R. Grau
Instituto de Ecología Regional (IER), Universidad Nacional de Tucumán (UNT)- Consejo Nacional de Investigaciones Científicas y Técnicas (CONICET), Yerba Buena, Tucumán, Argentina

N. Clerici (ed.), *Conservation of Andean Forests*,
https://doi.org/10.1007/978-3-031-80805-0_10

regulation and purification and serving as the source of major rivers that provide water resources for people (Célleri & Feyen, 2009), affecting the well-being of both local populations and those downstream. Additionally, the scenic beauty and varied landscapes attract tourists, contributing to local economies (Meisch, 2009).

2 Current Knowledge of Non-native Plants and Invasions in Andean Ecoregions

Numerous studies and systematic reviews have been conducted on non-native plants in the Andean region, underscoring a growing concern among researchers and environmentalists about their ecological impacts and the challenges they pose to biodiversity conservation of this region (Fuentes-Lillo & Pauchard, 2019; Fuentes-Lillo et al., 2023). Despite the importance of this topic, there are significant knowledge gaps and uncertainties surrounding different aspects related with non-native plants throughout the Andean region. For example, studies on non-native plants can show biases in certain countries or biogeographical regions (Fuentes-Lillo et al., 2023) so it is necessary to go further in understanding distribution patterns of non-native plants in Andean ecoregions. Current and growing availability lists of plant species and open databases like the Global Biodiversity Information Facility (GBIF) and Global Register of Introduced and Invasive Species (GRIIS) are useful tools for such purposes.

In this chapter, we conducted a systematic review using the Google Scholar database (https://scholar.google.com) to explore the distribution of richness and density of non-native plant species across Andean ecoregions, reasons of introduction, drivers of invasion, impacts and management. Additionally, this review aims to identify critical knowledge gaps and highlight key species that have been the focus of extensive research. We used combinations of several keywords: considering status: "exotic plants", OR "alien plants" OR "non-native plants"; OR "invasive plant" OR "plant invasion"; ecosystems: "alpine ecosystem", OR "Andean ecosystem" OR "mountain ecosystem", country: "Argentina" OR "Bolivia" OR "Colombia" OR "Chile" OR "Ecuador" OR "Peru" OR "Venezuela" and type of studies: "Impact" OR "drivers" OR "distribution" OR "dispersal" OR "uses" or "Management" OR "restoration" OR "social perception", in English and Spanish. The search was completed in June of 2023. The search resulted in 130 papers, the non-related publications (e.g. general revisions that did not mention plant species) were excluded by title, abstract or a careful reading of all text if necessary. After this first filter, we obtained 103 papers that matched our criteria. Studies were classified according to the Andean ecoregion and study topic: distribution, drivers of invasions (abiotic factors and species interactions), invasibility (the vulnerability of an ecosystem to invasion), invasiveness (the propensity of an introduced species to invade an ecosystem), impact, management, social perception and uses (Table S1). We mapped the richness and density of non-native plant species by Andean ecoregion. To do this we first obtained the geographical coordinates of each study site where non-native plant species were found according to the articles reviewed. In the cases that the studies did not report the georeferenced sites, but provided an accurate description of the study area, geographical coordinates were assigned using Google Earth Pro. Then, we utilized the Terrestrial Ecosystems of the World (TEOW) dataset developed by WWF-US (Olson et al., 2001) as the reference for ecoregions and we trimmed this global dataset to focus specifically on the Andes region, based on the delimitation outlined by Tovar et al. (2022). We estimated the density of non-native plant species in each ecoregion by dividing the richness in each ecoregions' area (km2) . The spatial distribution of the richness and density of non-native plants was analyzed using QGIS version 3.16 (QGIS Development Team, 2023).

We also explored the distribution of non-native plant species by Andean ecoregion using the databases GBIF and GRIIS. GBIF is an open database that provides species occurrence records across the globe with agreed standards. Although it has been questioned due to its biased data, this is the largest public database for georeferenced species occurrences data generally of high quality (Gaiji et al., 2013). First, we downloaded GBIF data including only records of vascular plants with geographic coordinates located in the Andean region. Then, we filtered records removing: (1) records with coordinates missing or failing to match the locality description, and (2) fossil records and records of unknown source(see Table S2 for a summary of all filters we used to ensure the replicability of this study). Second, we used GRIIS, a checklist of non-native species reviewed by a network of scientists and policymakers of each country, to classify the plant species as native and non-native species in the Andean region (see Table S2 for references to the downloaded datasets). Third, we systematized and integrated records from both databases using the bdc package (Ribeiro et al., 2022). This package allows standardized taxonomic names according to a taxonomic authority, harmonizing scientific names through matching algorithms with a reference list. Finally, we obtained the spatial distribution of richness and density of non-native plant species as we described above.

2.1 *Distribution of Richness and Density of Non-native Plants Across Andean Ecoregions*

According to the literature review, there are at least 219 non-native plant species in 20 of the 34 Andean ecoregions, but a little more than 60% of the total publications on non-native plants (including the invasive ones) were concentrated in only five Andean ecoregions (Fig. 1).

Of the total non-native plant species, 78% are herbaceous (171 species) and 22% are woody (48 species). These non-native species belong to 46 families (Table S1). We found that the families with the most non-native plant species in the Andes are Poaceae (67 species), followed by Fabaceae (21 species) and Asteraceae (21 species), as was previously reported by Fuentes-Lillo and Pauchard (2019). Most of the

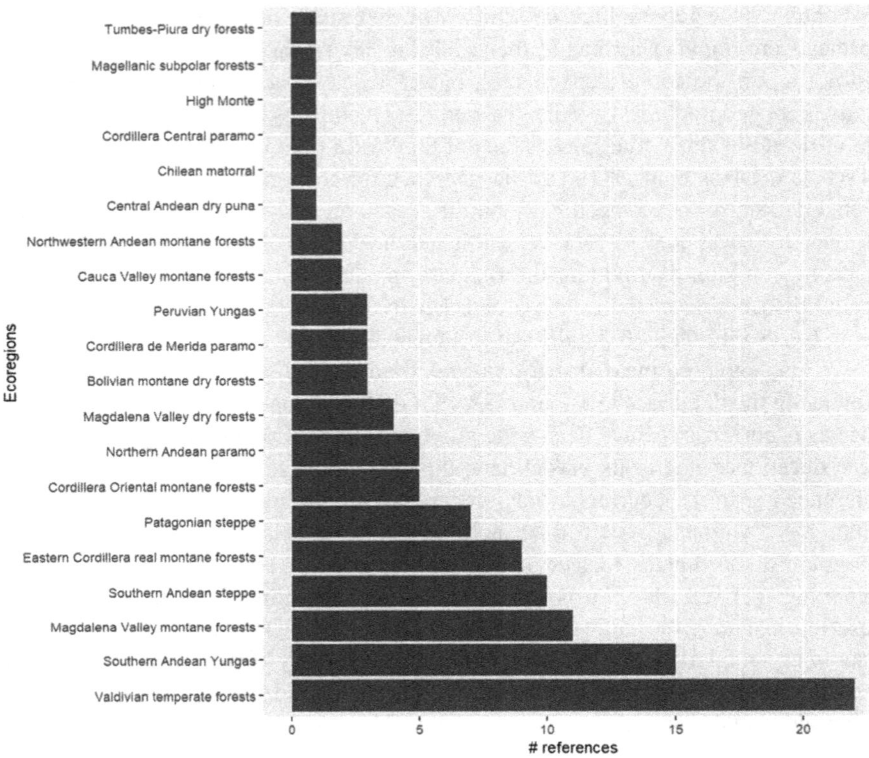

Fig. 1 .

families (67%) are represented in less than three Andean ecoregions and only the families Fabaceae (14 ecoregions), Poaceae (13 ecoregions), and Pinaceae (10 ecoregions) are represented in several Andean ecoregions. Our revision showed that the distribution of the recorded richness of non-native plants across Andean ecoregions exhibits a significant imbalance. The following ecoregions: Valdivian temperate forests (112), Southern Andean steppe (104) and Eastern Cordillera real montane forests (66) have the higher richness of non-native plants while the ecoregions: Cordillera Central páramo (1), Tumbes-Piura dry forests (2) and Cordillera de Merida páramo (3) presented the lower values of richness of these species (Table S1; Fig. 2A). The higher richness of non-native plants concentrated in ecoregions of the southern Andes may be given by the existing bias in sampling efforts, due to the concentration of studies in countries of the southern Andean ecoregions (Fuentes-Lillo & Pauchard, 2019; Fuentes-Lillo et al., 2023).

On the other hand, we found a total of 1082 non-native plant species in the Andes region according to the GBIF database (Table S3), being about seven times higher than the richness obtained from the literature review (Fig. 2B). There were records of non-native plants for almost of Andean ecoregions (33 ecoregions; Table S3), but the magnitude of richness of non-native plant species varied between ecoregions

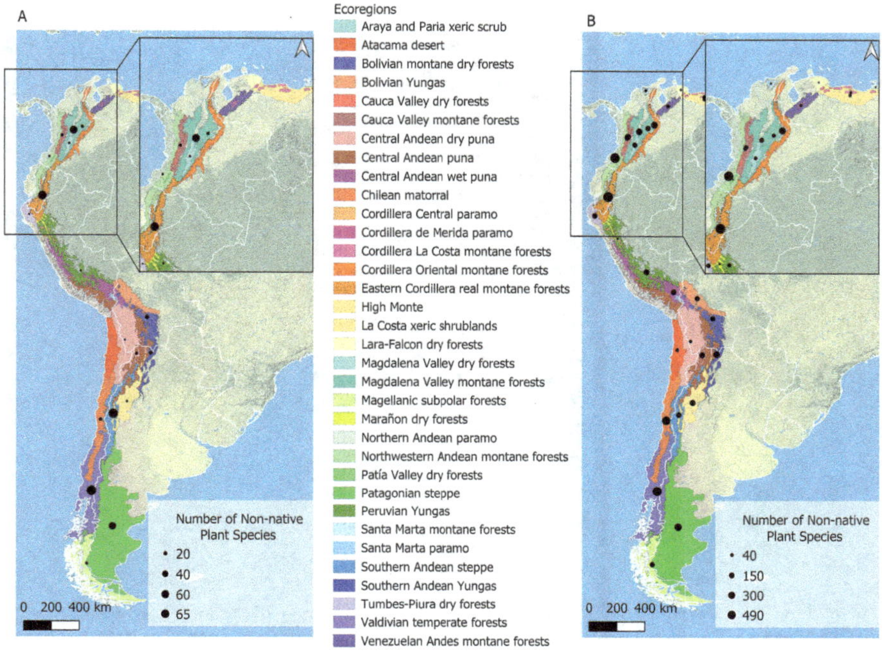

Fig. 2 .

(Fig. 2B). Eastern Cordillera Real Montane Forests, Northwestern Andean Montane Forests and Valdivian Temperate Forest were the ecoregions with the highest non-native plant species richness, while Patia Valley Dry Forest, Cordillera Central Paramo and Cordillera de Merida Paramo had the lowest non-native plant species richness (Fig. 2B). The species, *Cordia alliodora* (13,845) Boraginaceae and *Acacia decurrens* (12,930) Fabaceae, had the highest numbers of records across the Andean regions. Particularly, *Acacia decurrens* is classified as invasive non-native species in Colombia.

Similar to the richness pattern, the magnitude of the density of non-native plant species was lower according to the literature review (Fig. 3A) than the GBIF database (Fig. 3B). In both cases, the higher density of non-native plant species was concentrated in the North Andean ecoregions. However, the ecoregions Southern Andean steppe and Valdivian temperate forests from the south region also showed higher density of non-native plant species according to the literature review.

The results of our revision highlight that some non-native plant species received more attention in the literature of which different aspects of their invasion are recorded. In Table 1 we showed twelve of these most studied species in Andean ecoregions. It is important to note that except for *Ligustrum lucidum*, which has been highly studied but in a single Andean ecoregion, the rest of these most studied species are present in three or more Andean ecoregions. Most of the studies did not

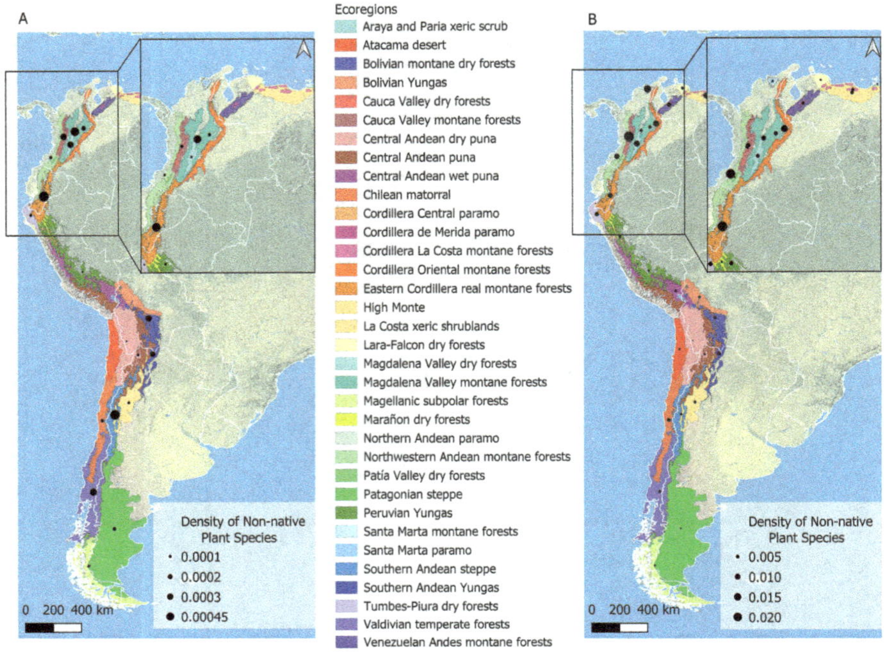

Fig. 3 .

specify the reason for the introduction of non-native plants in the Andes (83%). The most common reasons for the introduction of these species are commercial plantations, forage, ornamental and living fences.

To understand the distribution patterns of species, it is key to know their dispersion rate and the factors that influence or limit it (Zhang et al., 2023). The characteristic elevation gradient of the Andean region is an important factor that determines the distribution of non-native plants in Andean ecoregions. The harsh climate conditions increase with the altitude, making life difficult in general. This combined with the fact that there is higher population density and therefore higher human activities in low areas, could explain the higher richness and abundance of non-native plant species in lower altitude of Andean ecoregions (Pauchard & Alaback, 2004). However, even Andean ecosystems of high altitude are susceptible to plant invasion, particularly by plants native to Europe (Barros et al., 2015), where the history of human colonization is possibly important to explain the distribution of these non-native plants.

Considering the species traits, it was observed that growth form is an important factor influencing the rate of dispersal; herbaceous non-native plants tend to be dispersed faster than herbaceous native plants, while no difference was found in the dispersion rate of woody native vs. woody non-native plants (Zhang et al., 2023).

Table 1 Scientific name and growth form of twelve non-native plant species most studied in Andean ecoregions. The table indicates the total number of studies that include these species and the ecoregions where they were reported. For the complete list of non-native plant species recorded in the literature for the Andes ecoregions see Table S1

Species	Growth form	Number of studies	Andean ecoregions (Olson et al., 2001)
Taraxacum sect. Taraxacum	Perennial herb	25	Valdivian temperate forests,High Monte, Southern Andean steppe,Patagonian steppe, Central Andean dry puna, Magellanic subpolar forests, Eastern Cordillera real montane forests
Ligustrum lucidum	Tree	14	Southern Andean Yungas
Pinus patula	Tree	13	Northwestern Andean montane forests, Eastern Cordillera real montane forests,Magdalena Valley montane forests,Cordillera Central páramo,Peruvian Yungas, Northern Andean páramo
Rumex acetosella	Perennial herb	12	Southern Andean steppe,Central Andean dry puna, Valdivian temperate forests,Cordillera de Merida páramo, Eastern Cordillera real montane forests, Patagonian steppe
Trifolium repens	Perennial herb	11	Valdivian temperate forests,Southern Andean steppe,Magellanic subpolar forests, Eastern Cordillera real montane forests, Patagonian steppe
Ulex europaeus	Shrub	11	Magdalena Valley dry forests, Cordillera Oriental montane forests, Northern Andean páramo
Cerastium arvense	Perennial herb	10	High Monte, Southern Andean steppe, Patagonian steppe
Rosa rubiginosa	Shrub	10	Valdivian temperate forests, High Monte, Southern Andean steppe, Patagonian steppe
Cenchrus clandestinus	Perennial herb	8	Eastern Cordillera real montane forests, Central Andean dry puna, Cauca Valley montane forests, Magdalena Valley montane forests, Cordillera Oriental montane forests, Northern Andean páramo
Holcus lanatus	Perennial herb	8	Eastern Cordillera real montane forests, Cauca Valley montane forests, Magdalena Valley montane forests, Northern Andean páramo,Valdivian temperate forests, Patagonian steppe, Cordillera Oriental montane forests
Pinus radiata	Tree	8	Northwestern Andean montane forests, Magdalena Valley montane forests, Central Andean dry puna, Bolivian montane dry forests, Eastern Cordillera real montane forests, Valdivian temperate forests, Peruvian Yungas
Plantago lanceolata	Perennial herb	8	Southern Andean steppe, Valdivian temperate forests, Eastern Cordillera real montane forests, Patagonian steppe

This may contribute to explain the higher records for herbaceous non-native species in Andean ecoregions.

2.2 *Drivers of Plant Invasions in Andean Ecoregions*

Patterns of plant invasions in mountains are influenced by natural factors such as propagule availability, environmental conditions, biotic interactions and anthropogenic factors including different human activities (Pauchard et al., 2009; Quiroz et al., 2011; Alexander et al., 2016; Fuentes-Lillo et al., 2021; Joshi et al., 2022). Previous studies have emphasized the central role of human-induced factors, closely trailed by climate variables, as the primary influences determining the distribution of non-native plants in the Andes (Fuentes-Lillo & Pauchard, 2019; Fuentes-Lillo et al., 2021, 2023). The long history of human occupation in the Andean region has transformed and remodeled a considerable portion of its landscapes (Josse et al., 2011). This, combined with the region's diverse elevational and climatic gradients, renders it increasingly prone to the invasion of non-native plant species.

Disturbances associated with human activities can influence the distribution of non-native plants. Some examples of human activities that contribute to the introduction and spread of invasive non-native plants in the Andean region include agriculture, grazing, fire, urbanization, tourism and international trade (Pauchard & Alaback, 2004; Young, 2009; Barros et al., 2015; Montti et al., 2017). Increasing accessibility in the Andes due to infrastructure improvement and urban development leads to transport mechanisms as a source of long distance dispersal increasing propagule pressure in the most remote areas (Chapman et al., 2016). The creation and maintenance of roads is a disturbance factor removing the vegetation and changing the structure, chemistry and runoff of soil (Aschero et al., 2021). At the same time, roads act as vectors of dispersion of non-native plants, increasing the propagule pressure and the establishment of these species (Pauchard & Alaback, 2004; Mardones et al., 2015; Sandoya et al., 2017; Aschero et al., 2021). Thus, studies carried out in different Andean ecoregions found higher richness and abundance of non-native plants in sites closer to roads (e.g. Pauchard & Alaback, 2004; Fernández-Murillo et al., 2015; Sandoya et al., 2017; Aschero et al., 2021). Similarly, the disturbances in native plant communities and soils related with trails construction for recreational activities in Andean ecosystems such as hiking, climbing and sightseeing favor the dispersal and establishment of non-native plants (Barros et al., 2020). Livestock activities also contribute to the propagation of non-native plants in different Andean ecoregions, as the movement of these animals often acts as a vector for seed dispersal (Dacar et al., 2019; Aschero et al., 2021; Jimenez et al., 2021; Alvarez et al., 2022). Livestock has likely played a more significant role as a dispersal agent in the past when mules and horses were a key means of transport throughout the Andean region. Since livestock (and particularly live transport) is declining in many areas, their influence on the non-native plant invasion process is likely to be minor compared to other types of disturbance. Plant invasion in Andean ecoregions may result from different disturbances, as it was pointed out for the Páramo ecoregion, where plant invasions are a consequence of different disturbances including fire, livestock, agriculture and road construction (Ríos, 2013).

Different disturbances in the surrounding matrix and within protected areas in the Andean region facilitated the introduction, spread and invasion of non-native plants risking the effectiveness of these areas as conservation tools. Our revision revealed that nearly 50% of non-native plant species have been documented within protected areas. For example, in protected areas of High Monte and Southern Andean steppe in Argentina (Villavicencio, Cordon del Plata and Manzano), a heightened presence of non-native plants was observed in proximity to roads. Additionally, a positive correlation was identified between non-native herbivory disturbance and the distribution of non-native plants within these protected areas (Aschero et al., 2021). Similar findings were reported in protected areas of Valdivian temperate forests in Chile (Villarrica National Park, Lanín National Park (Pauchard & Alaback, 2004) and The Puyehue National Park (Mardones et al., 2015).

The distribution patterns of non-native plants in Andean ecosystems are also facing the effects of climate change (Fuentes-Lillo et al., 2023). Climate change could potentially reduce the climatic constraints of non-native plants distribution in mountain ecosystems (Petitpierre et al., 2016) and facilitate the establishment and spread of some non-native plants better adapted to the new conditions, with unknown impacts on biodiversity and ecosystem services (Hernández-Lambraño et al., 2017). Particularly, for the Andean region studies show projected increases in temperature and variable precipitation regimes depending on slope orientation (Báez et al., 2016; Tovar et al., 2022; Carilla et al., 2023b). Shifts in temperature and precipitation patterns may create more favorable conditions for the proliferation of non-native plant species, with upward migrations from some non-native plant species as long as they adapt to changing environmental conditions (Hernandez Lambraño et al., 2016). It is likely that this process will increase local diversity. For example, the Southern Andean Steppe, already has a high richness of non-native plants but the number of drought-adapted non-native plants may increase due to the expected increment of droughts (Molina-Montenegro et al., 2011). On the other hand, projected higher precipitations in Páramo might also reduce one of the existent barriers to the establishment of non-native plants in the region (Thomas et al., 2010) becoming an important, but little acknowledged, threat to the unique biodiversity of these high Andean ecosystems (Llambí et al., 2020).

2.3 *Ecological Consequences of Plant Invasions in Andean Ecoregions*

There are records of impacts of invasive non-native plants in around 50% of the Andean ecoregions, yet these impacts are linked to only a limited number of invasive non-native plant species. Studies focused on the negative impacts of plant invasions on native biodiversity, nutrient and water cycling and species interactions. While most studies assess the negative effects of non-native plants on native flora, limited attention has been given to their impact on other taxa (i.e. vertebrates, invertebrates, and microorganisms).

Evidence substantiating this includes *Pinus radiata* negatively affecting native species recruitment in Magdalena Valley montane forests (Cavelier & Santos, 1999), *Pinus contorta* impacting richness and cover in Valdivian temperate forests (Franzese et al., 2017), and *Pinus patula* reducing plant species diversity in the Northern Andean páramo (Van Wesenbeeck et al., 2003). Similarly, in the Southern Andean Yungas, *L. Lucidum* competes with native species, altering forest structure, slowing native tree growth (Malizia et al., 2017), and decreasing lianas abundance (Ceballos et al., 2015).

Regarding impacts on other taxa, it was documented the negative influence of *Ulex europaeus* on bird communities in Magdalena Valley montane forests (Amaya-Villarreal & Renjifo, 2010), and the adverse effects of *Pseudotsuga menziesii* and *Pinus* spp. on bird richness in the Patagonian steppe (Paritsis & Aizen, 2008). In relation to invertebrates, it was observed that *Pinus patula* decreased the richness and abundance of soil invertebrates in Magdalena Valley montane forests (León-Gamboa et al., 2010). Finally, it was reported changes in the abundance and composition of fungal decomposers in invaded forests by *L. lucidum* in the Southern Andean Yungas (Fernandez et al., 2017).

Concerning impacts on nutrient cycling, a study documented that commonly used non-native trees, such as *P. patula* and *Cupressus lusitanica* in the Andean region of Colombia, exhibit slower nutrient cycling compared to native forests and their forest have acidic soils with lower nutrient availability than native forest (Ramírez et al., 2014). These non-native species lead to the development of acidic soils with lower nutrient availability (Ramírez et al., 2014). Similarly, *P. patula* showed slower cycling of nutrients than native trees in the Eastern Cordillera real montane forests of Ecuador (Quichimbo et al., 2020). By contrast, *L. lucidum* and *Morus spp.* showed a high decomposition rate in Southern Andean Yungas, which could be translated into faster nutrient cycling at ecosystem level in invaded forests (Aragón et al., 2014).

Productivity in Andean ecoregions can be altered by non-native plants when, through different mechanisms, they use resources more efficiently than native plants. For example, in Yungas, *L. lucidum* is a perennial species with higher growth rate than native trees (Malizia et al., 2017). This may explain the higher productivity of *L. lucidum* invaded forest compared with native forest of Southern Andean Yungas (Zamora Nasca et al., 2014) and its higher carbon sequestration rate at local scale (Jimenez et al., 2021). This species also increases soil organic carbon and total nitrogen (Aragón et al., 2014). On the other hand, *P. patula* reduces organic matter, soil organic carbon, and total nitrogen in Eastern Cordillera real montane forests (Carrión-Paladines et al., 2022). A decrease in soil moisture and available water was reported in Southern Andean Yungas invaded by *L. lucidum* (Zamora Nasca et al., 2014) and in both, Eastern Cordillera real montane forests and Northwestern Andean montane forests invaded by *Pinus spp.* (Hofstede et al., 2002; Balthazar et al., 2015). Furthermore, plantations with *P. patula* and *P. radiata* have been found to significantly reduce the availability of water for irrigation, posing a substantial cost to downstream water users (Raboin & Posner, 2012).

The impacts on biotic interactions are focused on pollination. In this sense, it has been observed that the presence of pollen grains from *Pseudotsuga menziesii* and *Pinus spp.* in the native micropyles was associated with a strong increase in the percentage of seed abortion, indicating that these invasive non-native plants can interfere with the pollination and sexual reproduction of native wind-pollinated conifers in Valdivian temperate forests (Martyniuk et al., 2015). Additionally, results of another study in the same ecoregion suggested that non-native plants could facilitate the invasion of at least some non-native flower visitors to disturbed habitats (Morales & Aizen, 2002).

Invasive non-native plants can also alter the fire regime in the Andean region. It was documented that the invasive non-native *Pinus contorta* increases the intensity of forest fires in Valdivian temperate forests (Cóbar Carranza, 2013). Furthermore, its seeds are more resistant to high temperatures, so after high intensity fires caused by the invasion of this species, the germination of its seeds would be favored, increasing the invasion (Cóbar Carranza, 2013).

Although most of the impacts documented for the Andean ecoregions are negative, the direction and magnitude of the impact may vary with the density and traits of the non-native plant and characteristics of the invaded ecosystems. For example, the effect of the invasive non-native plant *Taraxacum sect. Taraxacum* on the pollination service of native plants in Valdivian temperate forests was positive or neutral at low densities while at higher densities its effect was negative (Muñoz & Cavieres, 2008). The direction of the impact can also vary significantly depending on the stakeholders involved. For example, species within the *Pinus* genus were introduced across diverse Andean ecoregions for plantation purposes, yielding favorable outcomes for the forestry sector and some landowners. However, substantial evidence points to their negative impacts on biodiversity and ecosystem services, as previously mentioned in this section.

2.4 Management of Plant Invasions in Andean Ecoregions

Mountains are few of the environments where proactive management is still possible (McDougall et al., 2011). Management strategies for non-native plants vary from no strategies, to focus on monetary valuable species to integrated programs. In line with this, for the Andes region there are regional management approaches such as the "Estrategia Regional de Biodiversidad para los Paises del Tropico Andino" (Banco Interamericano de Desarrollo, 2003). This strategy approved by Andean countries (Bolivia, Colombia, Ecuador Perú, and Venezuela) in 2002 has the goal of identifying and agreeing on joint priority actions for the conservation and sustainable use of biological biodiversity while promoting the socioeconomic development of the region. This involves a participative approach and the coordinated actions of these countries, which are key steps to minimize the impacts of invasive plants. Also, there are national approaches to developing a comprehensive checklist of non-native plant species by country, such as the Global Register of Introduced and Invasive Species organized by IUCN SSC Invasive Species Specialist Group (ISSG) and the Global Invasive Species Database (http://www.iucngisd.org/gisd/). These approaches facilitate effective prevention and management strategies of non-native plants by disseminating specialist's knowledge and experience to a broad audience. Applying risk assessment methods could be an effective option to prevent plant invasions in the Andean ecosystems. Despite that, we only found an example of application of this kind of method to assess non-native plant species that may potentially become invasive in a protected area of Valdivian temperate forests (Mardones et al., 2015).

Results of our review also evidenced only two studies about management strategies of invasive non-native plants in Andean ecoregions. These studies were carried out in Peruvian Yungas (Tito & de la Torre-Mayorga, 2016) and Magdalena Valley Dry Forests (Castro Gómez & Cervera Gómez, 2014). To advance in the development of management strategies for invasive non-native plants, there is an urgent need to improve the knowledge about the impacts of these species on biodiversity and ecosystem services (Fuentes-Lillo et al., 2023).

Also, it is necessary to understand the factors that influence the invasion process. Due to the key role that disturbances have been shown to play in the process of plant invasion in the Andes ecoregions, to conserve Andean biodiversity management actions should be focused on reducing disturbances, such as landscape fragmentation and edge effects (Rojas et al., 2011) or livestock (Aschero et al., 2021). In protected areas with higher touristic activities, where people are key dispersal vectors of non-native plants' propagule, management actions should include control of vehicles, clothes, camping equipment and reducing trampling damage by restricting visitor use to designated trails; with subsequent restoration of damaged areas (Barros & Pickering, 2014; Barros et al., 2015).

Management actions could benefit from the participation of people and take into account the perceptions of different stakeholders. For instance, results of a study showed that more than half the tourists interviewed were willing to pay a fee to contribute to the management of invasive non-native plants (Bravo-Vargas et al., 2019). Another study found differences in the perception among stakeholders (landowners, local governments, foresters, and nature conservationists) about the effect of plantations with non-native *Pinus spp.* in the Páramos of Ecuador (Quiroz Dahik et al., 2018). These types of studies highlight the importance of including the perceptions of different social actors in invasive plant management strategies to ensure their effectiveness and avoid potential conflicts of interest.

Future Perspectives and Research Recommendations We are achieving a better understanding of non-native and invasive species in Andean ecosystems. However, there are many biases towards certain species of non-native plants, specific Andean ecosystems, and topics of studies regarding these species, which make it difficult to obtain a balanced picture of the problem of plant invasions in this region. Therefore, we are still far from being able to indicate which ecosystems are most threatened by invasive non-native plants; and which invasions are detrimental, beneficial, or have mixed effects. There is a need to develop more field studies in those Andean ecoregions that received less attention to date, in order to identify the presence of non-native plants. Additionally, it is important to undertake more experimental studies in this region to evaluate the ecological and evolutionary mechanisms of non-native plants. On the other hand, to predict future plant invasions in the Andean region, coming studies should model the distribution of non-native species under different climate change scenarios, considering species traits such as their growth form and dispersion. Although management suggestions for non-native plants in the Andes are found in the literature, there are few examples of concrete actions. Given that economic resources in this region can be limited, preventive actions, such as risk assessments, may be the best option to avoid or reduce plant invasions in Andean ecosystems.

Competing Interests The authors have no conflicts of interest to declare that are relevant to the content of this chapter.

References

Alexander, J. M., Lembrechts, J. J., Cavieres, L. A., Daehler, C., Haider, S., Kueffer, C., et al. (2016). Plant invasions into mountains and alpine ecosystems: Current status and future challenges. *Alpine Botany, 126*, 89–103.

Alvarez, M. A., Barros, A. A., Vazquez, D. P., Bonjour, L. D. J., Lembrechts, J. J., Wedegärtner, R. E., & Aschero, V. (2022). Hiking and livestock favor non-native plants in the high Andes. *Biological Invasions, 24*(11), 3475–3488.

Amaya-Villarreal, Á. M., & Renjifo, L. M. (2010). Efecto del retamo espinoso (*Ulex europaeus*) sobre las aves de borde en un bosque altoandino. *Ornitología Colombiana, 10*, 11–25.

Aragón, R., Montti, L., Ayup, M. M., & Fernández, R. (2014). Exotic species as modifiers of ecosystem processes: Litter decomposition in native and invaded secondary forests of NW Argentina. *Acta Oecologica, 54*, 21–28.

Aschero, V., Barros, A., Bonjour, L., Mazzolari, A., Sosa, M. P., Cuevas, F., et al. (2021). *Non-native herbivores promote plant invasions away from mountain roads in the Andes*. https://doi.org/10.21203/rs.3.rs-979953/v1

Báez, S., Jaramillo, L., Cuesta, F., & Donoso, D. A. (2016). Effects of climate change on Andean biodiversity: A synthesis of studies published until 2015. *Neotropical Biodiversity, 2*(1), 181–194.

Balthazar, V., Vanacker, V., Molina, A., & Lambin, E. F. (2015). Impacts of forest cover change on ecosystem services in high Andean mountains. *Ecological Indicators, 48*, 63–75.

Banco Interamericano de Desarrollo. (2003). *Estrategia regional de biodiversidad para los países del trópico andino: recopilación de documentos técnicos*. BID.

Barros, A., & Pickering, C. M. (2014). Non-native plant invasion in relation to tourism use of Aconcagua Park, Argentina, the highest protected area in the Southern Hemisphere. *Mountain Research and Development, 34*(1), 13–26.

Barros, A., Monz, C., & Pickering, C. (2015). Is tourism damaging ecosystems in the Andes? Current knowledge and an agenda for future research. *Ambio, 44*, 82–98.

Barros, A., Aschero, V., Mazzolari, A., Cavieres, L. A., & Pickering, C. M. (2020). Going off trails: How dispersed visitor use affects alpine vegetation. *Journal of Environmental Management, 267*, 110546.

Bravo-Vargas, V., García, R. A., Pizarro, J. C., & Pauchard, A. (2019). Do people care about pine invasions? Visitor perceptions and willingness to pay for pine control in a protected area. *Journal of Environmental Management*, 229, 57–66.

Carilla, J., Aráoz, E., Acosta, O. O., Malizia, A., Malizia, M., Jimenez, Y., et al. (2023a). Long-term environmental and social monitoring in the Andes: State of the art, knowledge gaps, and priorities for an integrated agenda. *Mountain Research and Development, 43*(2), A1–A9.

Carilla, J., Aráoz, E., Foguet, J., Casagranda, E., Halloy, S., & Grau, A. (2023b). Hydroclimate and vegetation variability of high Andean ecosystems. *Frontiers in Plant Science, 13*, 1067096.

Carrión-Paladines, V., Benítez, Á., & García-Ruíz, R. (2022). Conversion of Andean montane forest to exotic forest plantation modifies soil physicochemical properties in the buffer zone of Ecuador's Podocarpus National Park. *Forest Ecosystems, 9*, 100076.

Castro Gómez, J. J., & Cervera Gómez, J. R. (2014). *Formulación de indicadores de sostenibilidad para la erradicación de la especie retamo espinoso (Ulex Europaeus) en las áreas afectadas de los cerros orientales de Bogotá*. Thesis (p 59).

Cavelier, J., & Santos, C. (1999). Efectos de plantaciones abandonadas de especies exóticas y nativas sobre la regeneración natural de un bosque montano en Colombia. *Revista de Biología Tropical, 47*(4), 775–784.

Ceballos, S. J., Malizia, A., & Chacoff, N. P. (2015). Influencia de la invasión de Ligustrum lucidum (Oleaceae) sobre la comunidad de lianas en la sierra de San Javier (Tucumán-Argentina). *Ecología Austral, 25*(1), 65–74.

Célleri, R., & Feyen, J. (2009). The hydrology of tropical Andean ecosystems: Importance, knowledge status, and perspectives. *Mountain Research and Development, 29*(4), 350–355.

Chapman, D. S., Makra, L., Albertini, R., Bonini, M., Páldy, A., Rodinkova, V., et al. (2016). Modelling the introduction and spread of non-native species: International trade and climate change drive ragweed invasion. *Global Change Biology, 22*(9), 3067–3079.

Cóbar Carranza, A. J. (2013). Cambios en el combustible en un bosque de *Araucaria araucana* y *Nothofagus antarctica* invadido por *Pinus contorta* Implicaciones en el régimen de incendios. Thesis (p 71).

Dacar, M. A., Dalmasso, A. D., Bobadilla, S. Y., & Cuevas, M. F. (2019). Rol del ganado doméstico en el establecimiento de la especie invasora Rosa Mosqueta (Rosa rubiginosa L.) en los Andes áridos, Argentina. *Mastozoología Neotropical, 26*, 331–339.

De la Cadena, M. (2010). Indigenous cosmopolitics in the Andes: Conceptual reflections beyond "politics". *Cultural Anthropology, 25*(2), 334–370.

Fernandez, R. D., Bulacio, N., Álvarez, A., Pajot, H., & Aragón, R. (2017). Fungal decomposers of leaf litter from an invaded and native mountain forest of NW Argentina. *Antonie van Leeuwenhoek, 110*, 1207–1218.

Fernández-Murillo, M., Rico, A., & Kindlmann, P. (2015). Exotic plants along roads near La Paz, Bolivia. *Weed Research, 55*(6), 565–573.

Fjeldså, J. (2007). The relationship between biodiversity and population centres: The high Andes region as an example. *Biodiversity and Conservation, 16*, 2739–2751.

Franzese, J., Urrutia, J., García, R. A., Taylor, K., & Pauchard, A. (2017). Pine invasion impacts on plant diversity in Patagonia: Invader size and invaded habitat matter. *Biological Invasions, 19*(3), 1015–1027.

Fuentes-Lillo, E., & Pauchard, A. (2019). Invasiones en montañas: ≐Cuánto hemos avanzado en los últimos 10 años y cuáles son los desafíos para los ecosistemas de los Andes? *Gayana. Botánica, 76*(2), 141–155.

Fuentes-Lillo, E., Lembrechts, J. J., Cavieres, L. A., Jiménez, A., Haider, S., Barros, A., & Pauchard, A. (2021). Anthropogenic factors overrule local abiotic variables in determining non-native plant invasions in mountains. *Biological Invasions, 23*, 3671–3686.

Fuentes-Lillo, E., Lembrechts, J. J., Barros, A., Aschero, V., Bustamante, R. O., Cavieres, L. A., et al. (2023). Going up the Andes: Patterns and drivers of non-native plant invasions across latitudinal and elevational gradients. *Biodiversity and Conservation, 32*(13), 4199–4219.

Gaiji, S., Chavan, V., Ariño, A. H., Otegui, J., Hobern, D., Sood, R., & Robles, E. (2013). Content assessment of the primary biodiversity data published through GBIF network: Status, challenges and potentials. *Biodiversity Informatics, 8*(2), 94–172.

González-Caro, S., Duque, Á., Feeley, K. J., Cabrera, E., Phillips, J., Ramirez, S., & Yepes, A. (2020). The legacy of biogeographic history on the composition and structure of Andean forests. *Ecology, 101*(10), e03131.

Hernández-Lambraño, R. E., González-Moreno, P., & Sánchez-Agudo, J. Á. (2017). Towards the top: Niche expansion of Taraxacum officinale and Ulex europaeus in mountain regions of South America. *Austral Ecology, 42*(5), 577–589.

Hofstede, R. G., Groenendijk, J. P., Coppus, R., Fehse, J. C., & Sevink, J. (2002). Impact of pine plantations on soils and vegetation in the Ecuadorian high Andes. *Mountain Research and Development, 22*(2), 159–168.

Jimenez, Y. G., Aráoz, E., Grau, H. R., & Paolini, L. (2021). Linking forest transition, plant invasion and forest succession theories: Socioeconomic drivers and composition of new subtropical andean forests. *Landscape Ecology, 36*, 1161–1176.

Joshi, S., Shrestha, B. B., Shrestha, L., Rashid, I., & Adkins, S. (2022). Plant invasions in mountains. In *Global plant invasions* (pp. 279–300). Springer International Publishing.

Josse, C., Cuesta, F., Navarro, G., Barrena, V., Becerra, M. T., Cabrera, E., et al. (2011). Physical geography and ecosystems in the tropical Andes. In *Climate change and biodiversity in the tropical* (pp. 152–169). MacArthur Foundation.

León-Gamboa, A. L., Ramos, C., & García, M. R. (2010). Efecto de plantaciones de pino en la artropofauna del suelo de un bosque Altoandino. *Revista de Biología Tropical, 58*(3), 1031–1048.

Llambí, L. D., Durbecq, A., Cáceres-Mago, K., Cáceres, A., Ramírez, L., Torres, E., & Méndez, Z. (2020). Interactions between nurse-plants and an exotic invader along a tropical alpine elevation gradient: Growth-form matters. *Alpine Botany, 130*, 59–73.

Malizia, A., Osinaga-Acosta, O., Powell, P. A., & Aragón, R. (2017). Invasion of Ligustrum lucidum (Oleaceae) in subtropical secondary forests of NW Argentina: Declining growth rates of abundant native tree species. *Journal of Vegetation Science, 28*(6), 1240–1249.

Mardones, D., Fuentes, N., & Saldaña, A. (2015). Comparación de la severidad de invasión asociada a caminos con diferentes niveles de perturbación en un bosque templado lluvioso del sur de Chile. *Gayana Botánica, 72*(2), 221–230.

Martyniuk, N. A., Morales, C. L., & Aizen, M. A. (2015). Invasive conifers reduce seed set of a native Andean cedar through heterospecific pollination competition. *Biological Invasions, 17*, 1055–1067.

McDougall, K. L., Khuroo, A. A., Loope, L. L., Parks, C. G., Pauchard, A., Reshi, Z. A., et al. (2011). Plant invasions in mountains: Global lessons for better management. *Mountain Research and Development, 31*(4), 380–387.

Meisch, L. A. (2009). Chapter seven. Tourism, the state, and the marketing of traditional Andean Artesanías: Problematic encounters, pitfalls, and competing interests. In *Cultural tourism in Latin America* (pp. 141–160). Brill.

Molina-Montenegro, M. A., Quiroz, C. L., Torres-Díaz, C., & Atala, C. (2011). Functional differences in response to drought in the invasive Taraxacum officinale from native and introduced alpine habitat ranges. *Plant Ecology & Diversity, 4*(1), 37–44.

Montti, L., Carrillo, V. P., Gutiérrez-Angonese, J., Gasparri, N. I., Aragón, R., & Grau, H. R. (2017). The role of bioclimatic features, landscape configuration and historical land use in the invasion of an Asian tree in subtropical Argentina. *Landscape Ecology, 32*, 2167–2185. https://doi.org/10.1007/s10980-017-0563-2

Morales, C. L., & Aizen, M. A. (2002). Does invasion of exotic plants promote invasion of exotic flower visitors? A case study from the temperate forests of the southern Andes. *Biological Invasions, 4*, 87–100.

Muñoz, A. A., & Cavieres, L. A. (2008). The presence of a showy invasive plant disrupts pollinator service and reproductive output in native alpine species only at high densities. *Journal of Ecology, 96*(3), 459–467.

Olson, D. M., Dinerstein, E., Wikramanayake, E. D., Burgess, N. D., Powell, G. V. N., Underwood, E. C., et al. (2001). Terrestrial ecoregions of the world: A new map of life on Earth (PDF, 1.1 M). *BioScience, 51*, 933–938.

Paritsis, J., & Aizen, M. A. (2008). Effects of exotic conifer plantations on the biodiversity of understory plants, epigeal beetles and birds in *Nothofagus dombeyi* forests. *Forest Ecology and Management, 255*(5–6), 1575–1583.

Pauchard, A., & Alaback, P. B. (2004). Influence of elevation, land use, and landscape context on patterns of alien plant invasions along roadsides in protected areas of South-Central Chile. *Conservation Biology, 18*(1), 238–248.

Pauchard, A., Kueffer, C., Dietz, H., Daehler, C. C., Alexander, J., Edwards, P. J., et al. (2009). Ain't no mountain high enough: Plant invasions reaching new elevations. *Frontiers in Ecology and the Environment, 7*(9), 479–486.

Pérez-Escobar, O. A., Zizka, A., Bermúdez, M. A., Meseguer, A. S., Condamine, F. L., Hoorn, C., et al. (2022). The Andes through time: Evolution and distribution of Andean floras. *Trends in Plant Science, 27*, 364–378.

Petitpierre, B., McDougall, K., Seipel, T., Broennimann, O., Guisan, A., & Kueffer, C. (2016). Will climate change increase the risk of plant invasions into mountains? *Ecological Applications, 26*(2), 530–544.

QGIS Development Team. (2023). *QGIS Geographic Information System*. Open Source Geospatial Foundation Project. http://qgis.osgeo.org

Quichimbo, P., Jiménez, L., Veintimilla, D., Potthast, K., Tischer, A., Günter, S., et al. (2020). Nutrient dynamics in an Andean forest region: A case study of exotic and native species plantations in southern Ecuador. *New Forests, 51*, 313–334.

Quiroz Dahik, C., Crespo, P., Stimm, B., Murtinho, F., Weber, M., & Hildebrandt, P. (2018). Contrasting stakeholders' perceptions of pine plantations in the páramo ecosystem of Ecuador. *Sustainability, 10*(6), 1707.

Quiroz, C. L., Cavieres, L. A., & Pauchard, A. (2011). Assessing the importance of disturbance, site conditions, and the biotic barrier for dandelion invasion in an Alpine habitat. *Biological Invasions, 13*, 2889–2899.

Raboin, M. L., & Posner, J. L. (2012). Pine or pasture? Estimated costs and benefits of land use change in the Peruvian Andes. *Mountain Research and Development, 32*(2), 158–169.

Radcliffe, S. A., & Laurie, N. (2006). Culture and development: Taking culture seriously in development for Andean indigenous people. *Environment and Planning D: Society and Space, 24*(2), 231–248.

Ramírez, J. A., León-Peláez, J. D., Craven, D., Herrera, D. A., Zapata, C. M., González-Hernández, M. I., et al. (2014). Effects on nutrient cycling of conifer restoration in a degraded tropical montane forest. *Plant and Soil, 378*, 215–226.

Ribeiro, B. R., Velazco, S. J. E., Guidoni-Martins, K., Tessarolo, G., Jardim, L., Bachman, S. P., & Loyola, R. (2022). bdc: A toolkit for standardizing, integrating and cleaning biodiversity data. *Methods in Ecology and Evolution, 13*(7), 1421–1428.

Ríos, O. V. (2013). Disturbios en los páramos andinos. In *Visión socioecosistémica de los páramos y la alta montaña colombiana: memorias del proceso de definición de criterios para la delimitación de páramos* (pp. 39–57). Instituto von Humboldt.

Rojas, I., Becerra, P., Gálvez, N., Laker, J., Bonacic, C., & Hester, A. (2011). Relationship between fragmentation, degradation and native and exotic species richness in an Andean temperate forest of Chile. *Gayana Botánica, 68*(2), 163.

Sandoya, V., Pauchard, A., & Cavieres, L. A. (2017). Natives and non-natives plants show different responses to elevation and disturbance on the tropical high Andes of Ecuador. *Ecology and Evolution, 7*(19), 7909–7919.

Särkinen, T., Pennington, R. T., Lavin, M., Simon, M. F., & Hughes, C. E. (2012). Evolutionary islands in the Andes: Persistence and isolation explain high endemism in Andean dry tropical forests. *Journal of Biogeography, 39*(5), 884–900.

Thomas, E., Damme, P., & Van Goetghebeur, P. (2010). Some factors determining species diversity of prepuna and puna vegetations in a Bolivian Andes region. *Plant Ecology and Evolution, 143*(1), 31–42.

Tito, R., & de la Torre-Mayorga, F. (2016). Effects of an aphid pest on an invasive plant in the Peruvian Andes. *Plant Ecology, 217*(7), 817–823.

Tovar, C., Carril, A. F., Gutiérrez, A. G., Ahrends, A., Fita, L., Zaninelli, P., et al. (2022). Understanding climate change impacts on biome and plant distributions in the Andes: Challenges and opportunities. *Journal of Biogeography, 49*(8), 1420–1442.

Van Wesenbeeck, B. K., van Mourik, T., Duivenvoorden, J. F., & Cleef, A. M. (2003). Strong effects of a plantation with Pinus patula on Andean subpáramo vegetation: A case study from Colombia. *Biological Conservation, 114*(2), 207–218.

Young, K. R. (2009). Andean land use and biodiversity: Humanized landscapes in a time of change. *Annals of the Missouri Botanical Garden, 96*(3), 492–507.

Zamora Nasca, L., Montti, L., Grau, R., & Paolini, L. (2014). Efectos de la invasión del ligustro, Ligustrum lucidum, en la dinámica hídrica de las Yungas del noroeste Argentino. *Bosque (Valdivia), 35*(2), 195–205.

Zhang, B., Hastings, A., Grosholz, E. D., & Zhai, L. (2023). The comparison of dispersal rate between invasive and native species varied by plant life form and functional traits. *Movement Ecology, 11*(1), 73.

The Role of Environmental Instruments in the Conservation of Andean Forests in Ecuador

Pablo Cuenca and Jin Kyoung Noh

1 Introduction

One of the paramount challenges confronting the global population today is acknowledging humanity's dependence on natural capital for a vast array of benefits crucial to our health, sustenance, and overall well-being (Díaz et al., 2015; Perschke et al., 2023). It is crucial for society to expand its comprehension of natural capital beyond merely associating it with the benefits provided by natural resources, or limiting it to environmental protection and sustainability, such as absorbing waste and pollution, or viewing it solely as the primary driver of economies (Jo et al., 2021). Thus, it must embrace the comprehensive and diverse aspects of natural capital, which include essential goods and services such as clean air, carbon sequestration, food production, pollination, and clean water. These rely directly and indirectly on biodiversity across its genetic, species, and ecosystem levels. In other words, achieving this broad spectrum of benefits in the "natural capital equation" necessitates the consideration of relationship or link between biodiversity and ecosystem services (Díaz et al. 2006, 2015). Moreover, biodiversity itself depends critically on forest ecosystems, which play a pivotal role in maintaining the balance between nature and human societies (with an estimated 80% of global biodiversity concentrated in forests) (FAO, 1993).

P. Cuenca (✉)
Tropical Ecosystems and Global Change Research Group, Universidad Regional Amazónica Ikiam, Road to Muyuna - Alto Tena km7, Tena, Napo, Ecuador

Global Change Laboratory, Universidad Regional Amazónica Ikiam, Road to Muyuna - Alto Tena km7, Tena, Napo, Ecuador
e-mail: pablo.cuenca@ikiam.edu.ec

J. K. Noh
Tropical Ecosystems and Global Change Research Group, Universidad Regional Amazónica Ikiam, Road to Muyuna - Alto Tena km7, Tena, Napo, Ecuador

N. Clerici (ed.), *Conservation of Andean Forests*,
https://doi.org/10.1007/978-3-031-80805-0_11

Increasingly, conservation organizations, institutions, and experts concur on the pivotal role of natural capital in generating economic and social benefits for society. The OECD (2015) underscores that achieving human well-being requires addressing four core pillars: economic health, political stability, legal robustness, and environmental health. From an economic perspective, the wealthiest countries worldwide prioritize the health of nature (natural capital) as a fundamental pillar for achieving societal well-being. Many scientists, institutions, and organizations view the OECD's approach as an urgent call to transition from a conventional sustainability paradigm—where environmental, social, and economic elements are in equilibrium—to a robust sustainability framework. This framework advocates for environmental considerations to encompass societal values, and for society to be integrated within economic frameworks. Essentially, it suggests that the rules governing human activities should prioritize impacts and benefits to nature (Díaz et al., 2006, 2015; Wu, 2013, 2021). For instance, Ecuador and Chile have formally incorporated this concept into their policy frameworks, promoting ecological and socio-ecological transitions to strengthen environmental governance.

Despite the growing recognition of natural capital's role in enhancing human well-being, rapid changes in land use, forest loss, and habitat fragmentation continue unabated (Cuenca & Echeverria, 2017; Kleemann et al., 2022b). Globally, it is reported that 488.2 million hectares of forest cover have been lost since the turn of the century (Institute, 2024). Additionally, landscape modification begins with deforestation, often driven by unregulated migration and unclear land tenure (Forman & Godron, 1986). This process escalates with landscape dissection, spurred by unplanned infrastructure development (e.g., roads, hydroelectric dams), followed by habitat fragmentation, marked by shifts in landscape context and matrix. The pattern continues with shrinking and depletion, reducing the size and eliminating remnants of forest patches, which threatens habitat specialist species with local extinction. Primary drivers of deforestation and fragmentation include extractive activities (e.g., gold, copper, petroleum), legal and illegal logging, expansion of agricultural frontiers, wildfires, and market failures (e.g., hunting bans, inadequate conservation incentives) (Forman & Godron, 1986; Turner, 2010; Turner & Gardner, 2015).

Given the urgent need to mitigate landscape modification, deforestation, and natural capital decline while ensuring the provision of ecosystem services (Han et al., 2019; Kim et al., 2014; Marshall & Jang, 2020; Yu & Han, 2016), governments worldwide have introduced conservation mechanisms. These include widely endorsed strategies such as protected areas (Cisneros et al., 2022; Kim et al., 2021; Wauchope et al., 2022), which currently amount to 295,520 globally (World Database on Protected Areas). Other effective instruments include payments for ecosystem services (PES), exemplified by Costa Rica's financial mechanism that promotes forest ecosystem conservation and combats land degradation (Arriagada et al., 2012), financial incentive mechanism like Ecuador's Socio Bosque Program (Cuenca et al., 2018), and extensive reforestation initiatives, such as Chile's Executive Decree 701 (Arriagada et al., 2016).

This chapter examines the impact of conservation instruments on forest loss, using propensity score matching, a widely employed methodology. It evaluates the strengths and weaknesses of this approach, its procedural aspects, and reviews various impact assessment studies. The culmination is a case study evaluating the effectiveness of protected areas in mitigating deforestation in the tropical Andean forests of Ecuador.

2 Quasi-Experimental Studies for Impact Assessment

Within the framework of negotiations such as REDD+ and the Zero Deforestation instrument implemented by the European Union, accurately estimating avoided deforestation is crucial (Pfaff et al., 2013; Ribas et al., 2020). This requires integrating spatially explicit information on socio-economic and biophysical factors, alongside employing robust impact assessment methodologies (Heilmayr et al., 2020; Soares-Filho et al., 2010; Yang et al., 2021).

The challenge in impact assessment lies in quantifying the effect of a program (or treatment) on a range of outcome variables across a group of individuals (Feng et al., 2021; Ribas et al., 2020). For instance, evaluating the impact of the Socio Bosque program on forest loss indicators (forest/non-forest) involves establishing the deforestation rate in the absence of the program to ascertain the attributable difference following its implementation (Feng et al., 2021; Soares-Filho et al., 2023). This difference is known as the treatment effect or program impact. Notably, it is impractical to observe both outcomes simultaneously for the same individual. The counterfactual outcome—the scenario for the participant if the program did not exist—is hypothetical and therefore unobservable. In impact evaluation literature, this hypothetical scenario is referred to as the counterfactual outcome (Bernal & Peña, 2012; Clemente et al., 2020; Gertler et al., 2011; Muñoz Brenes et al., 2018; Ribas et al., 2020).

Statistical methods used in impact assessment are grounded in the Roy-Rubin model (Rubin, 2006), where non-binary treatment is defined as Bi. For example, in conservation instruments, Bi could represent protected forest (Bi = 1) or non-protected forest (Bi = 0) for individual i. The outcome variables are denoted as Ei(Bi) for each individual i = 1 to N, where N denotes the total population. In simpler terms, Ei(1) signifies the outcome when the individual receives treatment, whereas Ei(0) represents the outcome when the individual does not. The treatment effect, or program impact for an individual i, is expressed as: Ti = Ei(1) – Ei(0).

As emphasized by Bernal and Peña (2012), Cuenca et al. (2018), Feng et al. (2021), Pfaff et al. (2013), and Ribas et al. (2020), the fundamental challenge in impact assessment is that researchers can only observe one potential outcome—the outcome variable if the individual receives treatment (Ei(1)) or the outcome variable if the individual does not (Ei(0))—but not both. This limitation means researchers lack the outcome with treatment if the individual was untreated, Ei(1) if Bi = 0, and similarly lack the outcome without treatment if the individual was treated, Ei(0) if

Bi = 1. It's important to note that the program impact estimated by the difference Ei(1) – Ei(0) pertains to a specific moment in time and does not equate to comparing the same individual at two distinct time points (pre- and post-intervention).

3 Strategies for Evaluating Conservation Instruments

A classical approach to assessing the impact of environmental instruments is through estimating avoided deforestation (Bruner et al., 2001; Oliveira et al., 2007; Wendland et al., 2015). Protected Areas (PAs), one of the most prevalent conservation policies (Assessment, 2005), are expected to influence deforestation patterns by preventing loss of woodlands through their protective measures (Cuenca et al., 2016; Kleemann et al., 2022a, b; Noh et al., 2019). Numerous studies assessed the impact of conservation policies globally and regionally by comparing deforestation rates at different scales in space and time (Bruner et al., 2001; Nagendra, 2008; Nelson & Chomitz, 2011). Some studies also employed indirect comparisons between protected and non-protected forests (Bruner et al., 2001; Joppa et al., 2008). However, results from these conventional methods may overestimate the deforestation prevented by protected areas, by not accounting for the non-random allocation of protection to forests in their analyses (Cuenca et al., 2016, 2018; Feng et al., 2021; Ribas et al., 2020).

Another challenge in impact evaluation strategy is that conservation instruments are not randomly assigned across landscapes, introducing experimental bias (Andam et al., 2008; Ribas et al., 2020). To address this lack of randomization, a widely used method in the scientific community is Propensity Score Matching (PSM), which is utilized in quasi-experimental studies (Cuenca et al., 2016; Feng et al., 2021; Ferraro et al., 2013; Liu & Ma, 2024; Ribas et al., 2020; Yu et al., 2024).

The basic concept of PSM is to find a "clone" of each treated individual in the control group and compare their outcome variables (Arriagada et al., 2016; Cuenca et al., 2018). Here, "clone" refers to an individual (or group of individuals) with the same observable characteristics X, such as proximity to roads, altitude, village proximity, temperature, precipitation, or any relevant biophysical or socioeconomic variables pertinent to the conservation instrument under analysis. Essentially, PSM aims to weigh observations in the control group so that the distribution of observable characteristics X closely matches that of the treatment group.

Impact assessment remains feasible for handling a large amount of data or variables due to the computational advancements in recent years (Achard et al., 2014; Guo et al., 2006; Joppa & Pfaff, 2010). Despite this computational progress, matching methods may only be effective when data or dimensions are not excessively large (Achard et al., 2014; Robalino et al., 2008). In other words, impact evaluation involves comparing the difference in the outcome variable (protected forest) for each cell, which consists of all possible values of X (protected forest/non-protected forest), generated from the weighted average effects in each cell of the program. If the datasets are extensive, data analysis will face challenges due to dimensionality (Bernal & Peña, 2012; Pfaff et al., 2009; Robalino et al., 2017).

In Propensity Score Matching (PSM), a key element is the ability to include variables that could bias the outcome and thus lead to reporting a causal conclusion. For example, in evaluating the impact of protected areas on deforestation, numerous variables can affect forest protection (distance to roads, distance to villages, distance to rivers, distance to forest, slope, altitude, etc.)- in other words, the observable characteristics of X (Cuenca et al., 2016; Ribas et al., 2020). This variety of variables complicates the analysis, and instead of matching based on a vector of observed characteristics X, individuals can be matched based on their estimated probability of participation in the conservation instrument, given their observable characteristics P(X), where: $P(X) = P(D = 1|X)$ (Guo & Fraser, 2014; Imbens, 2015; Wendland et al., 2015; Yang et al., 2019).

Our goal is to identify a replica(clone)of protected forests that matches the control group precisely, differing only in confounding variables. This allows us to match based solely on the probability P(X), which becomes a scalar (Imbens, 2015; Rosenbaum & Rubin, 1983; Sekhon, 2013). This method of matching based on probability has been widely demonstrated to prevent bias in estimators and to produce consistent effects (Imbens, 2015; Rosenbaum & Rubin, 1983).

As mentioned earlier, PSM is based on creating clones for treatment and control groups based on the participation probability of observable variables for each individual, in order to achieve very similar participation probabilities for both groups. In this process, it is important to analyze the common support region (Rosenbaum & Rubin, 1983). The rule of common support is that both treatment and control groups have a positive probability of being participants and non-participants in the conservation instrument. Taking the example of evaluating protected areas, it is not feasible to have a combination of cells that show a positive tendency to protect the forest and none that do not participate in forest protection. Therefore, the common support rule would only use cells in the control group analysis that assume the same propensity to participate as the treatment group (Gertler et al., 2011; Miteva et al., 2015).

An example of common support in a symmetrical curve is at extreme values to the right or left. If in the control group, the participation probabilities are very low, representing participation probability values far from the extremes, PSM eliminates control group participation probability values to perform the match (Bernal & Peña, 2012; Cuenca et al., 2018). This sounds simple; however, it could become inconvenient when there is a number of treated cells for which there are no similar control cells, meaning that the matching cells do not have similar confusion variables (Cuenca et al., 2016; Cuenca & Echeverria, 2017). This could lead to bias in impact evaluation before and after the implementation of a protected area because there is a set of treated cells that did not participate in the match, and therefore, the study conclusions will only apply to part of the data.

To clarify the concept of common support, let's adapt the example from Bernal and Peña (2012). Suppose we take a sample of 10 cells from the treatment group (protected forest) and 10 from the control group (non-protected forest), with the following estimated participation probabilities for the treatment group (T): T1(0.39); T2(0.45); T3(0.56); T4(0.63); T5(0.76); T6(0.78); T7(0.79); T8(0.81); T9(0.87), and T19(0.93), while for the Control group (C): C1(0.11); C2(0.30); C3(0.38); C4(0.42); C5(0.52); C6(0.55); C7(0.59); C8(0.62); C9(0.65); C10(0.79). Then, the common support zone is defined between the minimum value of the treatment group (T1, 0.39) and the maximum value of the control group (C10, 0.79). In this case, defining the common support zone results in the exclusion of three observations for both the treatment and control groups.

In practice, participation in the program may depend on both observable and unobservable variables or those not available in the database. The characteristics of each conservation instrument, particularly its targeting and agent response (e.g., Socio Bosque Program or Protected Areas), affect the significance of observable and non-observable factors in determining participation in the conservation instrument. Therefore, understanding the specific characteristics of each conservation instrument is crucial; for example, in the Socio Bosque program, observations related to forest loss are relevant due to its primary objective of reducing deforestation in the Tropical Andes. Using PSM in local context ensures a robust methodology.

4 Procedure for Applying Propensity Score Matching (PSM)

In many conservation instrument evaluation studies, the dependent variable is often binary (0 and 1). Therefore, the initial step involves establishing the model to use, typically a probit or logit model, as both yield comparable results (Gertler et al., 2011; Guo et al., 2006; Guo & Fraser, 2014; Miteva et al., 2012; Stuart et al., 2014).

Another critical consideration pertains to confounding variables. These variables influence deforestation positively or negatively and are directly linked to forest protection. For example, when assessing the effectiveness of protected areas (PAs) in preventing deforestation, it is imperative to study both direct and indirect drivers of deforestation. This thorough examination is necessary because PSM is highly sensitive to the selection of variables used to calculate matching probabilities (Cuenca et al., 2016, 2018; Pfaff et al., 2015; Ribas et al., 2020). Hence, it is crucial to include variables that remain constant over time or have been measured prior to participation in the conservation initiative, such as distance to main roads or forest area before PA implementation (Cuenca et al., 2018; Nolte, 2018; Ribas et al., 2020).

Subsequently, determining the common support is essential; as previously mentioned, this involves establishing the estimated participation probability for both control and treatment cells. To conclude the PSM process, the appropriate matching mechanism must be selected, whether nearest neighbor, stratification, maximum distance, or kernel. All these matching mechanisms aim to find identical or as similar as possible cells in their variables for both treatment and control groups. Therefore, the choice of matching mechanism or strategy in PSM is less pivotal in the analysis; what matters most is understanding whether different matching methods yield similar or divergent results and identifying the sources of any disparities to refine the model accordingly (Bernal & Peña, 2012; Miteva et al., 2012).

5 Empirical Evidence of Conservation Instrument Efficiency

Robust empirical evidence on the effectiveness of conservation instruments was scarce prior to 2008. Andam et al. (2008) conducted pioneering research in Costa Rica, which produced somewhat alarming findings at the time. They reported that traditional methods quantifying forest cover change before and after environmental policy, or other methods such as estimating deforestation inside and outside Costa Rican PAs, were overestimating the effectiveness of Costa Rica's protected area system. Furthermore, Andam et al. (2013) revealed that Costa Rica's national system of protected areas had a 49% effectiveness in preventing deforestation compared to robust methods like propensity score matching, which estimated an 11% impact of SNAP-Costa Rica on forest loss prevention. Since 2009, numerous studies on the impact of PAs, the vast majority employing PSM and Locally Weighted Scatterplot Smoothers (Nolte et al., 2013), have consistently demonstrated positive impacts of PAs in reducing deforestation ranging from 1–2% (Pfaff et al., 2013) to 24% (Gaveau et al., 2009).

Below, we summarize rigorous studies on the role of conservation instruments using the impact of protected areas on avoided deforestation as a proxy. Most of these investigations treat forest and non-forest as treatment variables at the pixel level. The results suggest that PAs effectively reduce deforestation (e.g., Chen et al., 2022; Zhang et al., 2022; Rahman & Islam, 2021; Feng et al., 2021; Cuenca et al., 2016; Andam et al., 2008). However, their impact on human welfare in many case studies is limited or nearly negligible (Clements et al., 2014). Research on ecosystem services using soil carbon as a proxy (Shi et al., 2020; Miteva et al., 2015) is also reported. Since Andam et al. (2008), methods have consistently focused on matching factual and counterfactual scenarios, spanning multiple scales from global to national levels (Table 1).

6 Case Study: Effectiveness of Protected Areas in the Tropical Andes on Forest Loss

6.1 Introduction

In recent years, human activities have escalated globally, impacting negatively biodiversity, ecosystem health, ecological processes, and intrinsic cultural values associated with wildlife (Ducros et al., 2023; Kleemann et al., 2022a, b). For instance, IPBES estimates that 1 million species are endangered (Brondízio et al., 2019).

Table 1 Protected area studies using rigorous empirical analysis

Study	Location	Unit of analysis	Sample size (protected/ unprotected)	PA type	Methods	Outcome
Chen et al. (2022)	China	Pixel	100% Pas of China	All PAs	Matching	1.37% pressure of human activities reduction.
Zhang et al. (2022)	China	Pixel	149 Pas of China	All PAs	Matching	93 protected areas showed effectiveness above zero in 2015 compared to 2000.
Rahman and Islam (2021)	Bangladesh	Pixel	18322.61 ha	IUCN II-IV	Matching	
Feng et al. (2021)	Global	Pixel	2975 Pas	All PAs	Matching	7.4% reduction in deforestation
Shi et al. (2020)	Global	Pixel	Global PAs from 1994 to 2015	All PAs	Matching	0.39% increase of carbon secuestration
Cuenca et al. (2016)	Ecuador	Pixel	14 million ha	IUCN Ia, Ib, II	Matching	5.6% reduction in deforestation
Hernandez et al. (2021)	Queensland, Australia	Pixel	13049.3 ha	IUCN I-II	Matching	10.5% reduction in deforestation
Clements et al. (2014)	Cambodia	Family	325/134 families	IUCN V	Mixed effects models, matching	PA intervention had limited impact on livelihood strategies
Graham et al. (2021)	Southeast Asian	Pixel	100% Pas from 2000 to 2008 of Southeast Asian	All PAs	Matching	9.67% of the average rate of forest cover loss avoided in METT protected areas
Black and Anthony (2022)	Cambodia	Pixel	46 Pas of Cambodia	All PAs	Matching	The probability of avoided deforestation was a 12.2–12.5% 2010–2012, ~9.6% and 10.3% between 2013 and 2015 between, and ~4.1% and 6.4% between 2016 and 2018.
Paiva et al. (2015)	Brazil	Pixel	2296.6/33500 ha	All PAs	Matching	Average avoided increase in anthropogenic land per km^2 of 15.49 hectares

Jones and Lewis (2015)	Ecuador & Russia	Pixel	60,000 + 51,500 ha	PES, IUCN I	Matching, regressions	Reduction of 0.4% and 0.03% in deforestation between 2011 and 2013 for Ecuador, and 1995 and 2010 for Russia.
Poor et al. (2019)	Sumatra, Indonesian	Pixel	83,000 + 127,698 + 136,000 ha	IUCN II, IV	Matching	4.19%, 12.8%, 26.36% reduction in deforestation of Tesso Nilo, Rimbang Baling, Bukit Tigapuluh respectively
Miteva et al. (2015)	Indonesia	Pixel	3931 villages	All pas except (IUCN V, VI)	Matching	The protected areas helped conserve mangroves, preventing the loss of approximately 14,100 ha, which translates to avoided emissions of approximately 13 million metric tons of blue carbon.
Andam et al. (2008)	Costa Rica	Pixel	100% Pas of Costa Rica	All PAs	Matching	10% reduction in deforestation
Miranda et al. (2016)	Peru	Pixel	29 national and regional protected areas	IUCN II	Matching	8% reduction in deforestation in the early 2000s.
Pfaff et al. (2017)	Mexico	Pixel	100% Pas of Mexico	All PAs	Matching	3.2% reduction in deforestation.

Matching: May include Propensity Score Matching, Mahalanobis covariate matching or difference-in-difference model
PES Payment for Environmental Services Program, *PAs* Protected Areas

New barriers often disrupt spatial connectivity between habitats, such as poorly planned infrastructure (e.g., roads, canals, hydroelectric projects) or agricultural fields fragmenting habitat patches, thereby creating physical barriers for wildlife movement (Díaz et al., 2019; Jaureguiberry et al., 2022).

Habitat fragmentation caused by multiple barriers has two primary effects: it divides habitat patches into smaller fragments and creates barriers between remaining patches, increasing isolation and rendering habitats unviable for many species (Cuenca & Echeverria, 2017; Noh et al., 2019, 2022). Habitat fragmentation is a crucial indicator for sustainable development and a key focus for many governmental and non-governmental environmental assessment methods.

Brondízio et al. (2019) highlight the global decline in ecosystem composition and functionality, which affects their ability to provide ecosystem services, potentially impacting human well-being (Wu, 2013). It is estimated that in recent years, more than half of all ecosystems have rapidly transitioned to other land uses, especially agriculture, shrimp farming, and urbanization; an example of this is reported in Ecuador, where 14 ecosystems are at risk of extinction (Noh et al., 2020).

Protected areas (PAs) are considered by decision-makers and conservation professionals as one of the most common biodiversity conservation policies at ecosystem, species, and genetic levels (Assessment, 2005; UNEP-WCMC, 2008). In recent years, the role of PAs has expanded to include maintaining food security, water supply, improving human health, and population well-being (Dudley et al., 2014; Nigel, 2008). Furthermore, PAs have become crucial for enhancing climate resilience (Juffe-Bignoli et al., 2014; Leberger et al., 2019; Rands et al., 2010; Scharlemann et al., 2010) and mitigating climate change effects (Soares-Filho et al., 2010, 2023). Due to their increasing relevance in addressing climate change impacts, PAs are considered essential landscape elements (Gaglio et al., 2017; Iwamura et al., 2013; Liu & Ma, 2024) and should be periodically evaluated to determine if they effectively protect their values and provide benefits to society (Ducros et al., 2023; Pessôa et al., 2023; Zhang et al., 2023).

One effective approach to assess the impact of forest conservation policies is estimating avoided deforestation (Black & Anthony, 2022; Kleemann et al., 2022a, b; Ribas et al., 2020; Wendland et al., 2015; Zhang et al., 2022). Protected areas (PAs), among the most common conservation strategies (Assessment, 2005), are expected to influence deforestation patterns by preventing forest logging through protection measures. Several studies globally and regionally assess conservation policy impacts by comparing deforestation rates over time and at landscape scales (Bruner et al., 2001; Nagendra, 2008; Nelson & Chomitz, 2011). Some studies also indirectly compare PAs with non-protected areas (Bruner et al., 2001; Joppa et al., 2008). However, conventional method-based studies may overestimate PA's avoided deforestation by neglecting the non-random assignment of forest protection (Andam et al., 2008; Ribas et al., 2020; Zhang et al., 2022).

To address this issue, it is crucial to integrate spatially explicit information on socio-economic and biophysical factors and utilize robust impact assessment methodologies. Propensity Score Matching (PSM) is one such widely used and robust method for evaluating the impact of forest conservation policies (Cisneros et al.,

2022; Cuenca et al., 2016; Cuenca & Echeverria, 2017; Hua et al., 2023; Ribas et al., 2020; Soares-Filho et al., 2023; Zhang et al., 2022).

This case study examines the avoided deforestation attributable to PAs in Ecuador. Ecuador was selected because its Andean tropical forests are classified as one of the most biodiverse globally per unit area, with 1250 plant species from 136 different families recorded in 1 km^2. These forests constitute a biodiversity hotspot that is disappearing due to rapid land use change to meet demands such as timber and non-timber forest products (Cuesta et al., 2017; Kleemann et al., 2022a, b; Myers et al., 2000).

6.2 *Methodology*

Variables

The estimation of the impact of protected areas (PAs) on avoided deforestation in the tropical Andean forests was conducted over an 18-year period, using historical deforestation baseline maps from 1990 and 2008 derived from Landsat 5 TM images with a resolution of 28.5 m.

Satellite image classifications categorized land into forest and non-forest areas to develop a binary map using GIS. Generally, land uses indicating productive activities, such as agricultural lands and plantations, were grouped as non-forest.

For impact assessment, three types of variables were defined: treatment variables, impact variables, and confounding variables. Treatment variables exclusively pertained to protected areas established between 1990 and 2008. The impact variable distinguished between protected and unprotected forests. In this study, PAs constituted the "treatment" group for comparison with other protected areas, using deforestation as a proxy.

Confounding variables are elements that could lead to a causal conclusion and have been operational since before the establishment of PAs. The literature on PA effectiveness considers essential confounding variables such as elevation, distance to roads, distance to settlements, distance to rivers, slope, distance to forest, temperature, and precipitation. These variables were primarily obtained from Ministry of Environment, Water, and Ecological Transition (MAATE) of Ecuador.

Propensity Score Matching Analysis

Simple impact estimates of protected and unprotected forests were determined using conventional methods to analyze differences and potential overestimations in measuring the impact of PAs on avoided deforestation.

Propensity Score Matching (PSM) estimated what would have happened if forest protection had not been implemented. PSM established the probability that a cell would be assigned protection, with 0 as the dependent variable for protected cells (treatment cells) and 1 for unprotected cells (control cells). Treatment cells were matched using the matching technique (Sekhon, 2008, 2013), ensuring that control cells had a similar probability of matching.

As mentioned earlier, PSM balances differences between treated groups based on a set of characteristics influencing the propensity for protection and impacting the outcome variable (avoided deforestation). PSM ensured that the covariance distribution of protected and unprotected cells was similar. To improve covariance balance, the *calipers* technique was used, defining a tolerance level to judge the quality of matches (Sekhon, 2008). Essentially, cells that did not match within the caliper were excluded from the sample, such as controls that did not match well with treatments.

Sensitivity Test

As previously mentioned, PSM assumes a limitation in accounting for unobserved confounding variables that affect both protection propensity and deforestation. If such unobserved confounding variables exist, differences observed between treatment and control cells may not be entirely controlled through observed confounding variables, potentially biasing the impact estimation.

Therefore, it is crucial to analyze matching robustness through a sensitivity test derived from the Wilcoxon rank-sum test, assuming each observation unit has a fixed value of an unobserved covariate (or a composite of unobserved covariates). A higher gamma (Γ) level at which the estimated effect of PAs on avoided deforestation remains unchanged provides more confidence that the conclusion regarding the impact estimation is unaffected by an unobserved difference.

6.3 *Results*

Covariates

Matching of confounding variables allowed for the comparison of PA effectiveness by blocking bias from assignment and controlling variables that could affect causal outcomes. For instance, comparing non-protected forest cells in 1990 with low altitude, near roads and settlements, to protected cells in 2008 with higher altitude and farther from roads and settlements, logically suggests that forest protection has functioned to prevent deforestation. In this example, a causal result indicates that avoided deforestation corresponds to confounding variables rather than the conservation instrument.

Covariate balance analysis showed that before matching, protected cells were located on steeper and higher altitude terrain compared to unprotected cells

Table 2 Covariate Balance between 1990 and 2008

Variables	Mean Value (Protected)	Mean Value (unprotected)	Differences in mean values	Raw eQQ difference
Slope (°)				
Unmatched	15.74	20.974	−41.44	5.2304
Matched	15.74	15.72	0.1568	0.29877
Altitude (m a.s.l.)				
Unmatched	1506.9	2320.8	−65.055	816.27
Matched	1506.9	1534.1	−2.1697	136.44
Distance to roads (km)				
Unmatched	14.278	14.887	−4.5869	3947
Matched	1.6615	1.6615	18.969	3665.2
Distance to secondary roads (log)				
Unmatched	3.7529	3.7529	−58.755	0.3372
Matched	4.089	3.8247	−12.546	0.1001
Distance to villages (log)				
Unmatched	3.7078	3.9871	−89.942	0.27981
Matched	3.7078	3.7307	−7.3522	0.027933
Distance to rivers (km)				
Unmatched	3.6903	3.6654	4.0825	0.047019
Matched	3.6903	3.6595	5.0558	0.071056
Distance to secondary rivers sec (km)				
Unmatched	2.6871	2.617	−11.985	0.07041
Matched	2.6871	2.689	−0.32095	0.039741

(Table 2). Comparisons of differences in covariate distribution between protected and unprotected cells revealed significant deviations in means and average distances between two functions of empirical quantiles where values greater than 0 indicate divergence between groups at some point in the empirical distribution.

Covariate analysis indicated that matching substantially improved covariate balance and eliminated biases in this analysis (Table 2). In other words, *matching* paired cells with similar characteristics of slope, altitude, distance to roads, settlements, and rivers. The only difference between treatment cells (protected forest) and control cells (unprotected forest) was the conservation treatment implemented in the forest.

Impact of Conservation Instruments

Conventional methods used to estimate the impact of conservation instruments (PAs) on avoided deforestation indicated that more than half of the protected cells would have been deforested by 2008 if forest protection had not been implemented after 1990.

The results demonstrate that 10% of protected cells would have been deforested by 2008 in the absence of protection ($P < 0.01$) (Table 3). When attempting to improve the balance of observational covariates through the calipers technique, bias was reduced, albeit at the cost of estimating avoided deforestation on a subsample that may not have been fully representative of the population of protected cells. Nevertheless, the estimation of avoided deforestation was 12% ($P < 0.01$), which was not significantly different from the estimation without calipers for the year 2008 (Table 3).

Unobserved Biases

Through sensitivity analysis, the presence of unobserved confounding variables in the study case was verified, which could have made the matching of cells less likely to be considered. The sensitivity test confirmed the existence of hidden biases in the allocation of protection, based on the gamma (Γ) indicator being equal to 1 in the absence of unobserved bias. A higher gamma level at which the protection effect on deforestation remains significantly zero indicates that avoided deforestation estimated is not merely a result of matching protected cells with unobserved non-protected cells. The sensitivity test indicated that the estimated avoided deforestation result of 12% for protected forest in 1990 showed significant differences ($P < 0.01$), even in the presence of moderate unobserved biases (Table 4).

Table 3 Estimation of Avoided Deforestation of Protected Areas for the period 1990–2008

Approaches	Protected native forest between 1990 and 2008
N° of protected cells (treatment)	1087
N° of controle cells	20,595
Matching approaches	
Matching de covariables	0.12
N° Matched control	(7939)
Matching con Calipers	−0.10
N° Matched control con calipers	(7939)
Post-matching	0.051
Conventional conservation science approach	
Differences in means	−0.6333
Differences of rates	0.69

Table 4 Unobserved Biases through Sensitivity Test with Critical p-value

Γ	Protected 2008 (never protected and forest in 1990 control)
1.0	< 0.0001
1.6	<0.0001
1.7	0.0003
1.8	0.0008
1.9	0.0018
2.0	0.0028
2.1	0.0064
2.2	0.0125
2.3	0.0174
2.4	0.0243
2.5	0.0418
2.6	0.0585
2.7	0.0791
2.8	0.1037
2.9	0.1322
3.0	0.1644
3.1	0.1999
3.2	0.2382
3.3	0.2789
3.4	0.3213
3.5	0.3648
3.6	0.4088
3.7	0.4526
3.8	0.4959
3.9	0.5362
4.0	0.5343

7 Current State of Environmental Instruments: What to Expect?

Despite the recent surge in the assessment of conservation instruments' impact, it is important to acknowledge that many governments, especially in tropical regions, still struggle to consistently incorporate impact assessment into their national and sectoral policies. This challenge largely stems from methodological gaps in knowledge, limited expertise in conservation instrument assessment, and inadequate planning. Consequently, funding allocation for post-implementation impact studies remains insufficient.

Nevertheless, international organizations are increasingly interested in understanding the impacts of projects or programs, and the reasons behind them. Under this pressure, guidelines have been established to implement impact assessments of their activities. Therefore, impact assessment presents a significant opportunity for tropical countries aiming to implement REDD+ initiatives. Solid results can attract more donors willing to invest in initiatives that positively affect both the environment and communities.

This chapter provided a framework review of impact assessment based on Rubin's causal model, focusing particularly on propensity score matching (PSM), outlining general steps for conducting impact assessment, and presenting a case study to illustrate its application. As demonstrated, a central element in estimating the role of conservation instruments is impact assessment, with PSM being one of the primary methods. It is essential to emphasize that PSM relies on achieving adequate covariate balance and can only mitigate biases through observed confounding factors. Therefore, sensitivity analysis is crucial to assess how conclusions might change if confounding factors were omitted. In other words, a robust sensitivity analysis can significantly enhance confidence in the results.

If we compare the evolution of conservation instruments, the most used proxy used to be deforestation. However, nowadays, many studies are assessing the impact of these instruments on governance and ecosystem health (Muñoz Brenes et al., 2018), ecosystem services (Yu et al., 2024; Zhang et al., 2023), carbon dynamics (Soares-Filho et al., 2023), forest fires (Pessôa et al., 2023), biodiversity (Wauchope et al., 2022), landscape fragmentation (Rahman & Islam, 2021), among others (Arriagada et al., 2016; Heilmayr et al., 2020). Unfortunately, the surge in these studies is primarily focused on Costa Rica, Brazil, China, and Indonesia, with very little research conducted in the Andean tropical forests.

8 Conclusion

To accurately estimate the impact of conservation instruments on the preservation of Andean tropical forests, employing robust methodologies is imperative. There is widespread consensus within the academic and professional communities that Propensity Score Analysis represents one of the most effective means to address non-randomness in the analysis of conservation policies. This approach enables us to mitigate potential causal effects by incorporating the spatial distribution of conservation instruments into our analyses. It is noteworthy that many conservation policies may demonstrate positive impacts on specific variables, such as deforestation, primarily due to their strategic placement away from deforestation hotspots rather than solely due to their implementation aimed at preventing forest loss, which may occur randomly across the landscape.

Therefore, to accurately assess the role of conservation instruments, we must transition from subjective evaluations to considering the before-and-after implementation of these instruments, such as the establishment of protected areas. Emphasizing the widespread adoption of Propensity Score Matching (PSM) is crucial, as it stands out as a primary tool in evaluating the impact of conservation instruments. It is essential to initiate with meticulous data preparation, typical in any analysis, involving scrutiny of data sources, review of spatial scale and resolution across different layers of information, and clear identification of treatment-affecting variables to select appropriate confounders, dependent (treatment), and independent variables.

Next, it is crucial to derive the propensity score using logistic or probit regressions. However, it is paramount to understand that the efficacy of Propensity Score Analysis hinges on grasping the concept of common support. In the core phase of propensity score analysis, selecting the appropriate propensity score method becomes pivotal. It is imperative to meticulously assess covariate balance and conduct sensitivity analyses for unobserved variables.

Now is the opportune moment to scale up the evaluation of environmental conservation instruments for forests through the generation of funding, shifting from management evaluation of conservation instruments to impact assessment. This requires allocating time, resources, and comprehending the theory of change that should underpin the implementation of environmental instruments for the conservation of Ecuador's Andean forests.

Acknowledgments We would like to thank the research project Impact Assessment of Forest and Land Conservation Mechanisms on Avoided Deforestation and Ecosystem Services—project code ECG-024-2024—funded by the Ikiam Amazon Regional University and the Jocotoco Foundation. We would also like to thank Shantall Ayala and other anonymous reviewers for their comments and support.

References

Achard, F., Beuchle, R., Mayaux, P., Stibig, H.-J., Bodart, C., Brink, A., Carboni, S., Desclée, B., Donnay, F., Eva, H. D., Lupi, A., Raši, R., Seliger, R., & Simonetti, D. (2014). Determination of tropical deforestation rates and related carbon losses from 1990 to 2010. *Global Change Biology, 20*(8), 2540–2554. https://doi.org/10.1111/gcb.12605

Andam, K. S., Ferraro, P. J., Pfaff, A., Sanchez-Azofeifa, G. A., & Robalino, J. A. (2008). Measuring the effectiveness of protected area networks in reducing deforestation. *Proceedings of the National Academy of Sciences of the United States of America, 105*(42), 16089–16094. https://doi.org/10.1073/pnas.0800437105

Andam, K. S., Ferraro, P. J., & Hanauer, M. M. (2013). The effects of protected area systems on ecosystem restoration: A quasi-experimental design to estimate the impact of Costa Rica's protected area system on forest regrowth. *Conservation Letters, 6*(5), 317–323. https://doi.org/10.1111/conl.12004

Arriagada, R. A., Ferraro, P. J., Sills, E. O., Pattanayak, S. K., & Cordero-Sancho, S. (2012). Do payments for environmental services affect forest cover? A farm-level evaluation from Costa Rica. *Land Economics, 88*(2), 382–399. http://le.uwpress.org/content/88/2/382.abstract

Arriagada, R. A., Echeverria, C. M., & Moya, D. E. (2016). Creating protected areas on public lands: Is there room for additional conservation? *PLoS One, 11*(2), e0148094. https://doi.org/10.1371/journal.pone.0148094

Assessment, M. E. (2005). *Ecosytems and human well-being: Policy responses*. Island Press.

Bernal, R., & Peña, X. (2012). *Guía práctica para la evaluación de impacto: Edición actualizada incluye casos chilenos*. Ediciones UC. https://books.google.com.co/books?id=95XyDwAAQBAJ

Black, B., & Anthony, B. P. (2022). Counterfactual assessment of protected area avoided deforestation in Cambodia: Trends in effectiveness, spillover effects and the influence of establishment date. *Global Ecology and Conservation, 38*, e02228. https://doi.org/10.1016/j.gecco.2022.e02228

Brondízio, E. S., Settele, J., Diaz, S., & Ngo, H. T. (2019). *Global assessment report on biodiversity and ecosystem services of the Intergovernmental Science-Policy Platform on Biodiversity and Ecosystem Services.*

Bruner, A. G., Gullison, R. E., Rice, R. E., & da Fonseca, G. A. (2001). Effectiveness of parks in protecting tropical biodiversity. *Science, 291*(5501), 125–128. https://doi.org/10.1126/science.291.5501.125

Chen, J., Shi, H., Wang, X., Zhang, Y., & Zhang, Z. (2022). Effectiveness of China's Protected Areas in Mitigating Human Activity Pressure. *International Journal of Environmental Research and Public Health, 19*(15), 9335. https://www.mdpi.com/1660-4601/19/15/9335

Cisneros, E., Börner, J., Pagiola, S., & Wunder, S. (2022). Impacts of conservation incentives in protected areas: The case of Bolsa Floresta, Brazil. *Journal of Environmental Economics and Management, 111*, 102572. https://doi.org/10.1016/j.jeem.2021.102572

Clements, T., Suon, S., Wilkie, D. S., & Milner-Gulland, E. J. (2014). Impacts of Protected Areas on Local Livelihoods in Cambodia. *World Development, 64*, S125–S134. https://doi.org/10.1016/j.worlddev.2014.03.008

Clemente, C. M. S., Espírito-Santo, M. M. D., & Leite, M. E. (2020). Estimates of deforestation avoided by protected areas: A case study in Brazilian tropical dry forests and Cerrado. *Landscape Research, 45*(4), 470–483. https://doi.org/10.1080/01426397.2020.1730773

Cuenca, P., & Echeverria, C. (2017). How do protected landscapes associated with high biodiversity and population levels change? *PLoS One, 12*(7), e0180537. https://doi.org/10.1371/journal.pone.0180537

Cuenca, P., Arriagada, R., & Echeverría, C. (2016). How much deforestation do protected areas avoid in tropical Andean landscapes? *Environmental Science & Policy, 56*, 56–66. https://doi.org/10.1016/j.envsci.2015.10.014

Cuenca, P., Robalino, J., Arriagada, R., & Echeverría, C. (2018). Are government incentives effective for avoided deforestation in the tropical Andean forest? *PLoS One, 13*(9), e0203545. https://doi.org/10.1371/journal.pone.0203545

Cuesta, F., Peralvo, M., Merino-Viteri, A., Bustamante, M., Baquero, F., Freile, J. F., Muriel, P., & Torres-Carvajal, O. (2017). Priority areas for biodiversity conservation in mainland Ecuador. *Neotropical Biodiversity, 3*(1), 93–106. https://doi.org/10.1080/23766808.2017.1295705

Díaz, S., Fargione, J., Chapin, F. S., III, & Tilman, D. (2006). Biodiversity loss threatens human well-being. *PLoS Biology, 4*(8), e277.

Díaz, S., Demissew, S., Joly, C., Lonsdale, W. M., & Larigauderie, A. (2015). A Rosetta Stone for nature's benefits to people. *PLoS Biology, 13*(1), e1002040. https://doi.org/10.1371/journal.pbio.1002040

Díaz, S., Settele, J., Brondízio, E. S., Ngo, H. T., Agard, J., Arneth, A., Balvanera, P., Brauman, K. A., Butchart, S. H. M., Chan, K. M. A., Garibaldi, L. A., Ichii, K., Liu, J., Subramanian, S. M., Midgley, G. F., Miloslavich, P., Molnár, Z., Obura, D., Pfaff, A., et al. (2019). Pervasive human-driven decline of life on Earth points to the need for transformative change. *Science, 366*(6471), eaax3100. https://doi.org/10.1126/science.aax3100

Ducros, D., Devillers, R., Messager, A., Suet, M., Wachoum, A. S., Deschamps, C., Breme, B. M., Petersen, I. K., Kayser, Y., Vincent-Martin, N., Djimasngar, M. B. N., Portier, B., Champagnon, J., Mondain-Monval, J.-Y., & Defos du Rau, P. (2023). Planning from scratch: A new modelling approach for designing protected areas in remote, data-poor regions. *Journal of Applied Ecology, 60*(9), 2018–2030. https://doi.org/10.1111/1365-2664.14470

Dudley, N., Groves, C., Redford, K. H., & Stolton, S. (2014). Where now for protected areas? Setting the stage for the 2014 World Parks Congress. *Oryx, 48*(04), 496–503. https://doi.org/10.1017/S0030605314000519

FAO. (1993). *Forest resources assessment 1990: Tropical country*. Food and Agriculture Organization of the United Nations.

Feng, Y., Wang, Y., Su, H., Pan, J., Sun, Y., Zhu, J., Fang, J., & Tang, Z. (2021). Assessing the effectiveness of global protected areas based on the difference in differences model. *Ecological Indicators, 130*, 108078. https://doi.org/10.1016/j.ecolind.2021.108078

Ferraro, P., Merlin, M. H., Daniela, A. M., Gustavo Javier, C.-B., Subhrendu, K. P., & Katharine, R. E. S. (2013). More strictly protected areas are not necessarily more protective: Evidence from Bolivia, Costa Rica, Indonesia, and Thailand. *Environmental Research Letters, 8*(2), 025011. http://stacks.iop.org/1748-9326/8/i=2/a=025011

Forman, R. T. T., & Godron, M. (1986). *Landscape ecology*. Wiley. http://books.google.es/books?id=ZvNEVs2MWqcC

Gaglio, M., Aschonitis, V. G., Mancuso, M. M., Reyes Puig, J. P., Moscoso, F., Castaldelli, G., & Fano, E. A. (2017). Changes in land use and ecosystem services in tropical forest areas: A case study in Andes mountains of Ecuador. *International Journal of Biodiversity Science, Ecosystem Services & Management, 13*(1), 264–279. https://doi.org/10.1080/21513732.2017.1345980

Gaveau, D. L. A., Epting, J., Lyne, O., Linkie, M., Kumara, I., Kanninen, M., & Leader-Williams, N. (2009). Evaluating whether protected areas reduce tropical deforestation in Sumatra. *Journal of Biogeography, 36*(11), 2165–2175. https://doi.org/10.1111/j.1365-2699.2009.02147.x

Gertler, P., Martinez, S., & World Bank. (2011). *Impact evaluation in practice*. World Bank. https://books.google.es/books?id=SV2nUAXnF3sC

Graham, V., Geldmann, J., Adams, V. M., Negret, P. J., Sinovas, P., & Chang, H.-C. (2021). Southeast Asian protected areas are effective in conserving forest cover and forest carbon stocks compared to unprotected areas. *Scientific Reports, 11*(1), 23760. https://doi.org/10.1038/s41598-021-03188-w

Guo, S., & Fraser, M. W. (2014). *Propensity score analysis: Statistical methods and applications*. SAGE Publications. https://books.google.es/books?id=5Y_MAwAAQBAJ

Guo, S., Barth, R. P., & Gibbons, C. (2006). Propensity score matching strategies for evaluating substance abuse services for child welfare clients. *Children and Youth Services Review, 28*(4), 357–383. https://doi.org/10.1016/j.childyouth.2005.04.012

Han, Y. W., Kang, W., Thorne, J., & Song, Y. (2019). Modeling the effects of landscape patterns of current forests on the habitat quality of historical remnants in a highly urbanized area. *Urban Forestry & Urban Greening, 41*, 354–363. https://doi.org/10.1016/j.ufug.2019.04.015

Heilmayr, R., Carlson, K. M., & Benedict, J. J. (2020). Deforestation spillovers from oil palm sustainability certification. *Environmental Research Letters, 15*(7), 075002. https://doi.org/10.1088/1748-9326/ab7f0c

Hernandez, S., Barnes, M. D., Duce, S., & Adams, V. M. (2021). The impact of strictly protected areas in a deforestation hotspot. *Conservation Science and Practice, 3*(9), e479. https://doi.org/10.1111/csp2.479

Hua, T., Zhao, W., Cherubini, F., Hu, X., & Pereira, P. (2023). Upgrading protected areas can improve or reverse the decline in conservation effectiveness: Evidence from the Tibetan Plateau, China. *Science of the Total Environment, 873*, 162345. https://doi.org/10.1016/j.scitotenv.2023.162345

Imbens, G. W. (2015). Matching methods in practice: Three examples. *Journal of Human Resources, 50*(2), 373–419. https://doi.org/10.3368/jhr.50.2.373

Institute, W. R. (2024). *Forest loss indicator.* Global Forest Review. Retrieved updated April 4, 2024 from Available online at https://research.wri.org/gfr/forest-extent-indicators/forest-loss

Iwamura, T., Guisan, A., Wilson, K. A., & Possingham, H. P. (2013). How robust are global conservation priorities to climate change? *Global Environmental Change, 23*(5), 1277–1284. https://doi.org/10.1016/j.gloenvcha.2013.07.016

Jaureguiberry, P., Titeux, N., Wiemers, M., Bowler, D. E., Coscieme, L., Golden, A. S., Guerra, C. A., Jacob, U., Takahashi, Y., Settele, J., Díaz, S., Molnár, Z., & Purvis, A. (2022). The direct drivers of recent global anthropogenic biodiversity loss. *Science Advances, 8*(45), eabm9982. https://doi.org/10.1126/sciadv.abm9982

Jo, J. H., Lee, C. B., Cho, H. J., & Lee, J. (2021). Estimation of citizens' willingness to pay for the implementation of payment for local forest ecosystem services: The case of taxes and donations. *Sustainability, 13*(11), 14, Article 6186. https://doi.org/10.3390/su13116186

Jones, K. W., & Lewis, D. J. (2015). Estimating the Counterfactual impact of conservation programs on land cover outcomes: The role of matching and panel regression Techniques. *PLoS One, 10*(10), e0141380. https://doi.org/10.1371/journal.pone.0141380

Joppa, L., & Pfaff, A. (2010). Reassessing the forest impacts of protection: The challenge of non-random location and a corrective method. *Annals of the New York Academy of Sciences, 1185*, 135–149. https://doi.org/10.1111/j.1749-6632.2009.05162.x

Joppa, L. N., Loarie, S. R., & Pimm, S. L. (2008). On the protection of "protected areas". *Proceedings of the National Academy of Sciences, 105*(18), 6673–6678. https://doi.org/10.1073/pnas.0802471105

Juffe-Bignoli, D., Burgess, N. D., Bingham, H., Belle, E. M. S., de Lima, M. G., Deguignet, M., Bertzky, B., Milam, A. N., Martinez-Lopez, J., Lewis, E., Eassom, A., Wicander, S., Geldmann, J., van Soesbergen, A., Arnell, A. P., O'Connor, B., Park, S., Shi, Y. N., Danks, F. S., et al. (2014). *Protected planet report 2014*. UNEP-WCMC. ISBN: 978-92-807-3416-4.

Kim, S. E., Kim, H., & Chae, Y. (2014). A new approach to measuring green growth: Application to the OECD and Korea. *Futures, 63*, 37–48. https://doi.org/10.1016/j.futures.2014.08.002

Kim, J. H., Kim, S. Y., & Yoo, S. H. (2021). South Koreans' willingness to pay for restoration of Gomsoman Tidal Flat. *Ocean & Coastal Management, 199*, 9, Article 105388. https://doi.org/10.1016/j.ocecoaman.2020.105388

Kleemann, J., Koo, H., Hensen, I., Mendieta-Leiva, G., Kahnt, B., Kurze, C., Inclan, D. J., Cuenca, P., Noh, J. K., Hoffmann, M. H., Factos, A., Lehnert, M., Lozano, P., & Fürst, C. (2022a). Priorities of action and research for the protection of biodiversity and ecosystem services in continental Ecuador. *Biological Conservation, 265*, 109404. https://doi.org/10.1016/j.biocon.2021.109404

Kleemann, J., Zamora, C., Villacis-Chiluisa, A. B., Cuenca, P., Koo, H., Noh, J. K., Fürst, C., & Thiel, M. (2022b). Deforestation in continental Ecuador with a focus on protected areas. *Land, 11*(2), 268. https://www.mdpi.com/2073-445X/11/2/268

Leberger, R., Rosa, I. M. D., Guerra, C. A., Wolf, F., & Pereira, H. M. (2019). Global patterns of forest loss across IUCN categories of protected areas. *Biological Conservation*, 108299. https://doi.org/10.1016/j.biocon.2019.108299

Liu, Y., & Ma, J. (2024). Significant early end of the growing season of forest vegetation inside China's protected areas. *iScience, 27*(1), 108652. https://doi.org/10.1016/j.isci.2023.108652

Marshall, T., & Jang, J. H. (2020). The role of social enterprises in urban sustainability: Insights from Anyang, South Korea. *Urban Science, 4*(3), 20, Article 42. https://doi.org/10.3390/urbansci4030042

Miranda, J. J., Corral, L., Blackman, A., Asner, G., & Lima, E. (2016). Effects of protected areas on forest cover change and local communities: Evidence from the Peruvian amazon. *World Development, 78*, 288–307. https://doi.org/10.1016/j.worlddev.2015.10.026

Miteva, D. A., Pattanayak, S. K., & Ferraro, P. J. (2012). Evaluation of biodiversity policy instruments: What works and what doesn't? *Oxford Review of Economic Policy, 28*(1), 69–92. https://doi.org/10.1093/oxrep/grs009

Miteva, D. A., Murray, B. C., & Pattanayak, S. K. (2015). Do protected areas reduce blue carbon emissions? A quasi-experimental evaluation of mangroves in Indonesia. *Ecological Economics, 119*, 127–135. https://doi.org/10.1016/j.ecolecon.2015.08.005

Muñoz Brenes, C. L., Jones, K. W., Schlesinger, P., Robalino, J., & Vierling, L. (2018). The impact of protected area governance and management capacity on ecosystem function in Central America. *PLoS One, 13*(10), e0205964. https://doi.org/10.1371/journal.pone.0205964

Myers, N., Mittermeier, R. A., Mittermeier, C. G., da Fonseca, G. A. B., & Kent, J. (2000). Biodiversity hotspots for conservation priorities. *Nature, 403*(6772), 853–858. https://doi.org/10.1038/35002501. http://www.nature.com/nature/journal/v403/n6772/suppinfo/403853a0_S1.html

Nagendra, H. (2008). Do parks work? Impact of protected areas on land cover clearing. *Ambio, 37*(5), 330–337.

Nelson, A., & Chomitz, K. M. (2011). Effectiveness of strict vs. multiple use protected areas in reducing tropical forest fires: A global analysis using matching methods. *PLoS One, 6*(8), e22722. https://doi.org/10.1371/journal.pone.0022722

Nigel, D. (Ed.). (2008). *Guidelines for applying protected area management categories*. IUCN.

Noh, J. K., Echeverra, C., Pauchard, A., & Cuenca, P. (2019). Extinction debt in a biodiversity hotspot: The case of the Chilean Winter Rainfall-Valdivian Forests. *Landscape and Ecological Engineering, 15*, 1–2.

Noh, J. K., Echeverria, C., Kleemann, J., Koo, H., Fürst, C., & Cuenca, P. (2020). Warning about conservation status of forest ecosystems in tropical Andes: National assessment based on IUCN criteria. *PLoS One, 15*(8), e0237877. https://doi.org/10.1371/journal.pone.0237877

Noh, J. K., Echeverria, C., Gaona, G., Kleemann, J., Koo, H., Fürst, C., & Cuenca, P. (2022). Forest ecosystem fragmentation in Ecuador: Challenges for sustainable land use in the Tropical Andean. *Land, 11*(2), 287. https://www.mdpi.com/2073-445X/11/2/287

Nolte, C. (2018). Buying forests for conservation: Contours of a global trend. *Current Opinion in Environmental Sustainability, 32*, 68–75. https://doi.org/10.1016/j.cosust.2018.05.003

Nolte, C., Agrawal, A., Silvius, K. M., & Soares-Filho, B. S. (2013). Governance regime and location influence avoided deforestation success of protected areas in the Brazilian Amazon. *Proceedings of the National Academy of Sciences, 110*(13), 4956–4961. https://doi.org/10.1073/pnas.1214786110

OECD. (2015). *Securing livelihoods for all*. https://doi.org/10.1787/9789264231894-en

Oliveira, P. J. C., Asner, G. P., Knapp, D. E., Almeyda, A., Galván-Gildemeister, R., Keene, S., Raybin, R. F., & Smith, R. C. (2007). Land-use allocation protects the Peruvian Amazon. *Science, 317*(5842), 1233–1236. https://doi.org/10.1126/science.1146324

Paiva, R. J. O., Brites, R. S., & Machado, R. B. (2015). The role of protected Areas in the Avoidance of Anthropogenic Conversion in a High Pressure Region: A Matching Method Analysis in the Core Region of the Brazilian Cerrado. *PLoS One, 10*(7), e0132582. https://doi.org/10.1371/journal.pone.0132582

Perschke, M. J., Harris, L. R., Sink, K. J., & Lombard, A. T. (2023). Ecological infrastructure as a framework for mapping ecosystem services for place-based conservation and management. *Journal for Nature Conservation, 73*, 126389. https://doi.org/10.1016/j.jnc.2023.126389

Pessôa, A. C. M., Morello R.S, T. F., Silva-Junior, C. H. L., Doblas, J., Carvalho, N. S., Aragão, L. E. O. C., & Anderson, L. O. (2023). Protected areas are effective on curbing fires in the Amazon. *Ecological Economics, 214*, 107983. https://doi.org/10.1016/j.ecolecon.2023.107983

Pfaff, A., Robalino, J., Sanchez-Azofeifa, G., Andam, K., & Ferraro, P. (2009). Park location affects forest protection: Land characteristics cause differences in park impacts across Costa Rica. *The BE Journal of Economic Analysis and Policy, 9*(2), 5.

Pfaff, A., Robalino, J., Lima, E., Sandoval, C., & Herrera, L. D. (2013). Governance, location and avoided deforestation from protected areas: Greater restrictions can have lower impact, due to differences in location. *World Development*. https://doi.org/10.1016/j.worlddev.2013.01.011

Pfaff, A., Robalino, J., Sandoval, C., & Herrera, D. (2015). Protected area types, strategies and impacts in Brazil's Amazon: Public protected area strategies do not yield a consistent ranking of protected area types by impact. *Philosophical Transactions of the Royal Society of London B: Biological Sciences, 370*(1681). https://doi.org/10.1098/rstb.2014.0273

Pfaff, A., Santiago-Ávila, F., & Joppa, L. (2017). Evolving protected-area impacts in Mexico: Political shifts as suggested by impact evaluations. *Forests, 8*(1), 17. https://www.mdpi.com/1999-4907/8/1/17

Poor, E. E., Frimpong, E., Imron, M. A., & Kelly, M. J. (2019). Protected area effectiveness in a sea of palm oil: A Sumatran case study. *Biological Conservation, 234*, 123–130. https://doi.org/10.1016/j.biocon.2019.03.018

Rahman, M. F., & Islam, K. (2021). Effectiveness of protected areas in reducing deforestation and forest fragmentation in Bangladesh. *Journal of Environmental Management, 280*, 111711. https://doi.org/10.1016/j.jenvman.2020.111711

Rands, M. R., Adams, W. M., Bennun, L., Butchart, S. H., Clements, A., Coomes, D., Entwistle, A., Hodge, I., Kapos, V., Scharlemann, J. P., Sutherland, W. J., & Vira, B. (2010). Biodiversity conservation: Challenges beyond 2010. *Science, 329*(5997), 1298–1303. https://doi.org/10.1126/science.1189138

Ribas, L. G. D. S., Pressey, R. L., Loyola, R., & Bini, L. M. (2020). A global comparative analysis of impact evaluation methods in estimating the effectiveness of protected areas. *Biological Conservation, 246*, 108595. https://doi.org/10.1016/j.biocon.2020.108595

Robalino, J., Alexander, P., Sánchez-Azofeifa, A., Alpízar, F., León, C., & Rodríguez, C. (2008). *Deforestation impacts of environmental services payments: Costa Rica's PSA Program 2000–2005.* http://www.rff.org/research/publications/deforestation-impacts-environmental-services-payments-costa-rica-s-psa-program

Robalino, J., Pfaff, A., & Villalobos-Fiatt, L. (2017). Heterogeneous local spillovers from protected areas in Costa Rica. *Journal of the Association of Environmental and Resource Economists.* https://doi.org/10.1086/692089

Rosenbaum, P. R., & Rubin, D. B. (1983). The central role of the propensity score in observational studies for causal effects. *Biometrika, 70*(1), 41–55. https://doi.org/10.2307/2335942

Rubin, D. B. (2006). *Matched sampling for causal effects.* Cambridge University Press. https://books.google.es/books?id=vgk9K8hJZ7kC

Scharlemann, J. P. W., Kapos, V., Campbell, A., Lysenko, I., Burgess, N. D., Hansen, M. C., Gibbs, H. K., Dickson, B., & Miles, L. (2010). Securing tropical forest carbon: The contribution of protected areas to REDD. *Oryx, 44*(03), 352–357. https://doi.org/10.1017/S0030605310000542

Sekhon, J. (2008). 271 The Neyman— Rubin model of causal inference and estimation via matching methods. In J. M. Box-Steffensmeier, H. E. Brady, & D. Collier (Eds.), The Oxford Handbook of Political Methodology (pp. 0). Oxford University Press. https://doi.org/10.1093/oxfordhb/9780199286546.003.0011

Sekhon, J. S. (2013). *Multivariate and propensity score matching with balance optimization.* R. S. Software.

Shi, H., Li, X., Liu, X., Wang, S., Liu, X., Zhang, H., Tang, D., & Li, T. (2020). Global protected areas boost the carbon sequestration capacity: Evidences from econometric causal analysis. Science of *The Total Environment, 715*, 137001. https://doi.org/10.1016/j.scitotenv.2020.137001

Soares-Filho, B., Moutinho, P., Nepstad, D., Anderson, A., Rodrigues, H., Garcia, R., Dietzsch, L., Merry, F., Bowman, M., Hissa, L., Silvestrini, R., & Maretti, C. (2010). Role of Brazilian Amazon protected areas in climate change mitigation. *National Academy of Sciences of the United States of America, 107*(24), 10821–10826. https://doi.org/10.1073/pnas.0913048107

Soares-Filho, B. S., Oliveira, U., Ferreira, M. N., Marques, F. F. C., de Oliveira, A. R., Silva, F. R., & Börner, J. (2023). Contribution of the Amazon protected areas program to forest conservation. *Biological Conservation, 279*, 109928. https://doi.org/10.1016/j.biocon.2023.109928

Stuart, E., Huskamp, H., Duckworth, K., Simmons, J., Song, Z., Chernew, M., & Barry, C. (2014). Using propensity scores in difference-in-differences models to estimate the effects of a policy change. *Health Services and Outcomes Research Methodology*, 1–17. https://doi.org/10.1007/s10742-014-0123-z

Turner, M. G. (2010). Disturbance and landscape dynamics in a changing world. *Ecology, 91*(10), 2833–2849. https://doi.org/10.1890/10-0097.1

Turner, M. G., & Gardner, R. H. (2015). *Landscape ecology in theory and practice: Pattern and process.* Springer. https://books.google.com.ec/books?id=S7zYCgAAQBAJ

UNEP-WCMC. (2008). *State of the world's protected areas: An annual review of global conservation progress.* UNEP-WCMC.

Wauchope, H. S., Jones, J. P. G., Geldmann, J., Simmons, B. I., Amano, T., Blanco, D. E., Fuller, R. A., Johnston, A., Langendoen, T., Mundkur, T., Nagy, S., & Sutherland, W. J. (2022). Protected areas have a mixed impact on waterbirds, but management helps. *Nature, 605*(7908), 103–107. https://doi.org/10.1038/s41586-022-04617-0

Wendland, K. J., Baumann, M., Lewis, D. J., Sieber, A., & Radeloff, V. C. (2015). Protected area effectiveness in European Russia: A postmatching panel data analysis. *Land Economics, 91*(1), 149–168. http://le.uwpress.org/content/91/1/149.abstract

Wu, J. (2013). Landscape sustainability science: Ecosystem services and human well-being in changing landscapes. *Landscape Ecology, 28*(6), 999–1023.

Wu, J. (2021). Landscape sustainability science (II): Core questions and key approaches. *Landscape Ecology, 36*(8), 2453–2485. https://doi.org/10.1007/s10980-021-01245-3

Yang, H., Choi, H. T., & Lim, H. (2019). Effects of forest thinning on the long-term runoff changes of coniferous Forest plantation. *Water, 11*(11), 16, Article 2301. https://doi.org/10.3390/w11112301

Yang, H., Viña, A., Winkler, J. A., Chung, M. G., Huang, Q., Dou, Y., McShea, W. J., Songer, M., Zhang, J., & Liu, J. (2021). A global assessment of the impact of individual protected areas on preventing forest loss. *Science of the Total Environment, 777*, 145995. https://doi.org/10.1016/j.scitotenv.2021.145995

Yu, D. D., & Han, S. J. (2016). Ecosystem service status and changes of degraded natural reserves – A study from the Changbai Mountain Natural Reserve, China. *Ecosystem Services, 20*, 56–65. https://doi.org/10.1016/j.ecoser.2016.06.009

Yu, C., Zhang, Z., Jeppesen, E., Gao, Y., Liu, Y., Liu, Y., Lu, Q., Wang, C., & Sun, X. (2024). Assessment of the effectiveness of China's protected areas in enhancing ecosystem services. *Ecosystem Services, 65*, 101588. https://doi.org/10.1016/j.ecoser.2023.101588

Zhang, Z., Tang, Y., Pan, H., Yao, C., & Zhang, T. (2022). Assessment of the ecological protection effectiveness of protected areas using propensity score matching: A case study in Sichuan, China. *International Journal of Environmental Research and Public Health, 19*(8), 4920. https://www.mdpi.com/1660-4601/19/8/4920

Zhang, X., Liao, L., Huang, Y., Fang, Q., Lan, S., & Chi, M. (2023). Conservation outcome assessment of Wuyishan protected areas based on InVEST and propensity score matching. *Global Ecology and Conservation, 45*, e02516. https://doi.org/10.1016/j.gecco.2023.e02516

Index

N. Clerici (ed.), *Conservation of Andean Forests*, https://doi.org/10.1007/978-3-031-80805-0

GPSR Compliance
The European Union's (EU) General Product Safety Regulation (GPSR) is a set of rules that requires consumer products to be safe and our obligations to ensure this.

If you have any concerns about our products, you can contact us on

ProductSafety@springernature.com

In case Publisher is established outside the EU, the EU authorized representative is:

Springer Nature Customer Service Center GmbH
Europaplatz 3
69115 Heidelberg, Germany

www.ingramcontent.com/pod-product-compliance
Lightning Source LLC
Chambersburg PA
CBHW050625290425
25869CB00003B/153
* 9 7 8 3 0 3 1 8 0 8 0 4 3 *